BASIC TECHNICAL PHYSICS

SECOND EDITION

PAUL E. TIPPENS
Professor of Physics
Southern College of Technology
Marietta, Georgia

GLENCOE

McGraw-Hill

New York, New York
Columbus, Ohio
Mission Hills, California
Peoria, Illinois

Technical Art and Production: York Production Services, Inc.

Library of Congress Cataloging-in-Publication Data

Tippens, Paul E.
  Basic technical physics.
  Includes index.
  1. Physics.   I. Title.
QC23.T62   1989   530   88-16459

Basic Technical Physics, Second Edition

Send all inquiries to:
Glencoe/McGraw-Hill
936 Eastwind Drive
Westerville, Ohio 43081

ISBN 0-07-065013-6

8 9 10 11 12 13 14 15   RRD-C   04 03 02 01 00 99 98 97 96

CONTENTS

* Asterisk indicates optional sections and other more difficult material.

PREFACE

Technology is changing so rapidly today that it is no longer practical to provide an educational program with a narrow job focus. The person trained as an electrician is suddenly faced with computers and robotics, which necessarily require a broader base in the principles of physics. Civil and mechanical technicians now use refined instrumentation, which requires greater understanding of optics and electrical phenomena. And many trade, industrial, and technical careers that once could be learned easily on the job now require advanced vocational training. Clearly, those who receive a broad preparation in the fundamentals of physics can better adjust to the changing demands of the modern world.

The dilemma facing postsecondary education today is to provide the underlying science base required by the job market while simultaneously providing the necessary training and skills. Some schools have decided to provide a strong one-semester applied physics course as early as possible to provide the background for the more specialized coursework. *Basic Technical Physics* has been written expressly for this purpose. It assumes that students have had rather weak preparation in mathematics, so it begins with a review of the mathematical principles needed in later chapters. (The mathematics can be taught either with right-triangle trigonometry or graphical approaches.) Those sections which require trigonometry or which are considered difficult are indicated by an asterisk (*) in the text. However, in keeping with modern demands, the concepts and problems of *Basic Technical Physics* are similar to those found in more advanced textbooks.

The first edition of *Basic Technical Physics* was well received and in a variety of educational settings. Most often it was used as a one-semester basic physics textbook, but in many cases it has been successfully adopted as a preparatory text for more advanced study. This second edition was prepared in such a manner as to maintain the flexibility of the first edition. A conscious effort was made to add only that material considered necessary based on user surveys. For example, a section was added on momentum, and the chapter on equilibrium was expanded and rewritten. However, there are still only 22 chapters, and a judicious selection of topics will allow for a strong one-semester course.

An ideal supplement for this text is available for those who have access to IBM and compatible personal computers. An independent series, *Tutorials in Physics*, has been prepared to enhance the learning process. These tutorials are interactive and take advantage of the motion, sound, and graphics capabilities of modern computers.

The questions and problems at the end of each chapter provide the necessary practice for understanding the principles of physics. These have been supplemented and replaced in many cases to improve their instructional value, but many of the favorite problems from the first edition remain. As always with such an undertaking, there may be a few errors in the text. The author earnestly solicits your assistance in eliminating these mistakes and welcomes your remarks and advice concerning the material and its presentation.

Paul E. Tippens

to

NENA AND SANDY

TECHNICAL MATHEMATICS

As a result of completing this chapter, you should be able to

1. Add, subtract, multiply, and divide signed numbers.
2. Solve simple formulas for any symbol appearing in the formula, then evaluate by substitution.
*3. Apply the rules for operations with exponents and radicals.
4. Express decimal numbers in scientific notation and perform the common mathematical operations with them.
5. Construct a graph from given data and interpret new information from the graph.
6. Apply the elementary rules of geometry to determine unknown angles in given situations.

Note to the Student   It is often frustrating to open a physics book and see that it begins with mathematics. Naturally, you want to learn only those things for which you see a definite need. You want to make measurements, to work with machines or engines, to get your hands on something, or, at the very least, to know that your time is not wasted. Depending on your previous experience, you may omit much of or all this first chapter at the discretion of your instructor. Keep in mind that fundamentals are

1

important and that some skills in mathematics are essential. You may have a thorough understanding of force, mass, energy, and electricity as concepts. But you may not be able to apply the concepts to your work because of weaknesses in basic mathematics. Mathematics is the language of physics.

In any industrial or technical occupation, we are concerned with physical measurements of some kind. It might be the length of a board, the area of a sheet of metal, the number of bolts to be ordered, the stress on an aircraft wing, or the pressure in an oil tank. The only way we can make sense out of such data is through the use of numbers and symbols. Mathematics provides the necessary tools for organizing the data and for predicting outcomes. For example, the formula $F = ma$ expresses the relationship between an applied force F and the acceleration a that force produces. The quantity m is a symbol which represents the *mass* of an object (a measure of the amount of matter it contains). Through appropriate mathematical steps, we can use formulas such as this one to predict future events. However, a general knowledge of algebra and geometry is required in most instances. This chapter provides you with a review of a few essential concepts in mathematics. Those objectives and chapter sections marked by an asterisk (*) are more difficult than other sections and may be omitted, if desired.

1-1 ■ SIGNED NUMBERS

Frequently, it is necessary to work with negative numbers as well as positive numbers. For example, a temperature of $-10°C$ means 10 degrees ''below'' a zero reference point, and 24°C refers to a temperature 24 degrees ''above'' zero (see Fig. 1-1). The numbers refer to the *magnitude* of the temperature, and the plus or minus sign refers to the *direction* from zero. The minus sign in $-10°C$ does not indicate a lack of tempera-

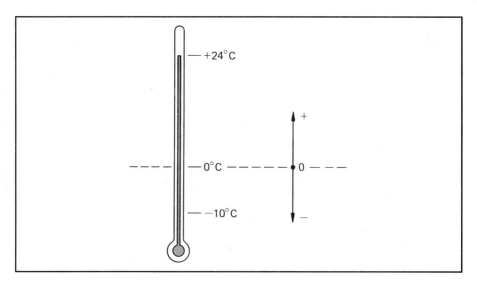

FIG. 1-1 Minus and plus signs often are used to indicate a direction from some zero reference point, as in the case of a thermometer.

ture; it means only that the temperature is less than zero. The number 10 in $-10°C$ describes only how far the temperature is from zero; the minus sign is necessary to indicate the direction from zero.

The value of a number without its sign is called its *absolute value*. In other words, if we disregard the signs of $+7$ and -7, the value is the same. Each number is 7 units from zero. The absolute value of a number is indicated by the vertical-bar symbol. The number $+7$ does not equal the number -7, but $|+7| = |-7|$. In performing arithmetic operations with signed numbers, the absolute values are used.

Plus and minus signs are also used to indicate arithmetic operations. Examples are

$7 + 5$ means "add the number $+5$ to the number $+7$"
$7 - 5$ means "subtract the number $+5$ from the number $+7$"

If we wish to indicate the addition or subtraction of negative numbers, parentheses are helpful:

$(+7) + (-5)$ means "add the number -5 to the number $+7$"
$(+7) - (-5)$ means "subtract the number -5 from the number $+7$"

When signed numbers are added, it is useful to recall the following rule:

Addition rule: *To add two numbers of like sign, we add the absolute values of the numbers and give the sum the common sign. To add two numbers of unlike sign, we find the difference of their absolute values and give the result the sign of the number of larger absolute value.*

Consider the following examples:

$$(+6) + (+2) = +(6 + 2) = +8$$
$$(-6) + (-2) = -(6 + 2) = -8$$
$$(+6) + (-2) = +(6 - 2) = +4$$
$$(-6) + (+2) = -(6 - 2) = -4$$

Whenever a number has another subtracted from it, we change the sign of the second number and then add it to the first, using the addition rule. In the expression $7 - 5$, the number $+5$ is to be subtracted from the number $+7$. The subtraction is accomplished by changing the $+5$ to -5 and then adding the two numbers of unlike sign: $(+7) + (-5) = +2$.

Subtraction rule: *To subtract one signed number b from another signed number a, we change the sign of b and then add it to a, using the addition rule.*

Consider the following examples:

$$(+8) - (+5) = 8 - 5 = 3$$
$$(+8) - (-5) = 8 + 5 = 13$$
$$(-8) - (+5) = -8 - 5 = -13$$
$$(-8) - (-5) = -8 + 5 = -3$$

EXAMPLE 1-1   The velocity of an object is considered positive when it is moving upward and negative when it is moving downward. What is the change in velocity of a ball if it strikes the floor at 60 feet per second (ft/s) and rebounds upward at 50 ft/s?

Solution   The initial velocity is −60 ft/s because the ball is traveling downward. Later, its velocity is +50 ft/s because it is traveling upward. We find the change in velocity by subtracting the initial velocity from the final velocity. Thus,

$$\text{Change in velocity} = \text{final velocity} - \text{initial velocity}$$
$$= (+50 \text{ ft/s}) - (-60 \text{ ft/s})$$
$$= 50 \text{ ft/s} + 60 \text{ ft/s} = 110 \text{ ft/s}$$

Without an understanding of signed numbers, we might have guessed that the change in speed was only 10 ft/s (60 − 50). A moment's thought, however, makes us realize that the speed must first decrease to zero (a change of 60 ft/s) and then attain a speed of 50 ft/s in the opposite direction (an additional change of 50 ft/s).

When two or more numbers are to be multiplied, each number is called a *factor* and the result is the *product*. We may now state the multiplication rule for signed numbers:

Multiplication rule: *If two factors have like signs, their product is positive. If two factors have unlike signs, their product is negative.*

Examples are

$$(+2)(+3) = +6 \qquad (-3)(-4) = +12$$
$$(-2)(+3) = -6 \qquad (-3)(+4) = -12$$

An extension of the multiplication rule is often helpful for products resulting from several factors. Rather than multiplying a series of factors two at a time, we might recall that

The product will be positive if all factors are positive or if there are an even number of negative factors. The product will be negative if there are an odd number of negative factors.

Consider the following examples:

$$(-2)(+2)(-3) = +12 \qquad \text{(two negative factors—even)}$$
$$(-2)(+4)(-3)(-2) = -48 \qquad \text{(three negative factors—odd)}$$
$$(-3)^3 = (-3)(-3)(-3) = -27 \qquad \text{(three negative factors—odd)}$$

Notice that in the last example a superscript 3 was used to indicate the number of times the number −3 was to be taken as a factor. The superscript 3 written in this manner is called an *exponent*.

When two numbers are to be divided, the number being divided is called the *dividend*. The number divided into the dividend is called the *divisor*. The result of division is called the *quotient*. The rule for dividing signed numbers is as follows:

Division rule: *The quotient of two numbers of like sign is positive, and the quotient of two numbers of unlike sign is negative.*

For example,

$$(+2) \div (+2) = +1 \qquad (-4) \div (-2) = +2$$

$$\frac{+4}{-2} = -2 \qquad \frac{-4}{+2} = -2$$

If either the numerator or denominator of a fraction contains two or more factors, the following rule is also useful:

The quotient is negative if the total number of negative factors is odd; otherwise, the quotient is positive.

For example,

$$\frac{(-4)(3)}{2} = -6 \qquad \text{odd}$$

$$\frac{(-2)(-2)(-3)}{(2)(-3)} = +2 \qquad \text{even}$$

You should practice the application of all the rules in this entire section. It is a serious mistake to assume that you understand these concepts without adequate proof. A major source of errors in physics problems can be traced to signed numbers.

EXERCISES 1-1

In Exercises 1 through 26, evaluate each expression by performing the indicated operations. Answers to exercises are in the back of the book.

1. $(+2) + (+5)$

2. $(-2) + (6)$

3. $(-4) - (-6)$

4. $(+6) - (+8)$

5. $(-3) - (+7)$

6. $(-15) - (+18)$

7. $(-4) - (+3) - (-2)$

8. $(-6) + (-7) - (+4)$

9. $(-2)(-3)$

10. $(-16)(+2)$

11. $(-6)(-3)(-2)$

12. $(-6)(+2)(-2)$

13. $(-3)(-4)(-2)(2)$

14. $(-6)(2)(3)(-4)$

15. $(-6) \div (-3)$

16. $(-14) \div (+7)$

17. $(+16) \div (-4)$

18. $(+18) \div (-6)$

19. $\dfrac{-4}{-2}$

20. $\dfrac{+16}{-4}$

21. $\dfrac{(-2)(-3)(-1)}{(-2)(-1)}$

22. $\dfrac{(-6)(+4)}{(-2)}$

23. $\dfrac{(-16)(4)}{2(-4)}$

24. $\dfrac{(-1)(-2)^2(12)}{(6)(2)}$

25. $(-2)(+4) - \dfrac{(-6)}{(+2)} - (-5)$

26. $(-2)(-2)^2 + \dfrac{(-3)(-2)(-8)}{(-4)(1)} - (-6)^3$

In Exercises 27 through 30, solve the given problem.

27. Distances above the ground are positive, and distances below the ground are negative. If an object is dropped from 20 ft above the ground into a 12-ft hole, what is the difference between the initial position and the final position?

28. In physics, work is measured in joules (J) and may be positive or negative depending on the direction of the force which does the work. What is the total work done if the works of the forces are 20, −40, and −12 J?

29. The temperature of a bolt is −12°C. (*a*) If the temperature increases by 6C°, what is the resultant temperature? (*b*) If the temperature decreases by 5C°, what is the resultant temperature? (*c*) If the temperature is multiplied by a factor of −3, what is the resultant temperature?

30. A metal expands when heated and contracts when cooled. Assume that the length of a rod changes by 2 millimeters (mm) for each 1C° change in temperature. What is the total change in the length of this rod as the temperature changes from −5 to −30°C?

1-2 ■ ALGEBRA REVIEW

Algebra is really a generalization of arithmetic in which letters are used to replace numbers. For example, we will learn that the space occupied by some objects (their volume) can be calculated by multiplying the length by the width by the height. By assigning letters to each of these measurements, we may establish a general *formula,* such as

$$\text{Volume} = \text{length} \times \text{width} \times \text{height}$$
$$V = l \cdot w \cdot h$$

The advantage of formulas is that they work for any number of situations. Given the length, width, and height of any rectangular solid, we may use the above formula to calculate its volume. If we wish to determine the volume of a rectangular metal block, we need only *substitute* the proper numbers into the formula.

EXAMPLE 1-2 Calculate the volume of a solid whose length is 6 inches (in.), whose width is 4 in., and whose height is 2 in.

Solution Substitution yields

$$V = lwh$$
$$= (6 \text{ in.})(4 \text{ in.})(2 \text{ in.})$$
$$= 48 \text{ (in.} \times \text{in.} \times \text{in.)} = 48 \text{ in.}^3$$

The treatment of units which results in volume being expressed as cubic inches is discussed later. For now, you should concentrate on the substitution of numbers.

When numbers are substituted for letters in a formula, it is very important to insert the proper sign of the number. Consider the following formula:

$$P = c^2 - ab$$

Suppose that $c = +2$, $a = -3$, and $b = +4$. Remember that plus and minus signs in formulas do not apply to any of the numbers which might be substituted. In this example, we have

$$
\begin{aligned}
P &= (c)^2 - (a)(b) \\
&= (+2)^2 - (-3)(+4) \\
&= 4 + 12 = 16
\end{aligned}
$$

It is easy to see how confusing a sign in the formula with the sign of a substituted number can result in an error.

It is frequently necessary to solve a formula or an equation for some letter that is only a part of the formula. For example, suppose we wanted to give a formula for the length of a rectangular solid in terms of its volume, height, and width. The letters in $V = lwh$ would need to be rearranged so that the letter l would appear by itself on the left side. Formula rearrangement is not difficult if we recall several rules for working with equations.

Basically, an equation is a mathematical statement that two expressions are equal. For example,

$$
2b + 4 = 3b - 1
$$

is an equation. In this case, it is evident that the letter b represents the *unknown* quantity. If we substitute $b = 5$ into both sides of this equation, we obtain $14 = 14$. Thus, $b = 5$ is the *solution* to the equation.

We can obtain solutions for equalities by performing the same operations on each side of an equation. Consider the equality $4 = 4$. If we add, subtract, multiply, or divide the number 2 into both sides, we do not change the fact that the two sides are equal. We *do* increase or decrease the magnitude of each side, but the equality remains. (You should verify this statement for the equality $4 = 4$.) Note also that taking the square or square root of each side does not disturb the equality. By performing a series of identical operations to each side of an equation, we can eventually obtain an equality with one letter by itself on the left side. In this case, we have *solved* the equation for that letter.

EXAMPLE 1-3   Solve the following equation for m:

$$
3m - 5 = m + 3
$$

Solution   The point is to get m by itself on one side of the equals sign and a number by itself on the other side. We first add $+5$ to both sides, then subtract m from both sides:

$$
\begin{aligned}
3m - 5 + 5 &= m + 3 + 5 \\
3m &= m + 8 \\
3m - m &= m + 8 - m \\
2m &= 8
\end{aligned}
$$

Finally, we divide both sides by 2:

$$\frac{2m}{2} = \frac{8}{2}$$
$$m = 4$$

To check this answer, we substitute $m = 4$ into the original equation and obtain $7 = 7$. This shows that $m = 4$ is the solution.

In formulas, the solution to an equation may also be expressed in terms of letters. For example, the literal equation

$$ax - 5b = c$$

might be solved for x in terms of a, b, and c. In cases such as this one, we decide in advance which letter is to be the "unknown." In our example, we will choose x. The remaining letters are treated as though they were known numbers. Adding $5b$ to both sides, we have

$$ax - 5b + 5b = c + 5b$$
$$ax = c + 5b$$

Now, we may divide both sides by a to obtain

$$\frac{ax}{a} = \frac{c + 5b}{a}$$
$$x = \frac{c + 5b}{a}$$

which is the solution for x. The values for a, b, and c in a given situation may be substituted to find a particular value for x.

EXAMPLE 1-4   The volume of a right circular cone is expressed by the formula

$$V = \frac{\pi r^2 h}{3}$$

What is the height of such a cone of radius $r = 3$ in. and $V = 81$ cubic inches (in.^{3})? (Assume that $\pi = 3.14$.)

Solution   First, we solve the formula for h in terms of r and V. Multiplying both sides by 3 gives

$$3V = \pi r^2 h$$

Dividing both sides by πr^2 yields

$$\frac{3V}{\pi r^2} = h \quad \text{or} \quad h = \frac{3V}{\pi r^2}$$

Substitution for V, π and r gives

$$h = \frac{3(81 \text{ in.}^3)}{(3.14)(3 \text{ in.})^2} = \frac{243 \text{ in.}^3}{28.26 \text{ in.}^2} = 8.60 \text{ in.}$$

The height of the cylinder is 8.60 in.

EXERCISES 1-2

In Exercises 1 through 16, determine the value for x when a = 2, b = −3, and c = −2.

1. $x = a + b + c$

2. $x = a - b - c$

3. $x = b + c - a$

4. $x = b(a - c)$

5. $x = \dfrac{b - c}{a}$

6. $x = \dfrac{a + b}{c}$

7. $x = b^2 - c^2$

8. $x = \dfrac{-b}{ac}$

9. $x = \dfrac{a}{bc}(a - c)$

10. $x = a^2 + b^2 + c^3$

11. $x = \sqrt{a^2 + b^2 + c^2}$

12. $x = ab(c - a)^2$

13. $2ax - b = c$

14. $ax + bx = 4c$

15. $3ax = \dfrac{2ab}{c}$

16. $\dfrac{4ac}{b} = \dfrac{2x}{b} - 16$

In Exercises 17 through 26, solve the equations for the unknown letter.

17. $5m - 16 = 3m - 4$

18. $3p = 7p - 16$

19. $4m = 2(m - 4)$

20. $3(m - 6) = 6$

21. $\dfrac{x}{3} = (4)(3)$

22. $\dfrac{p}{3} = \dfrac{2}{6}$

23. $\dfrac{96}{x} = 48$

24. $14 = 2(b - 7)$

25. $R^2 = (4)^2 + (3)^2$

26. $\dfrac{1}{2} = \dfrac{1}{P} + \dfrac{1}{6}$

In Exercises 27 through 40, solve the formulas for the letter indicated.

27. $V = IR$, R

28. $PV = nRT$, T

29. $F = ma$, a

30. $s = vt + d$, d

31. $F = \dfrac{mv^2}{R}$, R

32. $s = \frac{1}{2}at^2$, a

33. $2as = v_f^2 - v_o^2$, a

34. $C = \dfrac{Q^2}{2V}$, V

35. $\dfrac{1}{R} = \dfrac{1}{R_1} + \dfrac{1}{R_2}$, R

36. $MV = Ft$, t

37. $mv_2 - mv_1 = Ft$, v_2

38. $\dfrac{P_1 V_1}{T_1} = \dfrac{P_2 V_2}{T_2}$, T_2

39. $v = v_o + at$, a

40. $c^2 = a^2 + b^2$, b

*1-3 ■ EXPONENTS AND RADICALS

It is often necessary to multiply the same quantity a number of times. A shorthand method of indicating the number of times a quantity is taken as a factor uses a superscript number called an *exponent*. This notation works according to the following scheme:

For any number a:	For the number 2:
$a = a^1$	$2 = 2^1$
$a \times a = a^2$	$2 \times 2 = 2^2$
$a \times a \times a = a^3$	$2 \times 2 \times 2 = 2^3$
$a \times a \times a \times a = a^4$	$2 \times 2 \times 2 \times 2 = 2^4$

The powers of the number a are read as follows: a^2 is read "a squared," a^3 is read "a cubed," and a^4 is read "a to the fourth power." More generally, we speak of a^n as "a to the nth power." In these examples, the letter a is called the *base*, and the superscript numbers 1, 2, 3, 4, and n are called the *exponents*.

We will review several rules to follow in working with exponents.

Rule 1: *When two quantities of the same base are multiplied, their product is obtained by adding the exponents algebraically:*

$$(a^m)(a^n) = a^{m+n}$$

Multiplication Rule

Examples:

$$(2^4)(2^3) = 2^{4+3} = 2^7$$
$$y^8 y^6 = y^{14}$$
$$x^2 x^5 y^3 x^3 = x^{2+5+3} y^3 = x^{10} y^3$$

Rule 2: *When a is not zero, a negative exponent may be defined by either of the following expressions:*

$$a^{-n} = \dfrac{1}{a^n} \qquad \text{and} \qquad a^n = \dfrac{1}{a^{-n}}$$

Negative Exponent

Examples:

$$3^{-4} = \frac{1}{3^4} = \frac{1}{81} \qquad 10^2 = \frac{1}{10^{-2}}$$

$$a^{-5} = \frac{1}{a^5} \qquad \frac{x^{-3}y^2}{a^{-4}b^3} = \frac{a^4y^2}{x^3b^3}$$

Rule 3: *Any quantity raised to the power of zero is equal to 1:*

$$a^0 = 1$$

Zero Exponent

Examples:

$$x^3y^0 = x^3 \qquad (x^3y^2)^0 = 1$$

Rule 4: *The quotient of two nonzero quantities of the same base is found by taking the algebraic difference of their exponents:*

$$\frac{a^m}{a^n} = a^{m-n}$$

Division

Examples:

$$\frac{2^3}{2} = 2^{3-1} = 2^2 \qquad \frac{2^5}{2^7} = 2^{5-7} = 2^{-2} = \frac{1}{2^2}$$

$$\frac{a^{-3}}{a^{-5}} = a^{-3-(-5)} = a^{-3+5} = a^2$$

Rule 5: *When a quantity* a^m *is raised to the power* n, *the exponents are multiplied:*

$$(a^m)^n = a^{mn}$$

Power of a Power

Examples:

$$(2^2)^3 = 2^{2\cdot3} = 2^6 \qquad (2^{-3})^2 = 2^{-6} = \frac{1}{2^6}$$

$$(a^2)^4 = a^8 \qquad (a^2)^{-4} = a^{-8} = \frac{1}{a^8}$$

Rule 6: *The power of a product and of a quotient is obtained by applying the exponent to each of the factors:*

$$(ab)^n = a^n b^n$$

$$\left(\frac{a}{b}\right)^n = \frac{a^n}{b^n}$$

Examples:

$$(2 \cdot 3)^2 = 2^2 \cdot 3^2 = 4 \cdot 9 = 36$$
$$(ab)^3 = a^3 b^3$$
$$(ab^2)^3 = a^3 (b^2)^3 = a^3 b^6$$
$$\left(\frac{ax^3}{y^2}\right)^4 = \frac{a^4 x^{12}}{y^8}$$

If $a^n = b$, then not only is b equal to the nth power of a, but also by definition a is said to be the nth *root* of b. In general, this fact is expressed by using a *radical* ($\sqrt{}$):

$$\sqrt[n]{b} \qquad n\text{th root of } b$$

Consider the following statements:

$$2^2 = 4 \text{ means that 2 is the } square \ root \text{ of 4, or } \sqrt{4} = 2$$
$$2^3 = 8 \text{ means that 2 is the } cube \ root \text{ of 8, or } \sqrt[3]{8} = 2$$
$$2^5 = 32 \text{ means that 2 is the fifth root of 32, or } \sqrt[5]{32} = 2$$

A radical may also be expressed by using a fractional exponent. In general, we may write

$$\sqrt[n]{b} = b^{1/n}$$

For example,

$$\sqrt[3]{8} = 8^{1/3} \qquad \text{or} \qquad \sqrt{10} = 10^{1/2}$$

There are two additional rules to follow when you are working with radicals.

Rule 7: *The nth root of a product is equal to the product of the nth roots of each factor:*

$$\sqrt[n]{ab} = \sqrt[n]{a}\,\sqrt[n]{b}$$

Roots of a Product

Examples:

$$\sqrt{4 \cdot 16} = \sqrt{4} \sqrt{16} = 2 \cdot 4 = 8$$
$$\sqrt[5]{ab} = \sqrt[5]{a} \sqrt[5]{b}$$

Rule 8: *The roots of a power are found by using the definition of fractional exponents:*

$$\boxed{\sqrt[n]{a^m} = a^{m/n}}$$
Roots of Powers

Examples:

$$\sqrt[3]{2^9} = 2^{9/3} = 2^3 = 8$$
$$\sqrt{10^{-4}} = 10^{-4/2} = 10^{-2} = \frac{1}{10^2}$$
$$\sqrt{4 \times 10^8} = \sqrt{4} \sqrt{10^8} = 2(10)^{8/2} = 2 \times 10^4$$
$$\sqrt[3]{8 \times 10^{-6}} = \sqrt[3]{8}(10)^{-6/3} = 2 \times 10^{-2}$$

Most of the problems in this textbook will require only a limited understanding of the above rules. Squares, cubes, square roots, and cube roots are encountered most often. However, a good understanding of the rules for exponents and radicals is helpful.

*EXERCISES 1-3

In Exercises 1 through 22, simplify the given expressions, using the laws of exponents and radicals.

1. $2^5 \cdot 2^7$
2. $3^2 \cdot 2^3 \cdot 3^3$
3. $x^7 x^3$
4. $x^7 x^{-5} x^3$
5. $a^{-3} a^2$
6. $a^3 a^{-2} b^{-3} b$
7. $\dfrac{2^3}{2^5}$
8. $\dfrac{2a^3 b}{2ab^3}$
9. $\dfrac{2x^{17}}{x^{12}}$
10. $(ab)^{-2}$
11. $(m^{-3})^{-2}$

12. $(n^3 c^{-2})^{-2}$
13. $(4 \times 10^2)^3$
14. $(6 \times 10^{-2})^{-2}$
15. $\sqrt[3]{64}$
16. $\sqrt[4]{81}$
17. $\sqrt[3]{x^{15}}$
18. $\sqrt{a^4 b^6}$
19. $\sqrt{4 \times 10^4}$
20. $\sqrt[3]{8 \times 10^{-27}}$
21. $\sqrt[5]{32 a^{10}}$
22. $\sqrt{(x + 2)^2}$

1-4 ■ SCIENTIFIC NOTATION

In scientific work, one frequently encounters very large or very small numbers. For example, a machinist may measure the thickness of a thin metal sheet to be 0.00021 in. Similarly, an engineer may encounter an area of 1,200,000 ft^2 for an airport runway. It is convenient to be able to express these numbers, respectively, as 2.1×10^{-4} in. and 1.2×10^6 ft^2. Powers of 10 are used to keep track of the decimal point, and we are not forced to carry a large number of zeros along with our calculations. The expression of any number as a number between 1 and 10 times an integral power of 10 is called *scientific notation*.

Electronic calculators are often equipped with an $\boxed{EE}$ or $\boxed{EXP}$ button which allows even beginning students to use scientific notation in many computations. You are virtually certain to encounter scientific notation, even if your job does not require frequent use of numbers expressed in this form. Review the manual which comes with your calculator to learn how to work with powers of 10 on the calculator.

Consider the following multiples of 10 and examples of their use in scientific notation:

$$
\begin{array}{ll}
0.0001 = 10^{-4} & 2.34 \times 10^{-4} = 0.000234 \\
0.001 = 10^{-3} & 2.34 \times 10^{-3} = 0.00234 \\
0.01 = 10^{-2} & 2.34 \times 10^{-2} = 0.0234 \\
0.1 = 10^{-1} & 2.34 \times 10^{-1} = 0.234 \\
1 = 10^{0} & 2.34 \times 10^{0} = 2.34 \\
10 = 10^{1} & 2.34 \times 10^{1} = 23.4 \\
100 = 10^{2} & 2.34 \times 10^{2} = 234.0 \\
1000 = 10^{3} & 2.34 \times 10^{3} = 2340.0 \\
10,000 = 10^{4} & 2.34 \times 10^{4} = 23,400.0
\end{array}
$$

To write a number larger than 1 in scientific notation, you must determine the number of times the decimal point must be moved to the left in order to arrive at the shorthand notation. Examples are

$$
\begin{aligned}
467 &= 4\,6\,7. = 4.67 \times 10^2 \\
30 &= 3\,0. = 3.0 \times 10^1 \\
35,700 &= 3\,5\,7\,0\,0. = 3.57 \times 10^4
\end{aligned}
$$

Any decimal number less than 1 can be written as a number between 1 and 10 times a *negative* power of 10. The negative exponent in this case is the number of times the decimal point is moved to the right. This is always one more than the number of zeros that separate the first digit from the decimal. Examples are

$$
\begin{aligned}
0.24 &= 0\,2\,4 = 2.4 \times 10^{-1} \\
0.00327 &= 0.0\,0\,3\,2\,7 = 3.27 \times 10^{-3} \\
0.0000469 &= 0.0\,0\,0\,0\,4\,6\,9 = 4.69 \times 10^{-5}
\end{aligned}
$$

To transfer from scientific notation to decimal notation, the procedure is simply reversed.

By recalling the laws of exponents, scientific notation can be used in multiplication and division of very small or very large numbers. When two numbers are multiplied, the exponents of 10 are added. For example, 200×4000 may be written $(2 \times 10^2)(4 \times 10^3) = (2)(4) \times (10^2)(10^3) = 8 \times 10^5$. Other examples are

$$2200 \times 40 = (2.2 \times 10^3)(4 \times 10^1) = 8.8 \times 10^4$$
$$0.0002 \times 900 = (2.0 \times 10^{-4})(9.0 \times 10^2) = 18 \times 10^{-2}$$
$$1002 \times 3 = (1.002 \times 10^3)(3 \times 10^0) = 3.006 \times 10^3$$

Similarly, when one number is divided by another number, the exponent of 10 in the denominator is subtracted from the exponent of 10 in the numerator. Examples are

$$\frac{7000}{35} = \frac{7 \times 10^3}{3.5 \times 10^1} = \frac{7.0}{3.5} \times 10^{3-1} = 2.0 \times 10^2$$

$$\frac{1200}{0.003} = \frac{1.2 \times 10^3}{3.0 \times 10^{-3}} = \frac{1.2}{3.0} \times 10^{3-(-3)} = 0.4 \times 10^6$$

$$\frac{0.008}{400} = \frac{8 \times 10^{-3}}{4 \times 10^2} = \frac{8}{4} \times 10^{-3-2} = 2.0 \times 10^{-5}$$

When two numbers in scientific notation are added, care must be taken to adjust all numbers to be added so that they have identical powers of 10. Examples are

$$2000 + 400 = 2 \times 10^3 + 0.4 \times 10^3 = 2.4 \times 10^3$$
$$0.006 - 0.0008 = 6 \times 10^{-3} - 0.8 \times 10^{-3} = 5.2 \times 10^{-3}$$
$$4 \times 10^{-21} - 6 \times 10^{-20} = 0.4 \times 10^{-20} - 6 \times 10^{-20} = -5.6 \times 10^{-20}$$

Scientific calculators automatically make the necessary adjustments when adding and subtracting such numbers.

Scientific notation and powers of 10 take on an important and special significance when you work with metric units. In the next chapter, you will see that multiples of 10 are used to define a number of units in the metric system. For example, one kilometer is defined as a thousand (1×10^3) meters, and one millimeter is defined as one-thousandth (1×10^{-3}) of a meter.

EXERCISES 1-4

In Exercises 1 through 8, convert the decimal numbers to scientific notation.

1. 40,000
2. 67
3. 480
4. 497,000

5. 0.0021
6. 0.789
7. 0.087
8. 0.000967

In Exercises 9 through 16, convert the numbers to decimal notation.

9. 4×10^6

10. 4.67×10^3

11. 3.7×10^1

12. 1.4×10^5

13. 3.67×10^{-2}

14. 4×10^{-1}

15. 6×10^{-3}

16. 4.17×10^{-5}

In Exercises 17 through 40, simplify and express as a single number written in scientific notation.

17. $400 \times 20,000$

18. 37×2000

19. $(4 \times 10^{-3})(2 \times 10^5)$

20. $(3 \times 10^{-1})(6 \times 10^{-8})$

21. $(6.7 \times 10^3)(4.0 \times 10^5)$

22. $(3.7 \times 10^{-5})(200)$

23. $(4 \times 10^{-3})^2$

24. $(3 \times 10^6)^3$

25. $(6000)(3 \times 10^{-7})$

26. $(4)(300)(2 \times 10^{-2})$

27. $7000 \div (3.5 \times 10^{-3})$

28. $60 \div 30,000$

29. $(6 \times 10^{-5}) \div (3 \times 10^4)$

30. $(4 \times 10^{-7}) \div (7 \times 10^{-7})$

31. $\dfrac{4600}{0.02}$

32. $\dfrac{(1600)(4 \times 10^{-3})}{1 \times 10^{-2}}$

33. $4.0 \times 10^2 + 2 \times 10^3$

34. $6 \times 10^{-5} - 4 \times 10^{-6}$

35. $6 \times 10^{-3} - 0.075$

36. $0.0007 - 4 \times 10^{-3}$

37. $\dfrac{4 \times 10^{-6} + 2 \times 10^{-5}}{4 \times 10^{-2}}$

38. $\dfrac{6 \times 10^3 + 4 \times 10^2}{1 \times 10^{-3}}$

39. $\dfrac{600 - 3000}{0.0003}$

40. $(4 \times 10^{-3})^2 - 2 \times 10^{-5}$

1-5 ■ GRAPHS

Frequently, it is desirable to show the relationship between two quantities in the form of a graph. For example, we know that a car traveling at constant speed covers the same distances every minute (min) that it travels. We might record the distance traveled in feet at particular times as follows:

Distance, ft	200	400	600	800	1000
Time, min	1	2	3	4	5

Along the bottom of a sheet of graph paper, we might establish a time scale, maybe letting each division equal 1 min. On the left side of the paper, we can establish a distance scale. The scale selected should conveniently fill the graph paper. (This makes it easier to locate points on the graph.) Simple scale divisions are 1 division = 1, 2, or 5 times some power of 10. Good examples are 1 division = $1 \times 10^3 = 1000$, or 1 division = $2 \times 10^0 = 2$, or 1 division = $5 \times 10^{-2} = 0.05$. Awkward scale divisions such as 3 divisions = 100 ft should be avoided because they hinder the location of points. In our example, we let each division represent 100 ft. The data may then be plotted on the graph, as shown in Fig. 1-2. Each point on the horizontal line has a corresponding point on the vertical line. For example, the distance traveled after 3 min is 600 ft. Notice that when the points are connected, a straight line results.

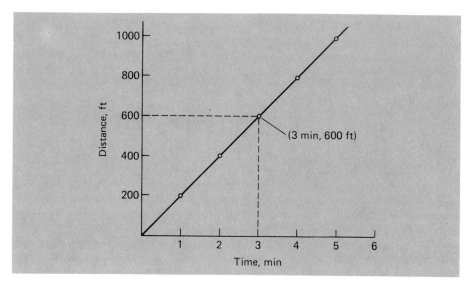

FIG. 1-2 A graph of distance as a function of time (a *direct* relationship).

When a graph of one quantity versus another results in a straight line, there is a *direct relationship*. In this example, the distance traveled is directly proportional to the time. When one quantity increases, the other increases by the same proportion. Doubling the time elapsed doubles the distance traveled.

Indirect or *inverse relationships,* in which an increase of one quantity results in a proportional decrease in another quantity, are also encountered. If we were to decrease the speed of an automobile, we would find that greater and greater time intervals were required to cover the same distance. Suppose we measure the time in seconds (s) required to travel a distance of 1 kilometer (km) [0.621 mile (mi)] at speeds of 20, 40, 60, 80, and 100 kilometers per hour (km/h). The following data are recorded:

Speed, km/h	20	40	60	80	100
Time, s	180	90	60	45	36

A graph of these data is shown in Fig. 1-3. Notice that the graph of an inverse relationship is not a straight line but a curved one.

A graph can be used to obtain information that was not available before the graph was constructed. For example, in Fig. 1-3, we can learn that a time of 120 s would be required to cover the distance if our speed were 30 km/h.

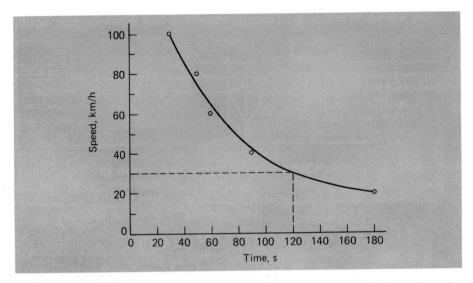

FIG. 1-3 A graph of the time required to cover a distance of 1 km as a function of speed (an *indirect* relationship).

EXERCISES 1-5

1. Plot a graph for the following data recorded for an object falling freely from rest.

Speed, ft/s	32	63	97	129	159	192	225
Time, s	1	2	3	4	5	6	7

What do you expect the speed to be after 4.5 s? How much time is required for the object to attain a speed of 100 ft/s?

2. The advance of a right-hand screw is proportional to the number of complete turns. For a particular screw, the following data are recorded:

Advance, in.	0.5	1.0	1.5	2.0	2.5	3.0
Number of turns	16	32	48	64	80	96

Plot a graph with the number of turns as the horizontal divisions and the advance in inches as the vertical divisions. What number of turns is required to advance the screw 2.75 in.?

3. Plot a graph showing the relationship between frequency and wavelength of electromagnetic waves. The following data are given:

Frequency, kilohertz (kHz)	150	200	300	500	600	900
Wavelength, meters (m)	2000	1500	1000	600	500	333

What are the wavelengths of electromagnetic waves with frequencies of 350 and 800 kHz?

4. The electric power loss in a resistor varies directly with the square of the current. In a particular experiment, the following data are recorded:

Current, amperes (A)	1.0	2.5	4.0	5.0	7.0	8.5
Power, watts (W)	1.0	6.5	16.2	25.8	50.2	72.0

Plot a graph and from the curve determine the power loss when the current is *(a)* 3.2 A and *(b)* 8.0 A.

1-6 ■ GEOMETRY

In this brief review, it is assumed that you already have an understanding of the concept of a point and a line. Other important concepts are reviewed only to the extent that they may be needed in physics problems. A lengthy review of the many possible theorems in this field is unnecessary. We will begin with angles and lines.

The *angle* between two straight lines is defined by drawing a circle with its center at the point of intersection (see Fig. 1-4*a*). The magnitude of angle *A* is proportional to the fraction of a complete circle which lies between the two lines. Angles are measured in *degrees*, as defined in Fig. 1-4*b*. One degree (°) is a portion of a circle equal to 1/360 of a full revolution (rev). Thus, there are 360° in 1 rev:

$$1° = \frac{1}{360} \text{ rev} \qquad 1 \text{ rev} = 360°$$

A special name is given to the angle which forms one-fourth of 1 rev, or 90°. It is called a *right angle* and is illustrated in Fig. 1-5*a*. When two straight lines meet so that the angle between them is a right angle, they are said to be *perpendicular*. Line *CA* in Fig. 1-5*b* is perpendicular to line *BD*. This may be written

$$CA \perp BD$$

where ⊥ means "is perpendicular to."

Two straight lines are said to be *parallel* if the extension of their sides will never intersect. In Fig. 1-6, line *AB* is parallel to line *CD*. We write

$$AB \parallel CD$$

where ‖ means "is parallel to."

The application of geometry usually requires only a few general rules. We will describe three of the most important.

Rule 1: *When two straight lines intersect, they form opposing angles which are equal (Fig. 1-7).*

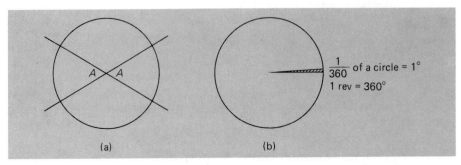

FIG. 1-4 An angle is a fraction of a complete circle. One degree is a portion of a circle equal to 1/360 of a full revolution.

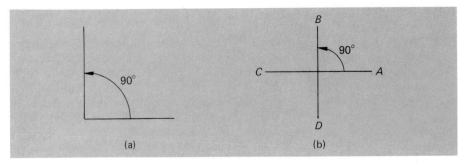

FIG. 1-5 (*a*) A right angle is one-fourth of a circle. (*b*) Lines which cross at right angles are said to be perpendicular.

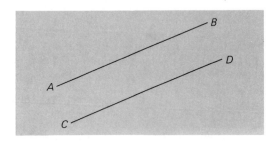

FIG. 1-6 Parallel lines extended indefinitely will never intersect (*AB* ∥ *CD*).

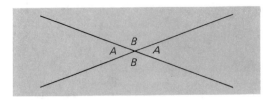

FIG. 1-7 When two straight lines intersect, the opposing angles are equal.

Rule 2: *When a straight line intersects (cuts across) two parallel lines, the alternate interior angles are equal (Fig. 1-8).*

Note from Fig. 1-8 that the angles *A* are on alternate sides of the intersecting line and are inside the two parallel lines. According to rule 2, these *alternate interior* angles are equal. (The remaining two interior angles are also equal.)

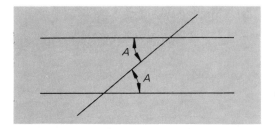

FIG. 1-8 When a straight line intersects two parallel lines, the alternate interior angles are equal.

EXAMPLE 1-5   Two studs at a construction site are braced by a cross member, as shown in Fig. 1-9. Determine angle *C*, using geometry.

Solution   Angle *A* is 60° by rule 1, angle *B* is 60° by rule 2, and angle *C* is 60° by rule 1. From this example, it is seen that *alternate exterior* angles are also equal, but no new rule is necessary.

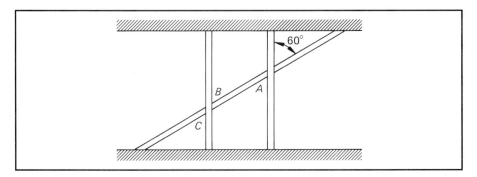

FIG. 1-9

A triangle is a plane closed figure with three sides. Figure 1-10 shows a general triangle with sides *a*, *b*, and *c* and angles *A*, *B*, and *C*. Such a triangle in which no two sides are equal and no two angles are equal is called a *scalene triangle*.

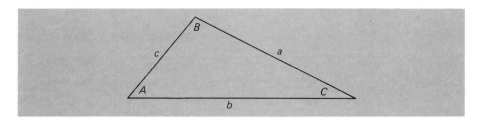

FIG. 1-10   A scalene triangle.

A special triangle of interest to us is a *right triangle,* as shown in Fig. 1-11. A right triangle is one in which one of the angles is equal to 90° (two of the sides are perpendicular). The side opposite the right angle joins the two perpendicular sides and is called the *hypotenuse.*

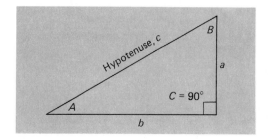

FIG. 1-11 A right triangle is a triangle in which one of the interior angles is a right angle.

Rule 3: *For any triangle, the sum of the interior angles is equal to 180°:*

$$A + B + C = 180°$$

Corollary: *For any right triangle (C = 90°), the sum of the two smaller angles is equal to 90°:*

$$A + B = 90°$$

EXAMPLE 1-6   Determine angles B and A in Fig. 1-12.

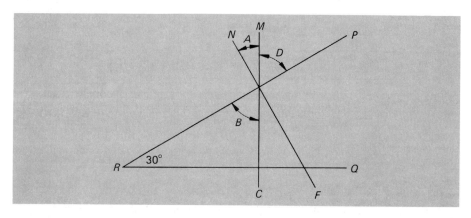

FIG. 1-12

Solution   Since $MC \perp RQ$, we have a right triangle in which one of the smaller angles is 30°. From rule 3,

$$30° + B = 90°$$

and

$$B = 90° - 30° = 60°$$

Since opposite angles are equal, D is also $60°$. Line $NF \perp RP$ so that $A + D = 90°$. Thus

$$A + 60° = 90°$$
$$A = 90° - 60° = 30°$$

The only additional rule you will need is the pythagorean theorem. This rule states that, for right triangles, the square of the hypotenuse is equal to the sum of the squares of the other two sides. We will discuss this rule and its application in Chap. 3.

EXERCISES 1-6

1. What are the magnitudes of the angles in Fig. 1-13, by your estimation?

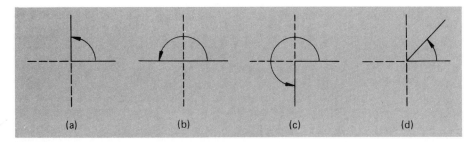

(a)   (b)   (c)   (d)

FIG. 1-13

2. Draw two lines AB and CD. Then draw a third line EF such that $AB \parallel EF$, $CD \perp AB$, and $CD \perp EF$.

3. Determine angles A and B for each of the arrangements drawn in Fig. 1-14.

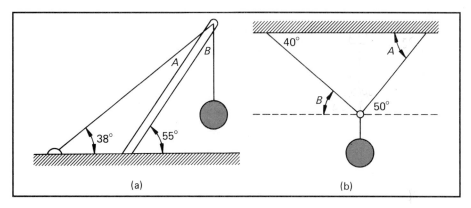

(a)   (b)

FIG. 1-14

4. Determine angles A and B in Fig. 1-15a and b.

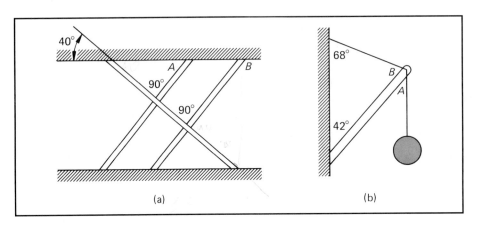

(a)

(b)

FIG. 1-15

Summary

This chapter was intended as a review of the mathematics necessary for industrial applications. Now that you have completed the unit, you should go back and review the chapter objectives. If you feel uneasy about any of the objectives, perhaps another review would be helpful. Remember the following points:

- When adding numbers of like sign, we add the absolute values of the numbers and give to the sum the common sign. To add numbers of unlike sign, we find the difference of their absolute values and affix the sign of the larger.
- To subtract a number b from a number a, we change the sign of number b and then add it to number a, using the addition rule.
- When we multiply or divide a group of signed numbers, the result is negative if the total number of negative factors is odd; otherwise, the result is positive.
- Formulas can be rearranged to solve for a particular unknown by performing similar operations (addition, subtraction, multiplication, division, etc.) to both sides of an equality.
- The following rules apply for exponents and radicals (optional):

$$a^m a^n = a^{m+n} \qquad a^{-n} = \frac{1}{a^n}$$

$$\frac{a^m}{a^n} = a^{m-n} \qquad (a^m)^n = a^{mn}$$

$$(ab)^n = a^n b^n \qquad \left(\frac{a}{b}\right)^n = \frac{a^n}{b^n}$$

$$\sqrt[n]{ab} = \sqrt[n]{a}\,\sqrt[n]{b} \qquad \sqrt[n]{a^m} = a^{m/n}$$

- Scientific notation uses positive or negative powers of 10 to express large or small numbers in shorthand notation.

■ Graphs are used to give a continuous description of the relationship between two variables, based on observed data.
■ When two straight lines intersect, they form opposing angles which are equal.
■ When a straight line intersects two parallel lines, the alternate interior angles are equal.
■ For any triangle, the sum of the interior angles is 180°; for a right triangle, the sum of the smaller angles is 90°.

Following each chapter summary are review questions and problems. These are in addition to the exercises within the chapter and give broad coverage to the general topics.

Questions

1-1. Define the following terms:

a. formula	**f.** parallel
b. scientific notation	**g.** perpendicular
c. exponent	**h.** triangle
d. base	**i.** right triangle
e. radical	

1-2. The sum of two numbers is always greater than their difference. Can you defend or refute this statement?

1-3. A negative number raised to an odd power will always be negative. Is this a true statement? Why?

1-4. What are two ways in which positive and negative signs are used? When substituting signed numbers into a formula that contains the operations of addition and subtraction, what precautions should you take?

1-5. When you transpose a term from one side of an equation to the other side, the sign is changed. Explain how this procedure works and why it works.

1-6. Cross multiplication is sometimes used in formula rearrangement where one fraction is equal to one fraction. For example,

$$\frac{a}{b} = \frac{c}{d} \qquad \text{becomes} \qquad ad = cb$$

Explain why this procedure works and discuss the dangers involved.

1-7. A common error in formula rearrangement is canceling *terms* instead of *factors*. The following are *not* permitted:

$$\frac{x+y}{x} \neq y \qquad \frac{x^2+y^2}{x+y} \neq \frac{x+y}{1}$$

Choose values of $x = 4$ and $y = 2$, then substitute to see why the above operations are not allowed.

Problems

1-1. A barometer reads 30.21 in. before a storm. It then drops 0.59 in. What is the later reading?

Answer 29.62 in.

1-2. A thermometer reads 29°C in a room. After a time in a freezer, it reads −15°C. What is the change in temperature?

1-3. The time required for a piece of metal to cool from 120 to 38°F after immersion in water is 20 min. Fifteen minutes before the metal was immersed in water, it was taken from a furnace. What was the time required for the entire process, and what was the change in temperature of the metal while it was in the water?

Answer 35 min, 82°F.

1-4. A length of wood is cut into six pieces, each 4¾ in. long. Each cut wastes ¹⁄₁₆ in. Determine the length of the original board.

1-5. The volume V of a cylinder is equal to 3.14 times the square of its radius r multiplied by its height h. Write a formula and solve for h.

$$Answer\ h = \frac{V}{3.14r^2}.$$

1-6. The centripetal force F_c is found by multiplying the mass m by the square of the velocity v and then dividing by the radius r of the circle. Write the formula and solve it for r.

1-7. Solve the equation $xb + cd = a(x + 2)$ for x, and find the value of x when $a = 2$, $b = -2$, $c = 3$, and $d = -1$.

$$Answer\ x = \frac{2a - cd}{b - a};\ x = -\frac{7}{4}.$$

1-8. Solve for b when $c^2 = a^2 + b^2$ and the values of c and a are 50 and 20, respectively.

*1-9. Given the formula $F = Gm_1m_2/R^2$ and the values $G = 6.67 \times 10^{-11}$, $m_1 = 4 \times 10^{-8}$, $m_2 = 3 \times 10^{-7}$, and $R = 4 \times 10^{-2}$, find the magnitude of F.

Answer 5×10^{-22}.

*1-10. After a rod has been heated, its new length L is given by

$$L = L_o + \alpha L_o(t_2 - t_1)$$

Assume that $L_o = 21.41$ cm, $\alpha = 2 \times 10^{-3}/°C$, $t_1 = 20°C$, and $t_2 = 100°C$. What is the value of L?

1-11. The voltage [in volts (V)] and the current [in milliamperes (mA)] for a particular resistance were recorded as follows:

Voltage, V	10	20	30	40	50	60
Current, mA	145	289	435	581	724	870

Plot a graph of these data. What would you expect the current to be for applied voltages of 26 and 48 V?

Answer 377 mA, 696 mA.

1-12. The pressure in a storage tank is dependent on the temperature. The following measurements are recorded:

Temperature, K	300	350	400	450	500	550
Pressure, pounds per square inch (lb/in.²)	400	467	535	598	668	733

Plot a graph of these data. What would you expect the pressure to be at the temperatures of 420 and 600 kelvins (K)?

1-13. Determine the unknown angles in Fig. 1-16*a* and *b*.

Answer (a) $A = 30°$, $B = 60°$, $C = 60°$;
(b) $A = 60°$, $B = 30°$, $C = 120°$, $D = 60°$.

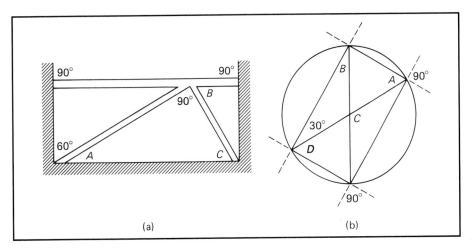

(a)

(b)

FIG. 1-16

2

TECHNICAL MEASUREMENT

As a result of completing this chapter, you should be able to

1. **Write the fundamental units for mass, length, and time in the International System (SI) and U.S. Customary System (USCS) units.**
2. **Define and apply the SI prefixes which indicate multiples of base units.**
3. **Convert from one unit to another unit for the same quantity when given the necessary definitions.**
4. **Calculate the surface area of a square, a rectangle, a triangle, and a circle when given the necessary dimensions.**
5. **Calculate the volume of a cube, a rectangular solid, and a cylinder when given the necessary dimensions.**
6. **Define mass and weight and discuss the use of the kilogram and the pound to describe objects.**

The application of physics, whether in the shop or in a large industrial plant, always requires measurements of some kind. An automobile mechanic might be required to measure the diameter, or *bore,* of an engine cylinder (Fig. 2-1). In air conditioning and refrigeration, the technician is concerned with the volume, pressure, and temperature of fluids. The television repairer depends on accurate instruments which measure elec-

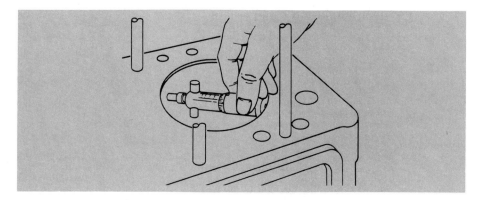

FIG. 2-1 Measuring the diameter, or the bore, of an engine cylinder.

tric current and voltage. In fact, it is difficult to imagine any occupation which does not routinely use numbers and units in some fashion. The accuracy of any of these measurements frequently determines the success or failure of a project.

2-1 ■ PHYSICAL QUANTITIES

The language of physics and technology is universal. Facts and laws must be expressed in an accurate and consistent manner if everyone is to mean exactly the same thing by the same term. For example, suppose an engine is said to have a piston displacement of 3.28 liters (L) (200 in.^{3}). Two questions must be answered if this statement is to be understood: (1) How is the *piston displacement* measured, and (2) what is the *liter*?

Piston displacement is the volume that the piston displaces, or "sweeps out," as it moves from the bottom of the cylinder to the top. It is really not a displacement in the usual sense of the word; it is a volume. A standard measure for volume that is easily recognized throughout the world is the liter. Therefore, when an engine has a label on it that reads "piston displacement = 3.28 L," all mechanics will give the same meaning to the label.

In the above example, the piston displacement (volume) is an example of a *physical quantity*. Notice that this quantity was defined by describing the procedure for its measurement. In physics, all quantities are defined in this manner. Other examples of physical quantities are length, weight, time, speed, force, and mass.

A physical quantity is measured by comparison with some known standard. For example, we might need to know the length of a metal bar. With appropriate instruments we might determine the length of the bar to be 12 ft. The bar does not contain 12 things called "feet"; it is merely compared with the length of some standard known as a "foot." The length could also be represented as 3.66 m or 4 yards (yd) if we used other known measures.

The *magnitude* of a physical quantity is given by a *number* and a *unit* of measure. Both are necessary because either the number or the unit by itself is meaningless. Except for pure numbers and fractions, it is necessary to include the unit with the number when you list the magnitude of any quantity.

> The **magnitude** of a physical quantity is completely specified by a number and a unit, for example, 20 m or 40 L.

Since there are many different measures for the same quantity, we need a way of keeping track of the exact size of particular units. To do this, it is necessary to establish standard measures for specific quantities. A standard is a permanent or an easily determined physical record of the size of a unit of measurement. For example, the standard for measuring electrical resistance, the *ohm* (Ω), might be defined by comparison with a standard resistor whose resistance is accurately known. Thus, a resistance of 20 Ω would be 20 times as great as that of a standard 1-Ω resistor.

Remember that every physical quantity is defined by telling how it is measured. Depending on the measuring device, each quantity can be expressed in a number of different units. For example, some distance units are *meters, kilometers, miles,* and *feet,* and some speed units are *meters per second, kilometers per hour, miles per hour,* and *feet per second.* Regardless of the units chosen, however, distance must be a *length* and speed must be *length* divided by *time.* Thus, *length* and *length/time* are the *dimensions* of the physical quantities of *distance* and *speed.*

Note that speed is defined in terms of two simpler quantities (length and time). Length and time, however, cannot be defined in lesser terms. Therefore, we say that length and time are *fundamental quantities* and that speed is not a fundamental quantity.

From this discussion, we are led to the notion that there may be a limit to the number of fundamental quantities. If we can reduce all physical measurements to a small number of fundamental measures with standard base units for each quantity, then there will be much less confusion in their application.

2-2 ■ INTERNATIONAL SYSTEM

The International System of units is called *Système International d'Unités* (SI) and is essentially the same as what we have come to know as the *metric system.* The International Committee on Weights and Measures has established seven fundamental quantities and has assigned official base units to each quantity. A summary of these quantities, their base units, and the symbols for the base units is given in Table 2-1.

TABLE 2-1 SI BASE UNITS FOR SEVEN FUNDAMENTAL QUANTITIES AND TWO SUPPLEMENTAL QUANTITIES

Quantity	Unit	Symbol
BASE UNITS		
Length	meter	m
Mass	kilogram	kg
Time	second	s
Electric current	ampere	A
Temperature	kelvin	K
Luminous intensity	candela	cd
Amount of substance	mole	mol
SUPPLEMENTAL UNITS		
Plane angle	radian	rad
Solid angle	steradian	sr

Each of the units listed in Table 2-1 has a specific measurable definition which can be duplicated anywhere in the world. Of these base units only one, the *kilogram,* is currently defined in terms of a single physical sample. This standard specimen is kept at the International Bureau of Weights and Measures in France. Copies of the original specimen have been made for use in other nations. The United States prototype is shown in Fig. 2-2. All other units are defined in terms of reproducible physical events and can be accurately determined at a number of locations throughout the world.

FIG. 2-2 The U.S. standard kilogram, a platinum-iridium cylinder housed in the National Bureau of Standards.

We can measure many quantities, such as volume, pressure, speed, and force, which are combinations of two or more fundamental quantities. However, no one has ever encountered a measurement that cannot be expressed in terms of length, mass, time, current, temperature, luminous intensity, and amount of substance. Combinations of these quantities are referred to as *derived quantities,* and they are measured in derived units. Several common derived units are listed in Table 2-2.

2-3 ■ U.S. CUSTOMARY SYSTEM UNITS

Unfortunately, the SI units are not fully implemented in many industrial applications. The United States is making progress toward the assimilation of SI units. However, wholesale conversions are costly, particularly in many mechanical and thermal applications, and total conversion to the International System will require some time. For this reason it is necessary to be familiar with older units for physical quantities. The USCS units for several important quantities are listed in Table 2-3.

TABLE 2-2 DERIVED UNITS FOR COMMON PHYSICAL QUANTITIES

Quantity	Derived units	Symbol	
Area	square meter	m^2	
Volume	cubic meter	m^3	
Frequency	hertz	Hz	s^{-1}
Mass density (density)	kilogram per cubic meter	kg/m^3	
Speed, velocity	meter per second	m/s	
Angular velocity	radian per second	rad/s	
Acceleration	meter per second squared	m/s^2	
Angular acceleration	radian per second squared	rad/s^2	
Force	newton	N	$kg \cdot m/s^2$
Pressure (mechanical stress)	pascal	Pa	N/m^2
Kinematic viscosity	square meter per second	m^2/s	
Dynamic viscosity	newton second per square meter	$N \cdot s/m^2$	
Work, energy, quantity of heat	joule	J	$N \cdot m$
Power	watt	W	J/s
Quantity of electricity	coulomb	C	
Potential difference, electromotive force	volt	V	J/C
Electric field strength	volt per meter	V/m	
Electric resistance	ohm	Ω	V/A
Capacitance	farad	F	C/V
Magnetic flux	weber	Wb	$V \cdot s$
Inductance	henry	H	$V \cdot s/A$
Magnetic flux density	tesla	T	Wb/m^2
Magnetic field strength	ampere per meter	A/m	
Magnetomotive force	ampere	A	
Luminous flux	lumen	lm	$cd \cdot sr$
Luminance	candela per square meter	cd/m^2	
Illuminance	lux	lx	lm/m^2
Wave number	1 per meter	m^{-1}	
Entropy	joule per kelvin	J/K	
Specific heat capacity	joule per kilogram kelvin	$J/(kg \cdot K)$	
Thermal conductivity	watt per meter kelvin	$W/(m \cdot K)$	
Radiant intensity	watt per steradian	W/sr	
Activity (of a radioactive source)	1 per second	s^{-1}	

TABLE 2-3 USCS UNITS

Quantity	SI unit	USCS unit
Length	meter (m)	foot (ft)
Mass	kilogram (kg)	slug (slug)
Time	second (s)	second (s)
Force (weight)	newton (N)	pound (lb)
Temperature	kelvin (K)	degree Rankine (°R)

It should be noted that even though the foot, slug, pound, and other units are frequently used in the United States, they have been redefined in terms of the SI standard units. Thus, all measurements are presently based on the same standards.

2-4 ■ MEASUREMENT OF LENGTH

The standard SI unit of length, the *meter* (m), was originally defined as one ten-millionth of the distance from the North Pole to the equator. For practical reasons, this distance was marked off on a standard platinum-iridium bar. In 1960, an atomic standard of length was adopted based on the wavelength of light from krypton-86. However, even this standard has now been revised in favor of a standard based on the speed of light. The 17th General Conference on Weights and Measures adopted the following definition in 1983:

> The **meter** is the length of path traveled by a light wave in a vacuum in a time interval of 1/299,792,458 second.

The result of such a definition is to define the speed of light to be exactly equal to:

$$c = 299,792,458 \text{ m/s (exactly)}$$

Of course, we do not need to know the above definition of the meter to make accurate measurements. Many tools, such as simple meter sticks and calipers, are calibrated to agree with the standard.

A distinct advantage of the metric system over other systems of units is the use of prefixes to indicate multiples of the base unit. Table 2-4 defines the accepted prefixes and demonstrates their use to indicate multiples and subdivisions of the meter.

TABLE 2-4 MULTIPLES AND SUBMULTIPLES FOR SI UNITS

Prefix	Symbol	Multiplier	Use
giga	G	$1,000,000,000 = 10^9$	1 gigameter (Gm)
mega	M	$1,000,000 = 10^6$	1 megameter (Mm)
kilo	k	$1000 = 10^3$	1 kilometer (km)
centi	c	$0.01 = 10^{-2}$	1 centimeter (cm)*
milli	m	$0.001 = 10^{-3}$	1 millimeter (mm)
micro	μ	$0.000001 = 10^{-6}$	1 micrometer (μm)
nano	n	$0.000000001 = 10^{-9}$	1 nanometer (nm)
—	Å	$0.0000000001 = 10^{-10}$	1 angstrom (Å)*

*The use of the centimeter and the angstrom is discouraged, but they are still widely used.

From the table, you can determine that

$$1 \text{ m} = 1000 \text{ mm}$$
$$= 100 \text{ cm}$$
$$1 \text{ km} = 1000 \text{ m}$$

The relationship between the millimeter and the *inch* can be seen in Fig. 2-3. By definition, 1 in. is equal to exactly 25.4 mm. This definition and other useful definitions are as follows:

$$1 \text{ in.} = 25.4 \text{ mm} \qquad 1 \text{ m} = 39.37 \text{ in.}$$
$$1 \text{ ft} = 0.3048 \text{ m} \qquad = 3.281 \text{ ft}$$
$$1 \text{ yd} = 0.914 \text{ m} \qquad = 1.094 \text{ yd}$$
$$1 \text{ mi} = 1.61 \text{ km} \qquad 1 \text{ km} = 0.621 \text{ mi}$$

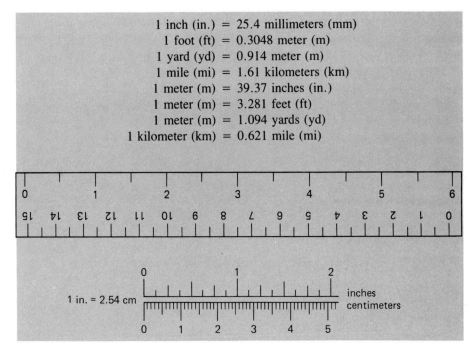

1 inch (in.) = 25.4 millimeters (mm)
1 foot (ft) = 0.3048 meter (m)
1 yard (yd) = 0.914 meter (m)
1 mile (mi) = 1.61 kilometers (km)
1 meter (m) = 39.37 inches (in.)
1 meter (m) = 3.281 feet (ft)
1 meter (m) = 1.094 yards (yd)
1 kilometer (km) = 0.621 mile (mi)

1 in. = 2.54 cm

FIG. 2-3 A comparison of the inch to the centimeter as a measure of length.

EXAMPLE 2-1   The symbol for the derived unit *volt* is V, as seen in Table 2-2. Electric current is measured in amperes (A). An electrician determines that the voltage across a resistor is 50 kilovolts and the current through the resistor is 30 milliamperes. Write the magnitudes of these quantities, using appropriate symbols, and give their meaning in terms of the base unit.

Solution   The prefix *kilo* indicates a multiplier of 1000, and the prefix *milli* indicates a multiplier of 0.001. Using appropriate symbols for prefixes and for the base units, we may write

$$50 \text{ kilovolts} = 50 \text{ kV} \qquad \text{or} \qquad 50,000 \text{ V}$$
$$30 \text{ milliamperes} = 30 \text{ mA} \qquad \text{or} \qquad 0.03 \text{ A}$$

In reporting data, it is preferable to use the prefix that will allow the number to be expressed in the range from 0.1 to 1000. For example, 7,430,000 m should be expressed as 7.43×10^6 m, and then it should be reported as 7.43 Mm. (See Sec. 1-4 on powers of 10.) It would not usually be desirable to write this measurement as 7430 km unless the distance were being compared with other distances measured in kilometers. In the case of the quantity 0.00064 A, it is proper to write either 0.64 mA or 640 μA. Normally, prefixes are chosen for multiples of 1000, such as milli (10^{-3}), kilo (10^3), or micro (10^{-6}). You should review the section on powers of 10 in Chap. 1 as an aid to writing SI units with the appropriate prefixes.

EXAMPLE 2-2   In an automobile engine, the piston clearance is the distance between the piston and the cylinder wall. A mechanic determines that the piston clearance is 0.002 in., or 0.051 mm. Express this measurement in the preferred manner.

Solution The proper prefix is the one which allows the number to be expressed between 0.1 and 1000. First, we note that

$$0.051 \text{ mm} = 0.051 \times 10^{-3} \text{ m}$$

since *milli* implies a multiple of 10^{-3}. Now

$$0.051 \times 10^{-3} \text{ m} = 51 \times 10^{-6} \text{ m}$$

The prefix *micro* (μ) indicates a multiple of 10^{-6}, and so the appropriate expression is 51 μm.

EXERCISES 2-1

1. Give the symbol and the meaning of the following units:
   a. millivolt
   b. microampere
   c. kilowatt
   d. megameter
   e. nanometer
   f. gigameter
2. Give the name and meaning of the units whose symbols are given below.
   a. cV
   b. Mg
   c. mA
   d. kW
   e. μJ
   f. nN
3. Express the following quantities in the proper SI form, using appropriate symbols.
   a. 40 millimeters
   b. 40,000 meters
   c. 6000 micrometers
   d. 0.0003 meter
   e. 0.0007 millimeter
   f. 0.02 kilometer

2-5 ■ *MEASURING INSTRUMENTS*

The choice of a measuring instrument is determined by the accuracy required and by the physical conditions surrounding the measurement. A basic choice for the mechanic or machinist is most often the steel rule, such as the one shown in Fig. 2-4. This rule is often accurate enough when you are measuring openly accessible lengths. Steel rules may be graduated as fine as thirty-seconds or even sixty-fourths of an inch. Metric rules are usually graduated in millimeters.

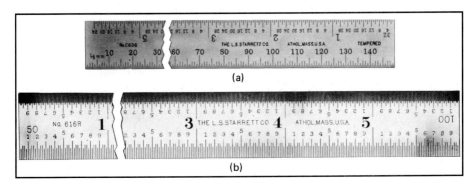

FIG. 2-4 Some 6-in. (15-cm) steel scales. (*a*) Scales 1/32 in. and 0.5 mm. (*The L. S. Starrett Company.*) (*b*) Scales 1/1000 and 1/50 in. (*The L. S. Starrett Company.*)

For the measurement of inside and outside diameters, calipers such as those shown in Fig. 2-5 may be used. The caliper itself cannot be read directly and therefore must be set to a steel rule or a standard-size gauge.

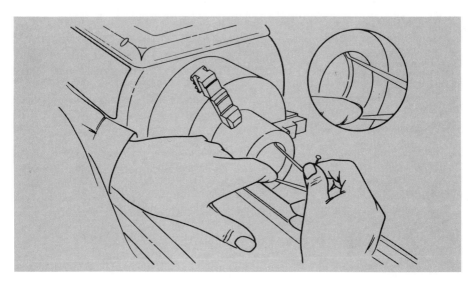

FIG. 2-5 Using calipers to measure an inside diameter.

The best accuracy possible with a steel rule is determined by the size of the smallest graduation and is of the order of 0.01 in., or 0.1 mm. For greater accuracy the machinist often uses a standard micrometer caliper, such as the one shown in Fig. 2-6, or a standard vernier caliper, as in Fig. 2-7. These instruments make use of sliding scales to record very accurate measurements. Micrometer calipers can measure to the nearest ten-thousandth of an inch (0.002 mm), and vernier calipers are used to measure within 0.001 in., or 0.02 mm.

The depth of blind holes, slots, and recesses is often measured with a micrometer depth gauge. Figure 2-8 shows such a gauge being used to measure the depth of a shoulder.

2-6 ■ UNIT CONVERSIONS

Because so many different units are required for a variety of jobs, it is often necessary to convert a measurement from one unit to another. For example, suppose that a machinist records the outside diameter of a pipe as 1³⁄₁₆ in. To order a fitting for the pipe, the machinist may need to know this diameter in millimeters. Such conversions can easily be accomplished by treating units algebraically and using the principle of cancellation.

In the above case, the machinist should first convert the fraction to a decimal:

$$1\tfrac{3}{16} \text{ in.} = 1.19 \text{ in.}$$

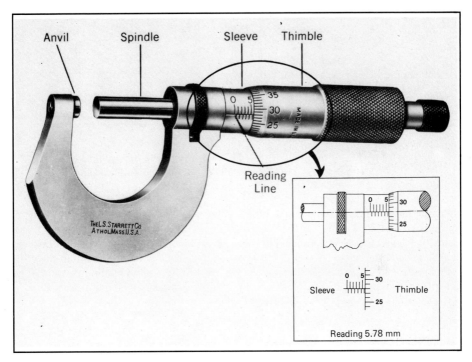

FIG. 2-6 A micrometer caliper, showing a reading of 0.250 in. (*The L. S. Starrett Company.*)

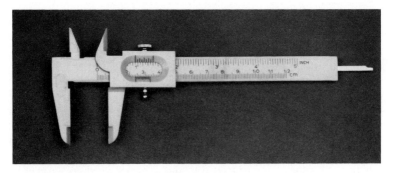

FIG. 2-7 The vernier caliper. (*The L. S. Starrett Company.*)

Next, the machinist should write down the quantity to be converted, giving both the number and the unit (1.19 in.). The definition which relates inches to millimeters is recalled:

$$1 \text{ in.} = 25.4 \text{ mm}$$

Since this statement is an equality, we can form two ratios, each equal to 1:

$$\frac{1 \text{ in.}}{25.4 \text{ mm}} = 1 \qquad \frac{25.4 \text{ mm}}{1 \text{ in.}} = 1$$

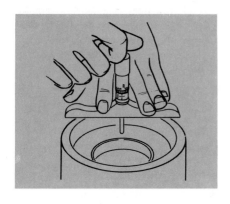

FIG. 2-8 Measuring the depth of a shoulder with a micrometer depth gauge.

Note that the number 1 does not equal the number 25.4, but the *length* of 1 in. is equal to the *length* of 25.4 mm. Thus, if we multiply some other length by either of these ratios, we will get a new number, but we will not change the length. Such ratios are called *conversion factors*. Either of the above conversion factors may be multiplied by 1.19 in. without changing the length represented. Multiplication by the first ratio does not give a meaningful result. Note that units are treated as algebraic quantities.

$$(1.19 \text{ in.}) \left(\frac{1 \text{ in.}}{25.4 \text{ mm}} \right) = \left(\frac{1.19}{25.4} \right) \left(\frac{\text{in.}^2}{\text{mm}} \right) \qquad Wrong!$$

Multiplication by the second ratio, however, gives the following result:

$$(1.19 \text{ in.}) \left(\frac{25.4 \text{ mm}}{1 \text{ in.}} \right) = \frac{(1.19)(25.4)}{1} \text{ mm} = 30.2 \text{ mm}$$

Therefore, the outside diameter of the pipe is 30.2 mm.

The following procedure is used in unit conversions:

1. Write down the quantity to be converted.
2. Define each unit appearing in the quantity to be converted in terms of the desired unit(s).
3. For each definition, form two conversion factors, one being the reciprocal of the other.
4. Multiply the quantity to be converted by those factors which will cancel all but the desired units.

EXAMPLE 2-3   A spool of copper wire contains a total length of 60 m according to the label. An electrician needs 200 ft of copper wire for the windings on a transformer. Does she have enough wire?

Solution   The information needed is the number of feet contained in 60 m of wire. First we should write down the quantity to be converted, then we should note the desired unit:

$$60 \text{ m} \rightarrow \text{ft}$$

If we can define meters in terms of feet (1 m = 3.281 ft), then only one conversion factor is necessary. However, for purposes of illustration, suppose we can only recall that 1 m = 39.37 in. and 12 in. = 1 ft. Possible conversion factors are

$$\frac{1 \text{ m}}{39.37 \text{ in.}}, \quad \frac{39.37 \text{ in.}}{1 \text{ m}} \quad \text{and} \quad \frac{12 \text{ in.}}{1 \text{ ft}}, \quad \frac{1 \text{ ft}}{12 \text{ in.}}$$

We must choose those factors which will cancel "m" and ultimately leave "ft" remaining. Thus,

$$60 \text{ m} \times \frac{39.37 \text{ in.}}{1 \text{ m}} \times \frac{1 \text{ ft}}{12 \text{ in.}} = \frac{(60)(39.37)}{12} \text{ ft}$$

and

$$60 \text{ m} = 197 \text{ ft}$$

If the job requires at least 200 ft of wire, the 60-m spool will not be sufficient.

Sometimes it is necessary to work with quantities which have multiple units. For example, *speed* is defined as *length* per unit *time* and may have units of *meters per second, feet per second,* or other units. The same algebraic procedure can help with conversion of multiple units.

EXAMPLE 2-4   Convert a speed of 60 mi/h to units of feet per second.

Solution   We recall two definitions which might result in four possible conversion factors:

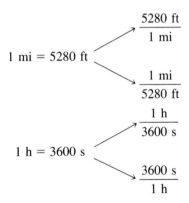

We write down the quantity to be converted, then choose the appropriate conversion factors:

$$60 \, \frac{\text{mi}}{\text{h}} \times \frac{5280 \text{ ft}}{1 \text{ mi}} \times \frac{1 \text{ h}}{3600 \text{ s}} = 88 \text{ ft/s}$$

If you could not recall that 3600 s = 1 h, then you could have used 60 min = 1 h and 1 min = 60 s to obtain the same result.

EXERCISES 2-2

1. Express each of the following length measurements in terms of the units given in parentheses:

    a. 400 mm (m)   **e.** 47 yd (ft)

    b. 20 in. (cm)   **f.** 0.012 in. (yd)

    c. 0.47 ft (mm)   **g.** 470,000 mm (in.)

    d. 0.028 km (m)   **h.** 86 Mm (km)

2. An electrician must install an underground cable from the highway to a home. If the home is located 1.2 mi into the woods, how many feet of cable will be needed?

3. The diameter of a groove on the top of a metal container is measured as 1.94 in. What must be the diameter in millimeters of an O ring designed to fit in this groove?

4. The installation of a television antenna requires 26 yd of antenna wire. If the wire costs 11¢/ft, what is the cost of the lead-in wire?

5. A gallon of white paint will cover a distance of 300 m down the centerline of a highway. How many gallons of paint are required to paint the centerline of a 12-km segment of highway?

6. Water flows from the end of a pipe at a speed of 20 ft/s. Convert this speed to meters per minute.

7. Given that 1 Btu = 252 cal and that heat is being lost at the rate of 500 Btu/h, how many calories (cal) of heat are lost in 1 min?

2-7 ■ MEASUREMENT OF AREA

Area is a measure of the surface of a geometric figure. When we find the area of a figure, we are finding the number of squares, one unit on a side, required to cover the surface. The rectangle in Fig. 2-9 has a length of 5 mm and a width of 3 mm. We say that this rectangle has an area of 15 square millimeters, written 15 mm². It contains 15 squares, each 1 mm on a side.

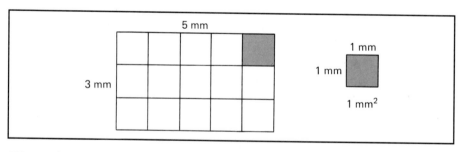

FIG. 2-9 Area is a measure of the surface occupied by a geometric figure. The area shown is 15 mm².

In general, the area of a rectangle is found by multiplying the length by the width (see Fig. 2-10). The formula is

$$A = lw$$

Rectangle

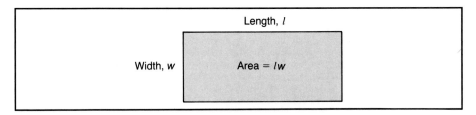

FIG. 2-10

where l and w must be measured in the same units of length and the area A is in square units of length. For example, consider a piece of metal 4 in. wide and 7 in. long. Its area is

$$A = lw = (4 \text{ in.})(7 \text{ in.})$$
$$= (4 \times 7)(\text{in.} \times \text{in.}) = 28 \text{ in.}^2$$

Since a square is a rectangle with all sides of equal length, its area is given by

$$A = s^2$$ *Square*

where s is the length of one side.

The area of a triangle is one-half the *base* of the triangle multiplied by the *height*, or *altitude* (see Fig. 2-11):

$$A = \tfrac{1}{2}bh$$ *Triangle*

The height h is the perpendicular line from a vertex (corner) to the opposite base b of the triangle.

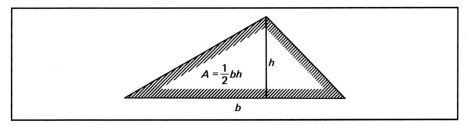

FIG. 2-11 The area of a triangle is one-half the base times the height.

EXAMPLE 2-5 Find the area of the piece of metal stock shown in Fig. 2-12.

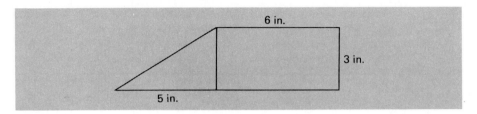

FIG. 2-12

Solution We divide the area of the piece of metal into two regular geometric figures, a triangle and a rectangle. The total area is the sum of these two areas. The rectangle has an area

$$A_1 = lw$$
$$= (6 \text{ in.})(3 \text{ in.}) = 18 \text{ in.}^2$$

Now, the triangle has a base of 5 in. and a height of 3 in. Its area is

$$A_2 = \tfrac{1}{2}bh$$
$$= \tfrac{1}{2}(5 \text{ in.})(3 \text{ in.}) = 7.5 \text{ in.}^2$$

Thus, the total area of the piece is

$$A = A_1 + A_2$$
$$= 18 \text{ in.}^2 + 7.5 \text{ in.}^2 = 25.5 \text{ in.}^2$$

The area of a circle (Fig. 2-13) is expressed in terms of the constant π and the radius of the circle:

$$A = \pi r^2 \qquad\qquad\qquad \textit{Circle}$$

Remember that the radius of a circle is the distance from the center of the circle to its edge. The constant π is approximately 3.1416.

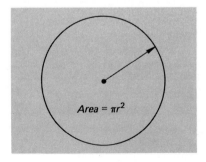

FIG. 2-13

EXAMPLE 2-6 Find the surface area of the face of the washer in Fig. 2-14.

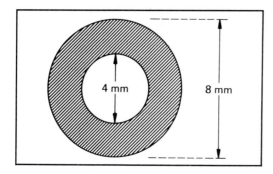

FIG. 2-14

Solution We find the area of the washer by subtracting the area of the hole A_1 from the area of the outside circle A_2. Note that the given measurements are diameters, not radii. Since the radius is one-half of the diameter, the radius of the hole is 2 mm, and the radius of the outside circle is 4 mm. First, we calculate the area of the hole:

$$A_1 = \pi r_1^2 = 3.1416(2 \text{ mm})^2$$
$$= (3.1416)(4 \text{ mm}^2) = 12.6 \text{ mm}^2$$

Now, the outside area is

$$A_2 = \pi r_2^2 = \pi(4 \text{ mm})^2$$
$$= \pi(16 \text{ mm}^2) = 50.3 \text{ mm}^2$$

The surface area of the face of the washer is therefore

$$A = A_2 - A_1$$
$$= 50.3 \text{ mm}^2 - 12.6 \text{ mm}^2 = 37.7 \text{ mm}^2$$

A term often heard in industrial work is the *cross-sectional area* or simply the *cross section*. If a geometric solid is cut by a thin plate, the surface area thus obtained is called the cross-sectional area. The cross section of a wire is the shaded area shown in Fig. 2-15.

It is often necessary to change the area measured in one unit to the corresponding value in another unit. The procedure for using conversion factors based on definitions of area units is the same as that previously discussed for length units. Several conversion factors for area measurements are given below:

$$1 \text{ ft}^2 = 144 \text{ in.}^2$$
$$1 \text{ yd}^2 = 9 \text{ ft}^2$$
$$1 \text{ cm}^2 = 100 \text{ mm}^2$$
$$1 \text{ m}^2 = 10{,}000 \text{ cm}^2$$
$$1 \text{ in.}^2 = 6.4516 \text{ cm}^2$$
$$1 \text{ m}^2 = 1550 \text{ in.}^2$$

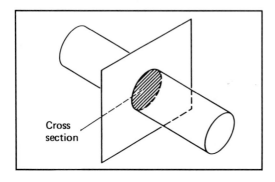

Cross
section

FIG. 2-15 Cross-sectional area.

EXAMPLE 2-7   How many square yards of carpet should be ordered to cover a living room floor that is 24 ft long and 18 ft wide?

Solution   The area is first calculated in square feet:

$$A = (24 \text{ ft})(18 \text{ ft}) = 432 \text{ ft}^2$$

Now, we convert 432 ft² to square yards. Since 1 yd² = 9 ft², we have

$$432 \text{ ft}^2 \times \frac{1 \text{ yd}^2}{9 \text{ ft}^2} = 48 \text{ yd}^2$$

Note that we could find the same answer as follows:

$$432 \text{ ft}^2 \times \frac{1 \text{ yd}}{3 \text{ ft}} \times \frac{1 \text{ yd}}{3 \text{ ft}} = 48 \text{ yd}^2$$

This technique is often useful if you can recall the length definitions but not the area definitions.

EXERCISES 2-3

1. Calculate the surface areas of the metal plates shown in Fig. 2-16.
2. Find the area of a circle whose radius is 2.87 cm.
3. A painter charges 25¢/ft² to paint the end wall of a house. The end of the house is a rectangle resting beneath a triangle under the roof. The left and right sides are each 9 ft high, and the base of the wall is 30 ft long. The highest corner of the wall (at the vertex of the triangle) is 16 ft from the ground. If there are two windows (each 6 ft × 2 ft), how much will the painter charge to paint the end wall?
4. Determine the area of the patterns in Fig. 2-17.
5. Convert 182 cm² to square inches.
6. Convert 48 ft² to square inches.
7. Convert 9.73 m² to square millimeters.
8. A heat loss of 50.0 Btu/h occurs for every square foot of glass. How much heat is lost in 1 h through a circular pane of glass 6 in. in diameter?

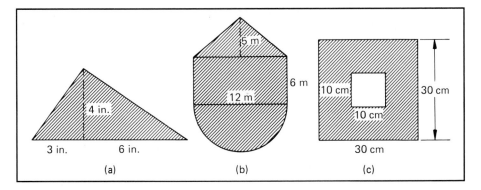

FIG. 2-16

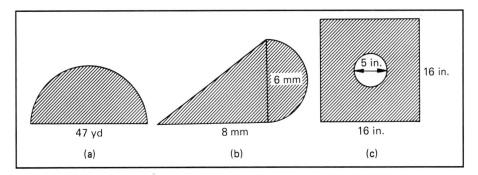

FIG. 2-17

2-8 ■ *MEASUREMENT OF VOLUME*

Volume can be thought of as the amount of space occupied by a particular solid figure or the space occupied by a fluid in a solid container. Finding the volume of a figure consists of finding the number of cubes, 1 unit on each edge, required to fill the figure. For example, the volume of the rectangular solid in Fig. 2-18 contains 30 cubes, each 1 in. on a side. The volume is said to be 30 *cubic inches*, written 30 cu in. or 30 in.^{3}.

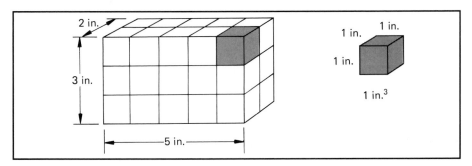

FIG. 2-18 Finding the volume of a solid consists of finding the number of unit cubes it contains. Here the volume is 30 in.^{3}.

In general, the volume of a rectangular solid of length l, height h, and width w is found from

$$V = lwh$$ *Rectangular Solid*

The units of l, w, and h must all be the same, and the volume V is in cubic units of length (refer to Fig. 2-19).

A rectangular solid in which all edges are equal in length is called a *cube*. In such a figure, l, w, and h are equal in length. The volume of a cube is

$$V = e^3$$ *Cube*

where e is the length of any side. For example, a cube 3 mm on each side has a volume of $(3 \text{ mm})^3$, or 27 mm^3.

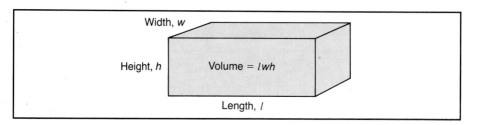

Width, w

Height, h   Volume = lwh

Length, l

FIG. 2-19

EXAMPLE 2-8   Find the volume of oil contained in a rectangular tank 50 cm long, 30 cm wide, and 25 cm deep.

Solution   The inside volume is calculated as though it were filled with a solid. Thus,

$$V = lwh = (50 \text{ cm})(30 \text{ cm})(25 \text{ cm})$$
$$= (50 \cdot 30 \cdot 25)(\text{cm} \cdot \text{cm} \cdot \text{cm}) = 37{,}500 \text{ cm}^3$$

A common metric unit for volume is the liter (L), which is defined as 1000 cm^3 (see Fig. 2-20). One liter, equal to 1.057 qt, is a convenient size for many applications. As mentioned earlier, the liter is often used to specify the volume piston displacement for engines. You should verify that the volume of 37,500 cm^3, calculated in Example 2-8, is equivalent to 37.5 L.

Two other commonly encountered solid shapes are the *right circular cylinder* and the *sphere*. These solid figures are shown in Fig. 2-21 along with the formulas for

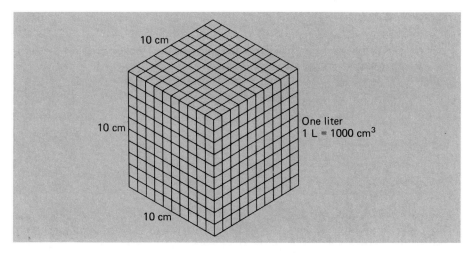

FIG. 2-20 The liter is equal to a volume of 1000 cm³.

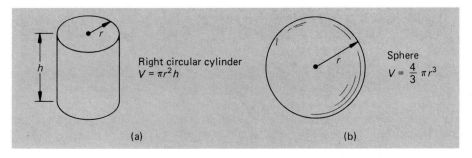

FIG. 2-21

computing their volume. For a cylinder, the volume is simply the area of the base (πr^2) times the height h:

$$V = \pi r^2 h \qquad\qquad \textit{Right Circular Cylinder}$$

For a sphere, the volume is given by

$$V = \tfrac{4}{3}\pi r^3 \qquad\qquad \textit{Sphere}$$

where r is the radius of the sphere. Remember that the radius is always equal to one-half of the diameter.

EXAMPLE 2-9   Find the total piston displacement (volume) of an eight-cylinder engine if each cylinder has a diameter of 4 in. and each piston moves a distance of 3½ in. from the bottom to the top of its stroke.

Solution   The radius of a cylinder is half its diameter, or 2 in. The height of each cylinder, for purposes of computing the volume displaced, is equal to the stroke of the piston, or 3.5 in. For each cylinder, we have

$$V = \pi r^2 h = (3.14)(2 \text{ in.})^2(3.5 \text{ in.}) = (3.14)(4 \text{ in.}^2)(3.5 \text{ in.}) = 44.0 \text{ in.}^3$$

Since there are eight such cylinders, the total volume displacement is

$$V_T = 8(44.0 \text{ in.}^3) = 352 \text{ in.}^3$$

In the metric system, this displacement is equivalent to 5.76 L.

TABLE 2-5 VOLUME MEASUREMENTS

1 ft^3 = 1728 in.^{3}	1 m^3 = 1,000,000 cm^3
1 yd^3 = 27 ft^3	1 ft^3 = 28.32 L
1 cm^3 = 1000 mm^3	1 L = 1.057 qt
1 L = 1000 cm^3	1 gal = 3.78 L

A summary of common volume units with appropriate definitions is given in Table 2-5. Unit conversions are accomplished in the usual manner. For example, 180 mm^3 is converted to cubic centimeters as follows:

$$180 \text{ mm}^3 \times \frac{1 \text{ cm}^3}{1000 \text{ mm}^3} = 0.18 \text{ cm}^3$$

EXERCISES 2-4

1. One U.S. gallon is a volume equivalent to 231 in.^{3}. Suppose a gasoline tank in an automobile is roughly equivalent to a rectangular solid 24 in. long, 18 in. wide, and 12 in. high. How many gallons will this tank hold?

2. An air-conditioning unit has a compressor with a 2-in. bore and a 3-in. stroke. What is the piston displacement (volume) (*a*) in cubic inches; (*b*) in liters?

3. What is the volume of gas contained in a spherical balloon 4 m in diameter?

4. What is the volume of a cube of side 0.43 m (*a*) in cubic meters; (*b*) in liters?

5. A concrete slab is to be poured to form the floor of a garage 20 ft wide and 15 ft deep. If the foundation is to be 4 in. thick, how many cubic yards of ready-mixed concrete should be ordered?

6. An oil storage tank is 4 m in diameter and 6 m high. How many liters of oil can be stored?

2-9 ■ MASS AND WEIGHT

Often it is necessary to describe the amount of material needed for a certain job. For example, how much copper is needed to produce 20 ft of 16-gauge electric wire? How much wax is needed to polish a car? What amount of gasoline will fill a given storage tank? In such cases, we are interested in a quantity of matter which is called the *mass* of an object. The internationally accepted unit of mass is the kilogram (kg). Thus, a 4-kg block of steel is recognized anywhere in the world as the same quantity of steel.

> One **kilogram** is that mass which is equivalent to the mass contained in the standard kilogram specimen located at the International Bureau of Weights and Measures in France.

We will see later that larger masses are more difficult to move, and we will be able to better define mass in terms of force and motion.

One important property of masses is their attraction to other masses. For example, any object near the earth will have a force exerted on it by the earth. This gravitational attraction that the earth has on an object is defined as the *weight* of the object. Unfortunately, the effects of gravity are not the same at different locations on the earth. Objects located at greater distances from the center of the earth will experience smaller attractions. In other words, the weight of an object on a mountaintop is slightly less than the weight of the same object at sea level. For this reason, the *weight* of objects should not be used as an indicator of quantity of material.

The fundamental unit of force in USCS units is the *pound,* and since weight is a force, the unit of weight must also be the pound. To avoid the problem of change in weight at different locations on the earth, objects are described in terms of their masses. The *pound-mass* (lb_m) is that quantity of matter which will have a weight of one pound when located at sea level on the earth. In other words, a box of nails weighing 20 lb at sea level would be labeled ''20 lb_m,'' and this would generally be understood to represent the same quantity of nails whatever its location. The pound-mass can be converted to kilograms by using the conversion

$$1 \ lb_m = 0.454 \ kg$$

Unfortunately, the subscript ''m'' is usually not included in descriptive material, and the term *weight* is used instead of *mass*. For example, you might read in a catalog ''Shipping weight: 45 lb.'' A more accurate label would be ''Shipping mass: 45 lb_m.'' But the numerical difference between 1 lb_m and 1 lb is very small near the earth's surface. You should remember that nearly all descriptions and pricing of objects by the pound are really based on the mass unit lb_m.

In physics, the term *pound-mass* is outdated and its use is not recommended. It is mentioned here only because, in the United States, many objects are described in this fashion. In a later chapter we discuss other units which are more appropriate for describing matter.

The units commonly used in industry to describe quantities of material are

USCS units	Metric units
1 pound-mass (lb_m) = fundamental unit	1 kilogram (kg) = fundamental unit
1 ounce (oz) = $\frac{1}{16}$ lb_m	1 gram (g) = 0.001 kg
1 ton = 2000 lb_m	1 milligram (mg) = 0.001 g
1 grain (gr) = $\frac{1}{7000}$ lb_m	1 metric ton (t) = 1000 kg

Mass is usually measured by comparing the mass of an object on a balance with objects of known mass (standards). The analytical balance and the triple-beam balance shown in Fig. 2-22 are often used to compare masses. Such instruments may be calibrated to indicate masses in a variety of units.

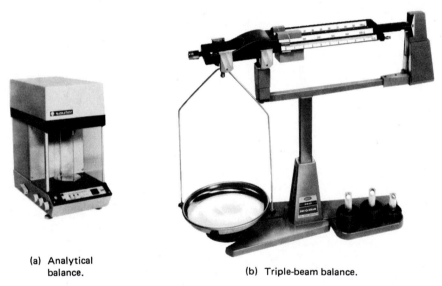

(a) Analytical
balance.   (b) Triple-beam balance.

FIG. 2-22 Two laboratory devices for determining masses. (*Central Scientific Co.*)

2-10 ■ UNIT ANALYSIS

When you are working with technical formulas, it is always helpful to substitute units as well as numbers. For example, the formula for speed v is

$$v = \frac{s}{t}$$

where s is the distance traveled in a time t. Thus, if a car travels 400 ft in 10 s, we find the speed as follows:

$$v = \frac{400 \text{ ft}}{10 \text{ s}} = 40 \text{ ft/s}$$

Notice that the units of velocity are feet per second, written "ft/s."

Whenever velocity appears in a formula, it must always have units of *length* divided by *time*. These are said to be the *dimensions* of velocity. There may be many different units for a given physical quantity, but the dimensions result from a definition, and they do not change.

In working with formulas, it will be useful to remember two rules concerning dimensions:

Rule 1: *If two quantities are to be added or subtracted, they must be of the same dimensions.*

Rule 2: *The quantities on both sides of an equals sign must be of the same dimensions.*

EXAMPLE 2-10 Show that the formula

$$s = v_0 t + \tfrac{1}{2}at^2$$

is dimensionally correct. The symbol s represents the distance traveled in a time t while the object accelerates at a rate a from an initial speed v_0. Assume that acceleration has units of meters per second squared (m/s^2).

Solution Since the units of a are given, the unit of length must be meters and the unit of time must be seconds. Ignoring the factor $\frac{1}{2}$, which has no dimensions, we substitute units for the quantities in the equation:

$$m = \frac{m}{\cancel{s}}(\cancel{s}) + \frac{m}{\cancel{s}^2}(\cancel{s})^2$$

Notice that both rule 1 and rule 2 are satisfied. Therefore, the equation is dimensionally correct.

The fact that an equation is dimensionally correct is a valuable check. Such an equation still may not be a *true* equality, but at least it is consistent dimensionally.

Summary

Technical measurements are essential to the application of physics. We have listed seven fundamental quantities and the SI units for each of these quantities. You should concentrate heavily on the three fundamental quantities of length, mass, and time, since you will encounter them more often than the others. The following points summarize this unit of study.

■ The SI prefixes used to express multiples and submultiples of the base units are given below:

$$\text{mega (M)} = 10^6 \qquad \text{milli (m)} = 10^{-3}$$
$$\text{kilo (k)} = 10^3 \qquad \text{micro } (\mu) = 10^{-6}$$
$$\text{centi (c)} = 10^{-2} \qquad \text{nano (n)} = 10^{-9}$$

■ To convert one unit to another,

1. Write down the quantity to be converted (number and unit).
2. Recall necessary definitions.
3. Form two conversion factors for each definition.
4. Multiply the quantity to be converted by those conversion factors which cancel all but the desired unit(s).

■ Surface areas may be calculated from the following formulas:

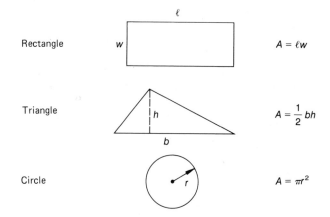

Rectangle $\qquad A = \ell w$

Triangle $\qquad A = \dfrac{1}{2}bh$

Circle $\qquad A = \pi r^2$

■ Volumes for common shapes may be calculated from the following formulas:

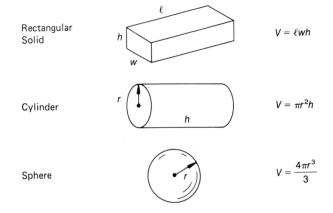

Rectangular Solid $\qquad V = \ell w h$

Cylinder $\qquad V = \pi r^2 h$

Sphere $\qquad V = \dfrac{4\pi r^3}{3}$

- **Weight** is the force (pull) of gravitational attraction that the earth exerts on all objects. Its magnitude is dependent on the location of an object.
- **Mass** is related to the amount of matter contained in an object. Its magnitude is *not* dependent on location.

Questions

2-1. Define the following terms:
- **a.** fundamental quantity
- **b.** length
- **c.** SI units
- **d.** USCS units
- **e.** conversion factor
- **f.** area
- **g.** cross section
- **h.** volume
- **i.** mass
- **j.** weight

2-2. Why are speed, area, and volume not considered fundamental quantities?

2-3. Form two triangles out of a rectangle by drawing a diagonal from an upper corner to a lower corner. Use this figure to show how the area of a triangle is one-half of its base times its height.

2-4. Give the symbol and the meaning of the following units:
- **a.** milliampere
- **b.** nanovolt
- **c.** megawatt

2-5. Express in proper SI form, using appropriate symbols:
- **a.** 1000 volts
- **b.** 2000 millimeters
- **c.** 0.020 kilometer

2-6. What are the dimensions of area, volume, and speed?

2-7. Accept or reject the following equations on the basis of unit analysis. Refer to Example 2-10.
- **a.** $s = v_0 t + \frac{1}{2} at$
- **b.** $2as = v_f^2 - v_0^2$
- **c.** $v_f = v_0 + at^2$
- **d.** $s = v_0 t + 4at^2$

2-8. The reciprocal of a fraction is a fraction with the numerator and denominator inverted. What is the meaning of the reciprocals of the following?
- **a.** mi/gal
- **b.** kg/m^3
- **c.** ft/s
- **d.** 2.54 cm/in.

Problems

2-1. Perform the following conversions:
- **a.** 29.4 cm to meters
- **b.** 80,600 mg to kilograms
- **c.** 350 mi/h to feet per second
- **d.** 600 ft to miles
- **e.** 1220 in. to meters
- **f.** 4400 ft^3/h to cubic meters per second
- **g.** 650 μm to meters
- **h.** 3.98×10^{-10} g to nanograms

Answer 0.294 m; 0.0806 kg; 513 ft/s; 0.114 mi; 31.0 m; 0.0346 m^3/s; 0.00065 m; 0.398 ng.

2-2. A large laboratory has a length of 36 ft and a width of 20 ft. What are these dimensions in meters?

2-3. Convert a speed of 40 km/h to centimeters per second.

Answer 1111 cm/s.

2-4. How many square kilometers are in a square mile?

2-5. The input piston for a hydraulic press has an area of 240 mm^2, and the output piston has an area of 4400 mm^2. Express these areas in square meters.

Answer 2.4×10^{-4} m^2, 4.40×10^{-3} m^2.

2-6. One "square" of roofing shingles covers 100 ft^2 of roof area. How many squares of shingles must be ordered to cover a square roof which is 68 ft on a side?

2-7. A cylindrical tank is 35 cm in diameter and 90 cm tall. Find the volume in cubic centimeters and in liters.

Answer 86,590 cm^3; 86.6 L.

2-8. How many cubic yards of concrete must be ordered to pour a 4-in.-thick slab that measures 24 ft by 32 ft?

2-9. The density of the earth is defined as its mass divided by its volume. Assume that the mass of the earth is 5.98×10^{24} kg and that its radius is 6.38×10^6 m. What is the density of the earth in kilograms per cubic meter?

Answer 5500 kg/m^3.

2-10. A chain-link fence costs $2.59 a linear foot. The distance around a certain lot is exactly 2 mi. Two gates, each 4 ft wide, must be installed at a cost of $100.00 per gate. What is the cost of fencing this property?

2-11. A Datsun engine has a displacement of 1595 cm^3 and a bore diameter of 83 mm. What are these measurements in cubic inches and in inches?

Answer 97.3 in.^{3}; 3.27 in.

2-12. A 50-ft length of extension cord has a shipping weight of 3 lb 2 oz. What is the mass in kilograms of a 28-ft section of this cord?

2-13. A Volkswagen engine has four cylinders, each with a bore diameter of 77 mm. If the piston stroke is 6.4 cm, what is the total piston displacement for the engine?

Answer 1190 cm^3.

2-14. A unit commonly used to measure cross sections of wire is the square mil, where 1 mil = 0.001 in. The diameter of 16-gauge electric wire is 50.82 mils. What is the cross section in square inches?

2-15. One coulomb (C) is a unit of charge equal to 6.25×10^{18} electrons. Electric current is measured in amperes (A), where one ampere is defined as one coulomb per second (C/s). If the current is 40 μA, how many electrons pass a given point in 1 min?

Answer 1.50×10^{16}.

2-16. Gasoline weighs about 5.6 lb/gal. How many kilograms of gasoline are contained in a liter?

2-17. The outside dimensions of a freezer are listed as 65⅞ × 29¾ × 30⅛ in. deep. What is the volume of space occupied by the freezer in cubic feet and in liters?

Answer 34.2 ft^3; 967 L.

2-18. Atmospheric pressure is 14.7 lb/in.^{2}. Express this pressure in tons per square foot.

2-19. How many liters of helium are required to completely fill a balloon 6 m in diameter?

Answer 113,000 L.

2-20. A woman weighs 130 lb at sea level and is 5 ft 9 in. tall. What are her mass in kilograms and her height in centimeters?

2-21. A steel beam 5.7 ft long has a cross section 8 in. × 4 in. What volume does the beam occupy in cubic feet and in cubic meters?

Answer 1.27 ft^3; 0.0359 m^3.

2-22. Highway signs indicating a speed limit of 55 mi/h are to become metric. What should be the new reading in kilometers per hour?

2-23. A nautical mile is about 6078 ft, and a knot is one nautical mile per hour. What is a speed of 20 knots in meters per minute?

Answer 618 m/min.

3

FORCE AND VECTORS

As a result of completing this chapter, you should be able to

1. Define a *vector* quantity and a *scalar* quantity and give examples of each.
2. Determine the components of a given vector.
3. Find the resultant of two or more vectors.
*4. Use trigonometry to determine the resultant of a number of vectors by the component method.

In many applications of physics it is necessary to indicate the direction as well as the magnitude of a quantity. The length of a rafter is determined partly by the angle it makes with the ceiling. The direction in which a conveyer belt moves is often just as important as the speed with which it moves. The effect of a pull of 12 lb at an angle with the floor is different from that of a pull of 12 lb parallel to the floor. Such physical quantities as *displacement, velocity,* and *force* are often encountered in industry. In this chapter, the concept of *vectors* is introduced as a method of using diagrams and mathematics to predict the effects of direction. Trigonometry is optional and required only for the later sections.

3-1 ▪ VECTOR AND SCALAR QUANTITIES

Some physical quantities can be totally described by a number and a unit. Only the physical magnitudes are of interest in an area of 12 cm^2, a volume of 40 ft^3, or a distance of 20 km. Such quantities are called *scalar* quantities.

*A **scalar** quantity is completely specified by its magnitude—a number and a unit. Examples are speed (15 mi/h), distance (12 km), and volume (200 cm³).*

Scalar quantities which are measured in the same units may be added or subtracted in the usual way. For example,

$$14 \text{ mm} + 13 \text{ mm} = 27 \text{ mm}$$
$$20 \text{ ft}^2 - 14 \text{ ft}^2 = 6 \text{ ft}^2$$

Some physical quantities, such as force and velocity, have direction as well as magnitude. In such cases, they are called *vector* quantities. The direction must be a part of any calculations involving such quantities.

*A **vector** quantity is completely specified by a magnitude and a direction. It consists of a number, a unit, and a direction. Examples are displacement (20 m, north) and velocity (40 mi/h, 30°north of west).*

The direction of a vector may be given by reference to conventional north, east, west, and south directions. Consider, for example, the vectors 20 m, west, and 40 m at 30°N of E, as shown in Fig. 3-1. The expression *north of east* indicates that the angle is formed by rotating a line northward from the easterly direction.

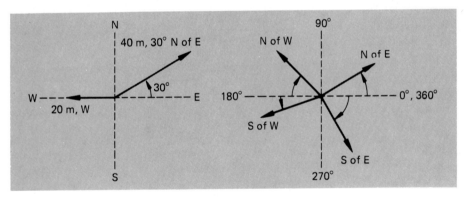

FIG. 3-1 Indicating the direction of a vector by reference to north (N), south (S), east (E), and west (W).

Another method of specifying direction which will be particularly useful later is to make reference to perpendicular lines called *axes*. These imaginary lines are usually chosen to be horizontal and vertical, but they may be oriented along other directions as long as the two lines remain perpendicular. An imaginary horizontal line is usually called the *x* axis, and an imaginary vertical line is called the *y* axis. (See Fig. 3-2.) Directions are given by angles measured counterclockwise from the positive *x* axis. The vectors 40 m at 60° and 50 m at 210° are shown in the figure.

Assume a person travels by car from Atlanta to St. Louis. The *displacement* from Atlanta can be represented by a line segment drawn to scale from Atlanta to St. Louis (see Fig. 3-3). An arrowhead is drawn on the St. Louis end to denote the direction. It

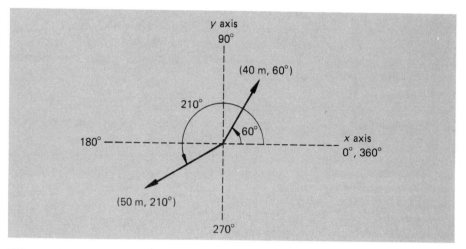

FIG. 3-2 Indicating the direction of a vector as an angle measured from the positive x axis.

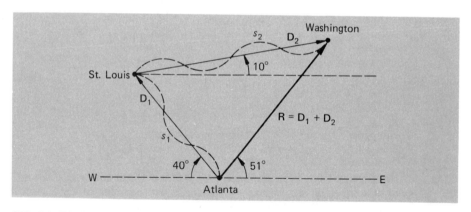

FIG. 3-3 Displacement is a vector quantity. Its direction is indicated by a solid arrow. Distance is a scalar quantity, indicated above by a dotted line.

is important to note that the displacement, represented by the vector D_1, is completely independent of the actual path or the mode of transportation. The odometer would show that the car had actually traveled a scalar distance s_1 of 541 mi. The magnitude of the displacement is only 472 mi.

Another important difference is that the vector displacement has a constant direction of 140° (or 40°N of W). However, the direction of the car at any instant on the trip is not important when we are considering the scalar distance.

Now, let us suppose that our traveler continues the drive to Washington. This time the vector displacement D_2 is 716 mi at a constant direction of 10°N of E. The corresponding ground distance s_2 is 793 mi. The total distance traveled for the entire trip from Atlanta is the arithmetic sum of the scalar quantities s_1 and s_2:

$$s_1 + s_2 = 541 \text{ mi} + 793 \text{ mi} = 1334 \text{ mi}$$

However, the *vector sum* of the two displacements $\mathbf{D}_1$ and $\mathbf{D}_2$ must take note of direction as well as magnitudes. The question now is not the distance traveled but the resulting displacement from Atlanta. This vector sum is represented in Fig. 3-3 by the symbol $\mathbf{R}$, where

$$\mathbf{R} = \mathbf{D}_1 + \mathbf{D}_2$$

Methods we will discuss in the next section will allow us to determine the magnitude and direction of $\mathbf{R}$. Using a ruler and a device for measuring angles, we would see that

$$\mathbf{R} = 545 \text{ mi}, \ 51°$$

Remember that in performing vector additions, both the magnitude and direction of the displacements must be considered. The additions are geometric instead of algebraic. It is possible for the magnitude of a vector sum to be less than the magnitude of either of the component displacements.

A vector is usually denoted in print by boldface type. For example, the symbol $\mathbf{D}_1$ denotes a displacement vector in Fig. 3-3. A vector can be denoted conveniently in handwriting by underscoring the letter or by putting an arrow over it. In print, the magnitude of a vector is usually indicated by italics; thus, D denotes the magnitude of the vector $\mathbf{D}$. A vector is often specified by a pair of numbers (R, θ). The first number and unit gives the magnitude, and the second number gives the angle measured counterclockwise from the positive x axis. For example,

$$\mathbf{R} = (R, \ \theta) = (200 \text{ km}, \ 114°)$$

Note that the magnitude R of a vector is always positive. A negative sign before the symbol of a vector merely reverses its direction; i.e., it interchanges the arrow tip without affecting the length. If $\mathbf{A} = (20 \text{ m}, \text{ east})$, then $-\mathbf{A}$ would be $(20 \text{ m}, \text{ west})$.

3-2 ■ ADDITION OF VECTORS BY GRAPHICAL METHODS

In this section we discuss two common graphical methods for finding the geometric sum of vectors. The *polygon method* is the more useful, since it can be readily applied to more than two vectors. The *parallelogram method* is useful for the addition of two vectors at a time. In each case the magnitude of a vector is indicated to scale by the length of a line segment. The direction is denoted by an arrow tip at the end of the line segment.

EXAMPLE 3-1   A ship travels 100 mi due north on the first day of a voyage, 60 mi northeast on the second day, and 120 mi due east on the third day. Find the resultant displacement by the polygon method.

Solution   A suitable scale might be 20 mi = 1 cm, as in Fig. 3-4. Using this scale, we note that

$$100 \text{ mi} = 100 \text{ mi} \times \frac{1 \text{ cm}}{20 \text{ mi}} = 5 \text{ cm}$$

$$60 \text{ mi} = 60 \text{ mi} \times \frac{1 \text{ cm}}{20 \text{ mi}} = 3 \text{ cm}$$

$$120 \text{ mi} = 120 \text{ mi} \times \frac{1 \text{ cm}}{20 \text{ mi}} = 6 \text{ cm}$$

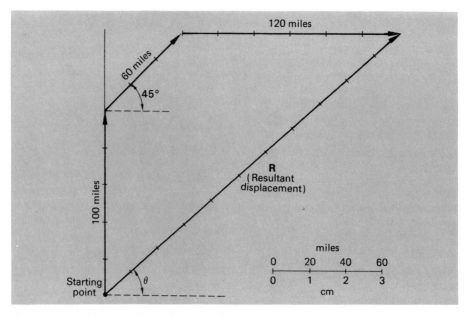

FIG. 3-4 The polygon method of vector addition.

By measuring with a ruler, we find from the scale diagram that the arrow for the resultant is 10.8 cm long. Therefore, the magnitude is

$$10.8 \text{ cm} = 10.8 \text{ cm} \times \frac{20 \text{ mi}}{1 \text{ cm}} = 216 \text{ mi}$$

Measuring the angle θ with a protractor shows the direction to be 41°. The resultant displacement is therefore

$$\mathbf{R} = (216 \text{ mi}, 41°)$$

Note that the order in which the vectors are added does not change the resultant in any way. We could have begun with any of the three distances traveled by the ship.

The polygon method can be summarized as follows:

1. Choose a scale and determine the length of the arrows which correspond to each vector.

2. Draw to scale an arrow representing the magnitude and direction of the first vector.

3. Draw the arrow of the second vector so that its tail is joined to the tip of the first vector.

4. Continue the process of joining tail to tip until the magnitude and direction of all vectors have been represented.

5. Draw the resultant vector with its tail at the origin (starting point) and its tip joined to the tip of the last vector.

6. Measure with ruler and protractor to determine the magnitude and direction of the resultant.

Graphical methods can be used to find the resultant of all kinds of vectors. They are not restricted to measuring displacement. In the next example we determine the resultant of two *forces* by the parallelogram method.

In the parallelogram method, which is useful only for two vectors at a time, these vectors are drawn to scale with their tails at a common origin. (See Fig. 3-5.) The two arrows then form two adjoining sides of a parallelogram. The other two sides are constructed by drawing parallel lines of equal length. The resultant is represented by the diagonal of the parallelogram included between the two vector arrows.

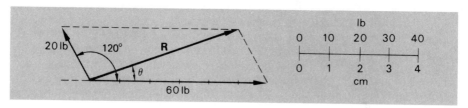

FIG. 3-5 The parallelogram method of vector addition.

EXAMPLE 3-2   A rope is wrapped around a telephone pole, making an angle of 120°. If one end is pulled with a force of 60 lb and the other with a force of 20 lb, what is the resultant force on the telephone pole?

Solution   Using a scale of 1 cm = 10 lb gives

$$60 \text{ lb} \times \frac{1 \text{ cm}}{10 \text{ lb}} = 6 \text{ cm}$$

$$20 \text{ lb} \times \frac{1 \text{ cm}}{10 \text{ lb}} = 2 \text{ cm}$$

A parallelogram is constructed in Fig. 3-5 by drawing the two forces to scale from a common origin with 120° between them. Completing the parallelogram allows the resultant to be drawn as a diagonal from the origin. Measurement of R and θ with a ruler and protractor yields values of 53 lb for the magnitude and 19° for the direction. Hence

$$\mathbf{R} = (53 \text{ lb, } 19°)$$

3-3 ■ *FORCE AND VECTORS*

A push or pull that tends to cause motion is called a *force*. A stretched spring exerts forces on the objects to which its ends are attached, compressed air exerts forces on the walls of its container, and a tractor exerts a force on the trailer it is pulling. Probably the most familiar force is the force of gravitational attraction exerted on every body by the earth. This force is called the *weight* of the body. A definite force exists even though there is no contact between the earth and the bodies it attracts. Weight as a vector quantity is directed toward the center of the earth.

The SI unit of force is the *newton* (N). Its relationship to the USCS unit the *pound* is

$$1 \text{ N} = 0.225 \text{ lb} \qquad 1 \text{ lb} = 4.45 \text{ N}$$

A 120-lb woman has a weight of 534 N. If the weight of a pipe wrench were 20 N, it would weigh about 4.5 lb in USCS units. Until all industries have converted completely to SI units, the pound will still be with us, and frequent conversions are necessary.

Two of the measurable effects of forces are (1) changing the dimensions or shape of a body and (2) changing a body's motion. If in the first case there is no resultant displacement of the body, the push or pull causing the change in shape is called a *static force*. If a force changes the motion of a body, it is called a *dynamic force*. Both types of forces are conveniently represented by vectors, as in Example 3-2.

The effectiveness of any force depends on the direction in which it acts. For example, it is easier to pull a sled along the ground with an inclined rope, as shown in Fig. 3-6, than to push it. In each case, the applied force is producing more than a single effect. That is, the pull on the cord is both lifting the sled and moving it forward. Similarly, pushing the sled would have the effect of adding to the weight of the sled. We are thus led to the idea of *components* of a force, i.e., the effective values of a force in directions other than that of the force itself. In Fig. 3-6, the force **F** can be replaced by its horizontal and vertical components $\mathbf{F}_x$ and $\mathbf{F}_y$.

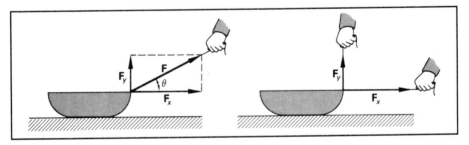

FIG. 3-6 The force F acting at an angle θ can be replaced by its horizontal and vertical components.

If a force is represented graphically by its magnitude and an angle (R, θ), its components along the x and y directions can be determined. A force **F** acting at an angle θ above the horizontal is drawn in Fig. 3-7. The meaning of the x and y components, $\mathbf{F}_x$

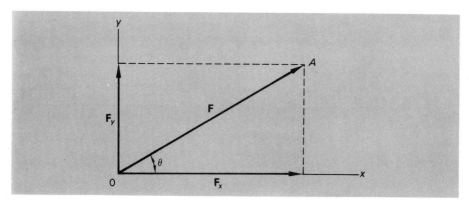

FIG. 3-7 Graphical representation of the x and y components of F.

and $\mathbf{F}_y$, can be seen in this diagram. The segment from O to the perpendicular dropped from A to the x axis is called the *x component* of $\mathbf{F}$ and is labeled $\mathbf{F}_x$. The segment from O to the perpendicular line from A to the y axis is called the *y component* of $\mathbf{F}$ and is labeled $\mathbf{F}_y$. By drawing the vectors to scale, we can determine the magnitudes of the components graphically. These two components acting together would have the same effect as the original force $\mathbf{F}$.

EXAMPLE 3-3 A lawn mower is pushed downward with a force of 40 lb at an angle of 50° with the horizontal. What is the magnitude of the horizontal effect of this force?

Solution A sketch is drawn, as in Fig. 3-8*a*, to translate the word problem to a picture. This approach often aids in understanding the problem. The 40-lb force is transmitted through the handle to the body of the lawn mower. A vector diagram is shown in Fig. 3-8*b*. A ruler and protractor are used to draw the diagram accurately. A scale of 1 cm = 10 lb is convenient for this example. The horizontal effect of the 40-lb force is the x component, as labeled in the figure. Measurement of this line segment gives

$$F_x = 2.57 \text{ cm}$$

Since 1 cm = 10 lb, we obtain

$$F_x = 2.57 \text{ cm} \, \frac{10 \text{ lb}}{1 \text{ cm}} = 25.7 \text{ lb}$$

Notice that the effective force is quite a bit less than the applied force. As an additional exercise, you should show that the magnitude of the *downward* component of the 40-lb force is $F_y = 30.6$ lb.

3-4 ■ RESULTANT FORCE

When two or more forces act at the same point on an object, they are said to be *concurrent forces*. The combined effect of such forces is called the *resultant force*.

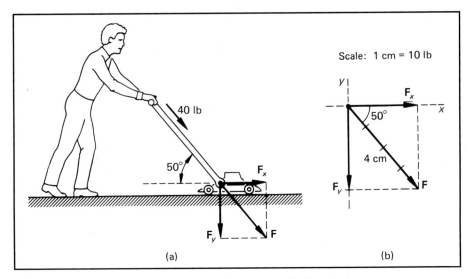

FIG. 3-8 Finding the components of a force by the graphical method.

The **resultant force** *is that single force which will produce the same effect in both magnitude and direction as two or more concurrent forces.*

Resultant forces may be calculated graphically by representing each concurrent force as a vector. The polygon or parallelogram method of vector addition will then give the resultant force.

Often forces act in the same line, either together or in opposition to each other. If two forces act on a single object in the same direction, the resultant force is equal to the sum of the magnitudes of the forces. The direction of the resultant will be the same as that of either force. Consider, for example, a 15-N force and a 20-N force acting in the same easterly direction. Their resultant is 35 N, east, as demonstrated in Fig. 3-9a.

If the above two forces act in opposite directions, the magnitude of the resultant force is equal to the *difference* of the magnitudes of the two forces, and it acts in the direction of the larger force. Suppose the 15-N force in our example were changed so that it pulled to the west. As seen in Fig. 3-9b, the resultant would be 5 N, east.

If forces act at an angle between 0 and 180° to each other, their resultant is the vector sum. The polygon method or the parallelogram method of vector addition may be used to find the resultant force. In Fig. 3-9c, our two forces of 15 and 20 N act at an angle of 60° with each other. The resultant force, calculated by the parallelogram method, is found to be 30.4 N at 34.7°.

EXERCISES 3-1

1. A boat sails 200 km west, then 500 km south. Represent each of these displacements as a vector, choosing a scale of 1 cm = 50 km. Construct a parallelogram, and using a ruler and protractor, determine the resultant displacement of the boat (include its direction).
2. A motorboat heads straight across a river at a speed of 20 mi/h. The river current flows parallel to the bank at a speed of 16 mi/h. Represent each of these velocities as a vector. (A possible scale is 1 in. = 4 mi/h.) Draw the river-speed vector first along a horizontal *x* axis

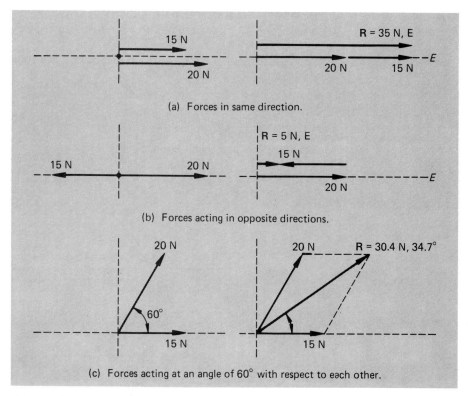

(a) Forces in same direction.

(b) Forces acting in opposite directions.

(c) Forces acting at an angle of 60° with respect to each other.

FIG. 3-9 The effect of direction on the resultant of two forces.

(along the river bank). Now, using the *polygon* method of vector addition, draw in the boat-speed vector. What is the resultant speed of the boat (vector sum)? What angle does the boat make with the river bank as it crosses?

3. What are the horizontal and vertical components of a 600-N force directed to the right and upward at an angle of 38° with the horizontal?

4. A woman walks 300 ft at an angle of 41°N of W. How far is she displaced west and how far north?

5. The wooden boom in Fig. 3-10 exerts a force **F** of 200 lb in the direction shown. What is the component of this force acting perpendicular to the wall?

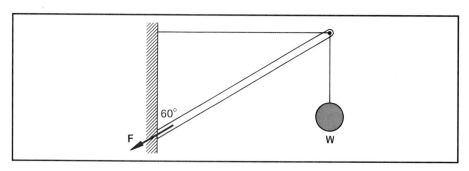

FIG. 3-10

6. In Fig. 3-11, a wooden rod is used to raise a window. What is the effective upward force if the force along the pole is 80 N?

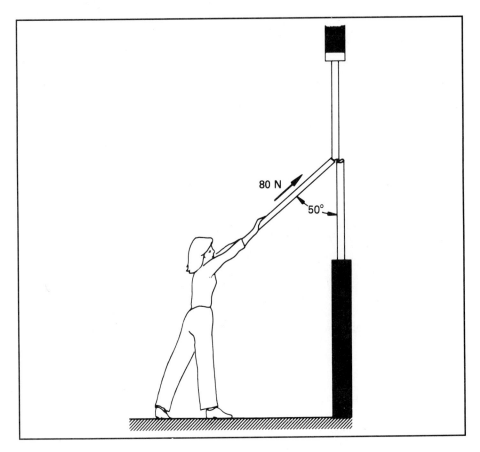

FIG. 3-11

7. Find the resultant (sum) of the following displacements: 400 km, east; 300 km, 30°; and 100 km, south.

8. Forces of 160 N at 0°, 80 N at 120°, and 40 N at 150° all act on the same object. Represent each of these forces as a vector drawn to scale. The resultant force is the vector sum. What are the magnitude and direction of the resultant force?

9. The forward thrust of an automobile is equal to 400 lb. Wind resistance is determined to be equivalent to a force of 20 lb. What is the resultant forward force?

*3-5 ■ TRIGONOMETRY

Often it is necessary to determine the components of forces or a resultant of two or more concurrent forces more accurately than graphs will allow. The graphical approaches are also time-consuming. By learning a few principles that apply to all right triangles, you can significantly improve your ability to work with vectors. Moreover, hand-held calculators make many of the calculations relatively simple.

First, let's review some of the things we already know about right triangles. We will follow the convention that uses Greek letters for angles and Roman letters for sides. Commonly used Greek symbols are

α alpha β beta γ gamma
θ theta ϕ phi δ delta

In the right triangle drawn as Fig. 3-12, the symbols R, x, and y refer to the side dimensions, and θ, ϕ, and 90° are the angles. From Chap. 1, we recall that the sum of the smaller angles in a right triangle is 90°:

$$\phi + \theta = 90° \qquad \textit{Right Triangle}$$

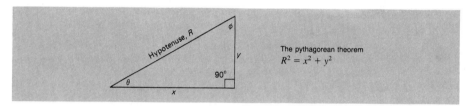

FIG. 3-12 The right triangle and the pythagorean theorem.

There is also a relationship between the sides, which is known as the pythagorean theorem:

Pythagorean theorem: *The square of the hypotenuse is equal to the sum of the squares of the other two sides.*

$$R^2 = x^2 + y^2 \qquad \textit{Pythagorean theorem} \quad (3\text{-}1)$$

The *hypotenuse* is defined as the longest side. It is conveniently located as that side directly opposite the right anngle—the line joining the two perpendicular sides.

EXAMPLE 3-4   What length of guy wire is needed to stretch from the top of a 40-ft telephone pole to a ground stake located 60 ft from the foot of the pole?

Solution   Draw a rough sketch, such as Fig. 3-13. Identify the length R as the hypotenuse of a right triangle; then from the pythagorean theorem

$$R^2 = (60 \text{ ft})^2 + (40 \text{ ft})^2$$
$$= 3600 \text{ ft}^2 + 1600 \text{ ft}^2 = 5200 \text{ ft}^2$$

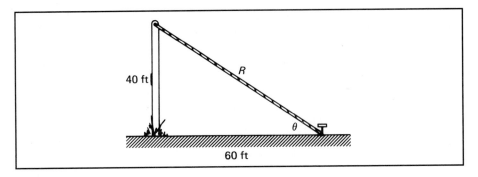

40 ft

R

θ

60 ft

FIG. 3-13

Taking the square root of both sides gives

$$R = \sqrt{5200 \text{ ft}^2} = 72.1 \text{ ft}$$

In general, to find the hypotenuse we could express the pythagorean theorem as

$$R = \sqrt{x^2 + y^2} \qquad \qquad \textit{Hypotenuse} \quad (3\text{-}2)$$

On some electronic calculators, the sequence of entries might be as follows:

$$x \quad \boxed{x^2} \quad \boxed{+} \quad y \quad \boxed{x^2} \quad \boxed{=} \quad \boxed{\sqrt{x}} \qquad (3\text{-}3)$$

In this instance, x and y are the values of the shorter sides, and the boxed symbols are operation keys on the calculator. You should verify the solution to the previous problem, using $x = 40$ and $y = 60$. (The input procedure varies depending on the make of calculator.)

Of course, the pythagorean theorem can also be used to find either of the shorter sides if the remaining sides are known. Solution for x or for y yields

$$x = \sqrt{R^2 - y^2} \qquad y = \sqrt{R^2 - x^2} \qquad (3\text{-}4)$$

Trigonometry is the branch of mathematics which takes advantage of the fact that similar triangles are proportional in size. In other words, for a given angle the ratio of any two sides is the same regardless of the overall dimensions of the triangle. For the three triangles in Fig. 3-14, the ratios of corresponding sides are equal as long as the angle is 37°. From Fig. 3-14, it is seen that

$$\frac{3}{4} = \frac{x_1}{y_1} = \frac{x_2}{y_2}$$

and

$$\frac{4}{5} = \frac{y_1}{R_1} = \frac{y_2}{R_2}$$

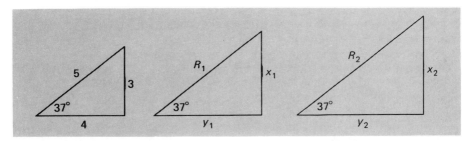

FIG. 3-14 All right triangles with the same internal angles are similar; that is, their sides are proportional.

Once an angle is identified in a right triangle, the sides *opposite* and *adjacent* to that angle may be labeled. The meanings of *opposite, adjacent,* and *hypotenuse* are given in Fig. 3-15. You should study this figure until you understand fully the meaning of these terms. Verify that the side opposite to θ is y and that the side adjacent to θ is x. Also notice that the sides described by *opposite* and *adjacent* change if we refer to angle ϕ.

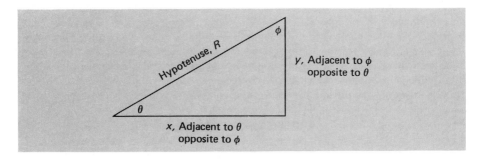

FIG. 3-15 Identification of sides in a right triangle.

In a right triangle, there are three side ratios that are very important. They are the *sine,* the *cosine,* and the *tangent,* defined as follows for angle θ:

$$\sin \theta = \frac{opp\ \theta}{hyp}$$

$$\cos \theta = \frac{adj\ \theta}{hyp} \tag{3-5}$$

$$\tan \theta = \frac{opp\ \theta}{adj}$$

To make sure that you understand these definitions, you should verify the following for the triangles in Fig. 3-16:

$$\sin \theta = \frac{9}{15} \qquad \cos \gamma = \frac{m}{H} \qquad \tan \alpha = \frac{y}{x}$$

$$\sin \alpha = \frac{y}{R} \qquad \cos \beta = \frac{n}{H} \qquad \tan \phi = \frac{12}{9}$$

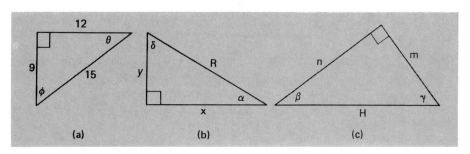

FIG. 3-16

You should first identify the right angle, then label the longest side (opposite the 90° angle) as the hypotenuse. Then, for a particular angle, the sides opposite and adjacent may be identified.

In Fig. 3-14, the tangent of 37° is ¾, or 0.75, for any triangle containing that angle. The constant values for the trigonometric functions of all angles between 0 and 90° have been calculated. They may be found from tables or, more conveniently, with a small calculator. Suppose we wish to know the cosine of 47°, for example. In a table of cosines we would see the number 0.682 adjacent to the angle 47°, and we would write

$$\cos 47° = 0.682$$

With some calculators, we would enter the angle 47, then strike the $\boxed{\cos}$ key to obtain the same result. From either a table or from your calculator, you should verify the following:

$$\tan 38° = 0.781 \qquad \cos 31° = 0.857$$
$$\sin 22° = 0.375 \qquad \tan 65° = 2.14$$

To find the angle whose tangent is 1.34 or to find the angle whose sine is 0.45, we would reverse the above process. On a calculator, for example, we would enter the number 1.34, then we would look for one of the following sequences, depending on the calculator: $\boxed{\text{INV}}$ $\boxed{\tan}$, $\boxed{\text{ARC}}$ $\boxed{\tan}$, or $\boxed{\tan^{-1}}$. Any of these will give the angle whose tangent is the entered value. In the above examples, we find that

$$\tan \theta = 1.34 \qquad \theta = 53.3°$$
$$\sin \theta = 0.45 \qquad \theta = 26.7°$$

You should now be able to apply trigonometry to find unknown angles or sides in a right triangle. The following procedure will be helpful:

APPLICATION OF TRIGONOMETRY

1. Draw the right triangle from the stated conditions of the problem. (Label all sides and angles with either their known values or a symbol for an unknown value.)
2. Isolate an angle for study; if an angle is known, choose that one.
3. Label each side according to whether its relation to the chosen angle is opp, adj, or hyp.
4. Decide which side or angle is to be found.
5. Recall the definitions of the trigonometric functions:

$$\sin = \frac{opp}{hyp} \qquad \cos = \frac{adj}{hyp} \qquad \tan = \frac{opp}{adj}$$

6. Choose that trigonometric function which involves (a) the unknown quantity and (b) no other unknown quantity.
7. Write the trigonometric equation and solve for the unknown.

EXAMPLE 3-5   What is the length of the rope segment x in Fig. 3-17?

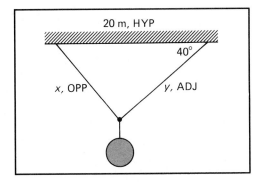

FIG. 3-17

Solution   The sketch is already drawn, so we proceed with step 2. The 40° angle is selected for reference, and *opp, adj,* and *hyp* are labeled on the figure. The decision is made to solve for x (*opp*). Since the sine function involves *opp* and *hyp,* we choose that function and write the equation

$$\sin 40° = \frac{x}{20 \text{ m}}$$

Solving for x by multiplying both sides by 20 m, we obtain

$$x = (20 \text{ m})(\sin 40°)$$

If we use tables, we determine that sin 40° = 0.643, and then

$$x = (20 \text{ m})(0.643) = 12.9 \text{ m}$$

On some calculators, we would calculate x as follows:

$$(20 \text{ m})(\sin 40°) = 20 \quad \boxed{\times} \quad 40 \quad \boxed{\sin} \quad \boxed{=} \quad = 12.9 \text{ m}$$

It is easy to see what an advantage the calculator is in the application of trigonometry. As an additional exercise, you might show that $y = 15.3$ m.

EXAMPLE 3-6   Refer to Fig. 3-13. What is the angle θ that the guy wire makes with the ground?

Solution   Note the location of the angle θ; then label *opp*, *adj*, and *hyp*. For the angle θ, we may write

$$\tan \theta = \frac{opp}{adj} = \frac{40 \text{ ft}}{60 \text{ ft}}$$

The tangent function was the only one that involved the two *known* sides. If tables are to be used, then θ is the angle whose tangent is $^{40}\!/_{60}$, or 0.667. In this case, we find that

$$\tan \theta = 0.667 \quad \text{and} \quad \theta = 33.7°$$

The sequence on some calculators might be as follows:

$$40 \quad \boxed{\div} \quad 60 \quad \boxed{=} \quad \boxed{\text{INV}} \quad \boxed{\text{tan}}$$

Of course, the procedure varies. Study your manual and work with your calculator until you understand the process thoroughly.

*EXERCISES 3-2

In Exercises 1 through 18, use your calculator or tables to evaluate each example.

1. $\sin 67°$
2. $\cos 48°$
3. $\tan 59°$
4. $\sin 34°$
5. $\cos 29°$
6. $\tan 15°$
7. $20 \cos 15°$
8. $400 \sin 21°$

9. $600 \tan 24°$
10. $170 \cos 79°$
11. $240 \sin 78°$
12. $1400 \tan 60°$
13. $\dfrac{200}{\sin 17°}$
14. $\dfrac{300}{\sin 60°}$

15. $\dfrac{167}{\cos 78°}$
16. $\dfrac{256}{\cos 16°}$
17. $\dfrac{670}{\tan 17°}$
18. $\dfrac{2000}{\tan 51°}$

In Exercises 19 through 27, determine the unknown angles, using your tables or a calculator.

19. $\sin \theta = 0.811$
20. $\sin \theta = 0.111$
21. $\tan \theta = 1.2$
22. $\tan \theta = 0.511$

23. $\cos \beta = 0.228$
24. $\cos \theta = 0.81$
25. $\cos \theta = \dfrac{400}{500}$

26. $\tan \theta = \dfrac{16}{4}$
27. $\sin \phi = \dfrac{140}{270}$

In Exercises 28 through 35, solve the triangles for the unknown angles and sides.

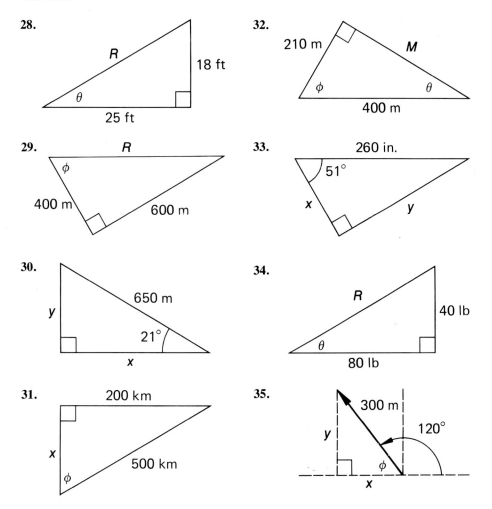

28.

R

18 ft

θ

25 ft

29.

R

φ

400 m

600 m

30.

650 m

y

21°

x

31.

200 km

x

500 km

φ

32.

210 m

M

φ

θ

400 m

33.

260 in.

51°

x

y

34.

R

40 lb

θ

80 lb

35.

300 m

y

120°

φ

x

In Exercises 36 through 42, solve the given problems.

36. A man walks 200 km north and then 400 km east. Use the pythagorean theorem to determine his resultant displacement.

37. A triangular piece of metal has perpendicular sides of 40 and 68 mm. What is the length of the longest edge?

38. The length of a guy wire is 200 ft, and the cable stretches from a ground stake to the top of a 90-ft tower. What is the distance from the base of the vertical tower to the ground stake? What angle does the cable make with the ground?

39. A ladder 22 ft long leans against the wall of a house. If the top of the ladder rests at a point on the wall 16 ft above the ground, what angle does the ladder make with the wall?

40. A loading platform is 4.86 m above the ground. What length ramp is needed if it must make an angle of 20° with the ground?

41. A force of 600 lb acts in an upward direction of 38°. Represent this force as a vector, and determine its horizontal and vertical components.

42. A force of 20 N acts perpendicular to another force of 13 N. If the forces act together on the same object, find the resultant force, using trigonometry instead of graphs.

*3-6 ■ TRIGONOMETRY AND VECTORS

Trigonometric methods can improve your accuracy and speed in determining the resultant vector or in finding the components of a vector. In most cases, it is helpful to use imaginary x and y axes when you are working with vectors in an analytical way. Any vector can then be drawn with its tail at the center of these imaginary lines. Components of the vector might be seen as effects along the x or y axis.

EXAMPLE 3-7 What are the x and y components of a force of 200 lb at an angle of 60°?

Solution A diagram is drawn placing the tail of the 200-lb vector at the center of the x and y axes (see Fig. 3-18).

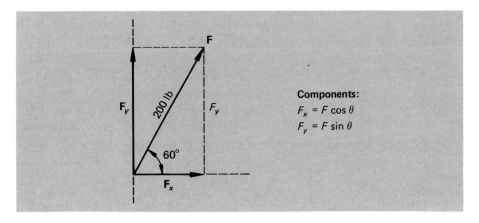

FIG. 3-18 Using trigonometry to find the x and y components of a vector.

We first compute the x component, F_x, by noting that it is the side adjacent. Thus,

$$\cos 60° = \frac{F_x}{200 \text{ lb}}$$

or

$$F_x = (200 \text{ lb})(\cos 60°) = 100 \text{ lb}$$

For purposes of calculation, we recognize that the side opposite to 60° is equal in length to F_y. Then we may write

$$\sin 60° = \frac{F_y}{200 \text{ lb}}$$

or

$$F_y = (200 \text{ lb})(\sin 60°) = 173.2 \text{ lb}$$

In general, we may write the x and y components of a vector in terms of its magnitude F and direction θ:

$$F_x = F \cos \theta$$
$$F_y = F \sin \theta$$

Components of a Vector   (3-6)

where θ is the angle between the positive x axis and the vector, measured in a counterclockwise direction.

The sign of a given component can be determined from a vector diagram. The four possibilities are shown in Fig. 3-19. The magnitude of the component can be found by using the small reference angle ϕ when angle θ is greater than 90°.

EXAMPLE 3-8   Find the x and y components of a 400-N force at an angle of 220° from the positive x axis.

Solution   Refer to Fig. 3-19c, which describes this problem when $\theta = 220°$. The small angle ϕ is found by reference to 180°:

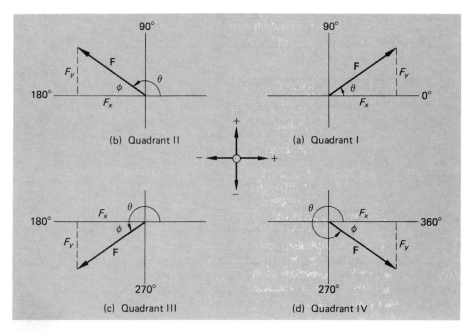

FIG. 3-19 (a) In the first quadrant, angle θ is between 0 and 90°; both F_x and F_y are positive. (b) In the second quadrant, angle θ is between 90 and 180°; F_x is negative, and F_y is positive. (c) In the third quadrant, angle θ is between 180 and 270°; F_x and F_y are negative. (d) In the fourth quadrant, angle θ is between 270 and 360°; F_x is positive, and F_y is negative.

$$\phi = 220° - 180° = 40°$$

From the figure, both F_x and F_y will be negative:

$$F_x = -|F \cos \phi| = -(400 \text{ N})(\cos 40°)$$
$$= -(400 \text{ N})(0.766) = -306 \text{ N}$$
$$F_y = -|F \sin \phi| = -(400 \text{ N})(\sin 40°)$$
$$= -(400 \text{ N})(0.643) = -257 \text{ N}$$

Note that the signs were determined from the figure. With many electronic calculators, both the magnitude and the sign of F_x and F_y can be found directly from Eq. (3-6) by using $\theta = 220°$. You should verify this fact.

Trigonometry is also useful in calculating the resultant force. In the special case for two forces F_x and F_y at right angles to each other, as in Fig. 3-20, the resultant (R, θ) may be found from

$$R = \sqrt{F_x^2 + F_y^2} \qquad \tan \theta = \frac{F_y}{F_x} \qquad (3\text{-}7)$$

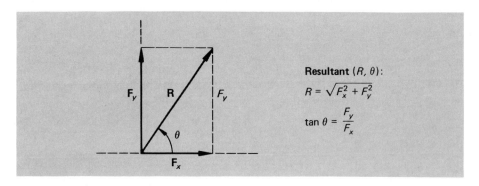

FIG. 3-20 Resultant of perpendicular vectors.

If either F_x or F_y is negative, it is usually easier to determine the small angle ϕ, as described in Fig. 3-19. The sign (or direction) of the forces F_x and F_y determines which of the four quadrants is used. Then, Eq. (3-7) becomes

$$\tan \phi = \left| \frac{F_y}{F_x} \right|$$

Only the absolute values of F_x and F_y are needed. If desired, the angle θ from the positive x axis may be determined. In any case, the direction must be clearly identified.

EXAMPLE 3-9　What is the resultant of a 5-N force directed horizontally to the right and a 12-N force directed vertically downward?

Solution   Label the two forces $F_x = 5$ N and $F_y = -12$ N (downward). Draw a diagram for the situation described by Fig. 3-19*d*. The magnitude of the resultant **R** is found from Eq. (3-7):

$$R = \sqrt{F_x^2 + F_y^2} = \sqrt{(5 \text{ N})^2 + (-12 \text{ N})^2}$$
$$= \sqrt{25 \text{ N}^2 + 144 \text{ N}^2} = \sqrt{169 \text{ N}^2} = 13 \text{ N}$$

To find the direction of R, we first find the small angle ϕ:

$$\tan \phi = \left| \frac{-12 \text{ N}}{5 \text{ N}} \right| = 2.4$$

$$\phi = 67.4° \text{ below positive } x \text{ axis}$$

The angle θ measured counterclockwise from the positive x axis is

$$\theta = 360° - 67.4° = 292.6°$$

The resultant force is 13 N at 292.6°.

*3-7 ■ COMPONENT METHOD OF VECTOR ADDITION

Some of you may require a more advanced treatment of vectors which allows you to find the resultant of forces not at right angles to each other. For the sake of completeness, this section is included. You should consult your instructor before proceeding to determine what is required.

For the more general case in which not all vectors are completely specified along a particular axis, the *component method* of vector addition may be used. Consider the vectors **A, B,** and **C** in Fig. 3-21 on page 78. The resultant **R** is the vector sum **A** + **B** + **C**. However, **A** and **B** are not along an axis and cannot be added in the usual manner. The following procedure may be used:

1. Draw each vector from the center of imaginary x and y axes.
2. Find the x and y components of each vector.
3. Find the x component of the resultant by adding the x components of all vectors. (Components to the right are positive, and those to the left are negative.)

$$R_x = A_x + B_x + C_x \qquad (3\text{-}8)$$

4. Find the y component of the resultant by adding the y components of all vectors. (Upward components are positive, and downward components are negative.)

$$R_y = A_y + B_y + C_y \qquad (3\text{-}9)$$

5. Find the magnitude and direction of the resultant from its perpendicular components R_x and R_y.

$$R = \sqrt{R_x^2 + R_y^2} \qquad \tan \theta = \frac{R_y}{R_x} \qquad (3\text{-}10)$$

The above steps are shown graphically in Fig. 3-21.

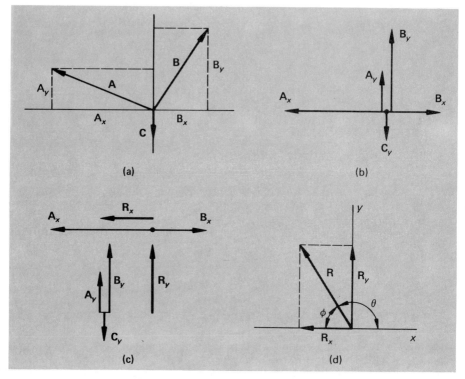

FIG. 3-21 Component method of vector addition.

EXAMPLE 3-10 Three ropes are tied to a stake, and the following forces are exerted: **A** = 20 lb, east; **B** = 30 lb, 30°N of west; and **C** = 40 lb, 52°S of west. Determine the resultant force.

Solution Follow the steps described above.

1. Draw a figure representing each force (Fig. 3-22a). Two things should be noticed from the figure: (a) all angles are determined from the x axis, and (b) the components of each vector are labeled opposite and adjacent to known angles.

2. Find the x and y components of each vector. (Refer to Fig. 3-22b, c, and d.) Note that the force A has no y component. Care must be taken to obtain the correct sign for each component. For example, B_x, C_x, and C_y are each negative. The results are listed in Table 3-1.

3. Add the x components to obtain R_x:

$$R_x = A_x + B_x + C_x$$
$$= 20 \text{ lb} - 26 \text{ lb} - 24.6 \text{ lb} = -30.6 \text{ lb}$$

4. Add the y components to obtain R_y:

$$R_y = A_y + B_y + C_y$$
$$= 0 + 15 \text{ lb} - 31.5 = -16.5 \text{ lb}$$

TABLE 3-1 A TABLE OF COMPONENTS

Force	ϕ_x	x Component	y Component
$A = 20$ lb	$0°$	$A_x = 20$ lb	$A_y = 0$
$B = 30$ lb	$30°$	$B_x = -(30 \text{ lb})(\cos 30°)$ $= -26.0$ lb	$B_y = (30 \text{ lb})(\sin 30°)$ $= 15$ lb
$C = 40$ lb	$52°$	$C_x = -(40 \text{ lb})(\cos 52°)$ $= -24.6$ lb	$C_y = (-40 \text{ lb})(\sin 52°)$ $= -31.5$ lb
		$R_x = \Sigma F_x = -30.6$ lb	$R_y = \Sigma F_y = -16.5$ lb

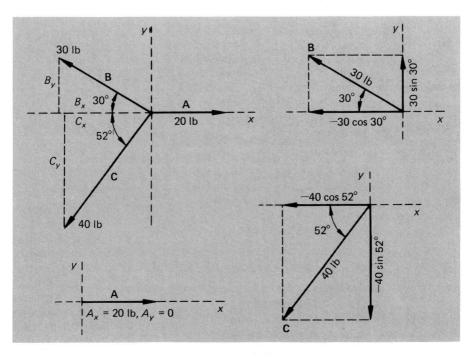

FIG. 3-22 Calculating the x and y components of all vectors.

5. Now we find R and θ from R_x and R_y (see Fig. 3-23 on page 22):

$$R = \sqrt{R_x^2 + R_y^2} = \sqrt{(-30.6)^2 + (-16.5)^2}$$
$$= \sqrt{936.4 + 272.2} = 34.8 \text{ lb}$$

$$\tan \phi = \left| \frac{R_y}{R_x} \right| = \left| \frac{-16.5}{-30.6} \right| = 0.539$$

$$\phi = 28.3°\text{S of W (or } 208.3°)$$

Thus, the resultant force is 34.8 lb at 28.3°S of E.

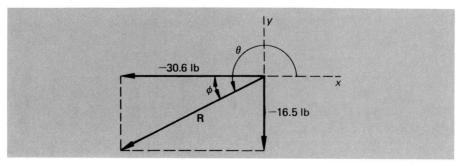

FIG. 3-23

*EXERCISES 3-3

In Exercises 1 through 10, determine the reference angle φ from the given polar angle θ.

1. 41°	**6.** 316°
2. 115°	**7.** 215°
3. 165°	**8.** 200°
4. 133°	**9.** 290°
5. 330°	**10.** 351°

In Exercises 11 through 18, determine the polar angle θ from the given reference angle φ.

11. 60°N of W	**15.** 47°S of W
12. 75°N of W	**16.** 53°S of E
13. 70°S of E	**17.** 10°N of E
14. 35°S of W	**18.** 51°S of W

*In Exercises 19 through 28, determine the x and y components of each vector (**R**, θ). Make sure the signs are correct for each component.*

19. (400 m, 50°)	**24.** (200 N, 46°N of W)
20. (60 lb, 15°)	**25.** (24 lb, 63°S of E)
21. (200 N, 120°)	**26.** (500 km, 30°N of E)
22. (650 mi/h, 220°)	**27.** (1600 mi, 140°)
23. (160 m, 300°)	**28.** (18 m/s, 210°)

*In Exercises 29 through 38, determine the resultant (**R**, θ) given the components (**R**x, **R**y). Use the sign of the component or its direction to place the resultant in the correct quadrant.*

29. (−300 ft, +400 ft)	**34.** (−100 m, 600 m)
30. (20 m, −40 m)	**35.** (60 lb, east; 24 lb, north)
31. (140 m, 400 m)	**36.** (200 lb, south; 80 lb, east)
32. (−600 lb, −400 lb)	**37.** (400 mi, south; 350 mi, west)
33. (48 N, −16 N)	**38.** (18 N, west; 24 N, north)

In Exercises 39 through 47, solve the problems.

39. A force of 90 N acts to the left and downward at an angle of 40° with the horizontal. What is the effective downward force and what component is acting to the left?

40. A 600-lb pull acts in a direction 28°S of E. What is the eastward component? What is the southward component?

41. A block of metal weighing 20 lb is resting on an inclined plane sloped at an angle of 32° with the horizontal floor. What is the component of the weight down the plane? What is the component of the weight perpendicular to the plane?

42. What are the x and y components of a force of 160 N at 120° from the positive x axis?

43. What are the x and y components of a vector **A** = 400 lb, 42°S of E?

44. Find the force on the docking cable of a boat on which the wind acts in a northerly direction with a force of 200 lb and the tide acts in an easterly direction with a force of 300 lb.

45. A rope is wrapped around a pole so that a force of 80 lb acts on one end and a force of 60 lb acts on the other. If the angle between the two forces is 120°, what is the resultant force? What angle does the resultant force make with the 60-lb force?

46. A man pushes with a force directed along a lawn mower handle which makes an angle of 50° with the ground. What must be the magnitude of his force in order to produce a horizontal force of 40 lb?

47. Determine the resultant force for the forces in Fig. 3-24.

48. What are the magnitude and direction of the resultant force in Fig. 3-25?

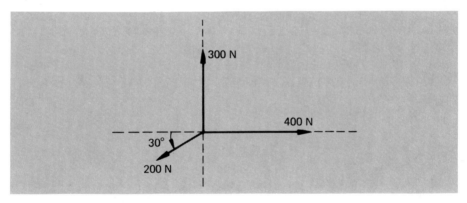

FIG. 3-24

Summary

In this chapter, we introduced the concepts of scalar and vector quantities. The fact that some quantities, called vectors, have direction as well as magnitude means that they may not be added in the usual manner. Both graphical and analytical methods for the treatment of vectors were presented as summarized below:

   ■ The **polygon method** of vector addition: The resultant vector is found by drawing each vector to scale, placing the tail of one vector to the tip of another until all vectors are drawn. The **resultant** is the straight line drawn from the starting point to the tip of the last vector (Fig. 3-26).

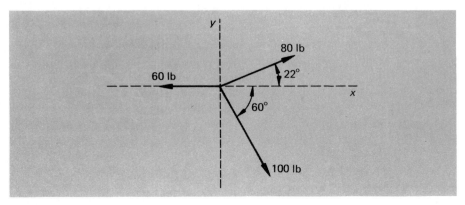

FIG. 3-25

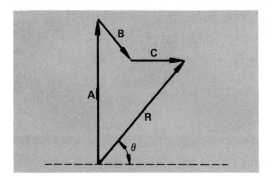

FIG. 3-26

■ The **parallelogram method** of vector addition: The resultant of two vectors is the diagonal of a parallelogram formed with the two vectors as adjacent sides. The direction is away from the common origin of the two vectors (Fig. 3-27).

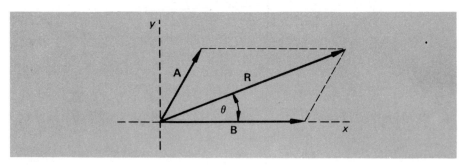

FIG. 3-27

*■ The x and y components of a vector (R, θ) (Fig. 3-28):

$$R_x = R \cos \theta$$

$$R_y = R \sin \theta$$

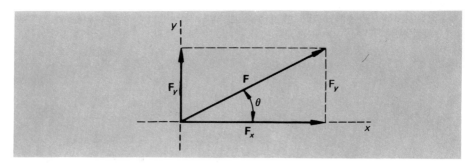

FIG. 3-28

*■ The resultant of two perpendicular vectors (R_x, R_y):

$$R = \sqrt{R_x^2 + R_y^2} \qquad\qquad \tan\theta = \frac{R_y}{R_x}$$

*■ The **component method** of vector addition:

$$R_x = A_x + B_x + C_x + \ldots \qquad\qquad R_y = A_y + B_y + C_y + \ldots$$

$$R = \sqrt{R_x^2 + R_y^2} \qquad\qquad \tan\theta = \frac{R_y}{R_x}$$

Questions

3-1. Define the following terms:

a. scalar   f. sine

b. components   g. force

c. vector   h. cosine

d. resultant   i. concurrent forces

e. displacement   j. tangent

3-2. Distinguish between vector and scalar quantities. Give two examples of each.

3-3. Explain the difference between adding scalars and adding vectors. Is it possible for the magnitude of a vector sum to be less than that of either of the vectors which are a part of the sum?

3-4. When three or more vectors are added, which graphical method is preferred? Show how one can use the parallelogram method to find the resultant of three or more vectors (by adding two at a time).

3-5. What are the minimum and maximum possible resultants of two forces of 10 and 7 N acting on the same object?

*3-6. Look up a section on rectangular and polar coordinates in a mathematics book. What are the similarities between components of a vector and the rectangular and polar coordinates of a point?

3-7. Can a resultant vector ever be smaller than its x and y components? Explain and point out how this question differs from Question 3.

3-8. Write the pythagorean theorem and solve the resulting equation explicitly for each parameter in the equation.

Problems

*3-1. A team of surveyors marks off distances represented by the vectors $D_1 = (80 \text{ m}, 60°)$, $D_2 = (40 \text{ m}, 0°)$, and $D_3 = (20 \text{ m}, 330°)$. Use the polygon method of vector addition to find the resultant displacement.

Answer (114 m, 31.3°N of E).

3-2. Find the resultant of the following forces by the polygon method: $A = (450 \text{ N}, 150°)$, $B = (200 \text{ N}, 270°)$, and $C = (300 \text{ N}, 135°)$.

3-3. A woman walks 16 m east and then 24 m in a direction 70°N of E. Use the parallelogram method to find her resultant displacement.

Answer 33.1 m, 43.0°.

3-4. An engine is held from the ceiling by two ropes. Rope A exerts a 60-lb force, and rope B exerts a 90-lb force. If the angle between the two ropes is 30°, what is the resultant force as determined by the parallelogram method?

3-5. Find the horizontal and vertical components of the following forces by graphical methods: (*a*) $F_1 = (260 \text{ lb}, 60°)$ and (*b*) $F_2 = (320 \text{ lb}, 210°)$.

Answer (a) $F_{1x} = 130$ lb, $F_{1y} = 225$ lb; (b) $F_{2x} = -277$ lb, $F_{2y} = -160$ lb.

3-6. Find the x and y components of the vector $R = (670 \text{ m}, 330°)$. Use graphical methods.

3-7. In removing a nail, a force of 260 lb is applied by a hammer in the direction shown in Fig. 3-29. What are the horizontal and vertical components of this force?

Answer $F_v = 251$ lb, $F_H = 67.3$ lb.

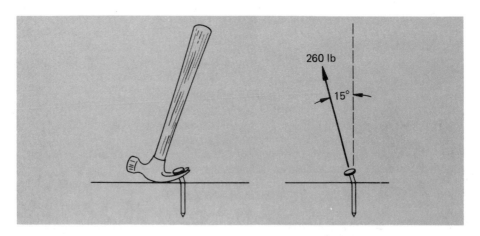

260 lb

15°

FIG. 3-29

3-8. What is the effective force acting to push the cart up the plane in Fig. 3-30?

3-9. A chain is wrapped around the bumper of an automobile, and forces of 400 and 250 N are exerted at right angles to each other. What is the resultant force on the bumper?

Answer 472 N.

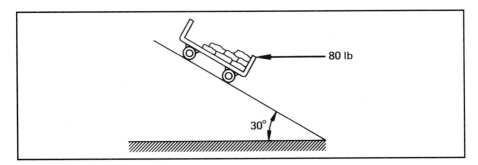

FIG. 3-30

3-10. A rope is tied to the edge of a large box so that it makes an angle of 43° with the horizontal. What force along the rope will produce an effective horizontal force of 70 N?

3-11. An upward force of 32 N is sufficient to lift a window. What force must be exerted along a pole which makes an angle of 30° with the wall in order to give the necessary upward lift?

Answer 37.0 N.

3-12. To go from Chicago to Miami, a plane must fly about 1800 km at an angle of 60°S of east. How far south of Chicago is Miami? How far east?

3-13. Determine the resultant force on the bolt in Fig. 3-31.

Answer 69.6 lb, 154.1°.

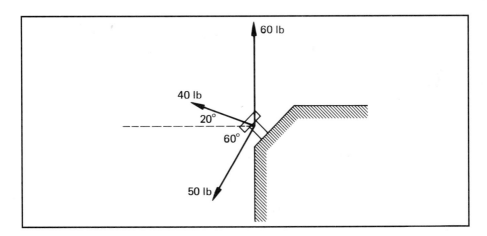

FIG. 3-31

3-14. Three boats exert forces on a mooring hook. What is the resultant force on the hook if boat A exerts a force of 420 N, boat B exerts a force of 150 N, and boat C exerts a force of 500 N? (Refer to Fig. 3-32.)

3-15. A boat heads directly across a stream with a velocity of 20 km/h. The current perpendicular to the bank is at 15 km/h. What are the resulting direction and velocity of the boat?

Answer 25 km/h, 53.1° with bank.

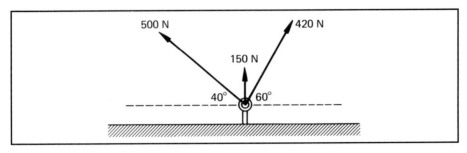

FIG. 3-32

3-16. A cable is wrapped around a steel beam so that it will not slip. What pull at an angle of 40° with the beam is required to produce an effective pull of 200 N along the beam?

3-17. A rope is wrapped around a stake so that an angle of 60° exists between the two rope segments. If the tension in each rope segment is 40 N, what is the resulting force on the stake?

Answer 69.3 N.

3-18. A picture is hung from a nail so that the two equal wire segments make an angle of 70°. What is the resultant force on the nail if the tension in each segment is 20 lb?

3-19. A bee flies directly west at a velocity of 12 ft/s. If a crosswind blows southward at 6 ft/s, what are the resulting velocity and direction of the bee?

Answer 13.4 m/s; 26.6°S of W.

3-20. A river 20 m wide flows northward at a speed of 7 m/s. What must be the velocity and the heading of a boat if it is to move directly from the west bank to the east bank in 5 s?

3-21. In Fig. 3-33, what force F at what angle θ is needed to pull the car directly east with a resultant force of 400 lb? (*Hint:* Set up the sum of components as usual, then solve for F_x and F_y.)

Answer 223 lb; 17.9°.

3-22. Determine the resultant of the following forces by the component method: $\mathbf{A} =$ (200 lb, 30°), $\mathbf{B} =$ (300 lb, 330°), and $\mathbf{C} =$ (400 lb, 250°).

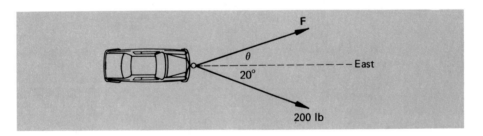

FIG. 3-33

4

EQUILIBRIUM
AND FRICTION

As a result of completing this chapter, you should be able to

1. **Demonstrate by example or experiment your understanding of Newton's first and third laws of motion.**
2. **State the first condition for equilibrium, give a physical example, and demonstrate graphically that the first condition is satisfied.**
3. **Construct a free-body diagram representing all forces acting on an object which is in equilibrium.**
4. **Solve for unknown forces by applying the first condition for equilibrium.**
5. **Apply your understanding of kinetic and static friction to the solution of equilibrium problems.**
*6. **Use trigonometry and vector components in the solution of equilibrium problems.**

Forces may act in such a manner as to cause motion or to prevent motion. Large bridges must be designed so that the overall effect of forces is to prevent motion. Every truss, girder, beam, and cable must be in *equilibrium*. In other words, the resultant forces acting at any point on the entire structure must be balanced. Shelves, chain hoists, hooks, lifting cables, and even large buildings must be constructed so that the effects of forces are controlled and understood. In this chapter we will continue our study of forces by studying objects at rest. The friction force which is so essential for equilibrium in many applications will also be introduced in this chapter as a natural extension of our work with all forces.

4-1 ■ NEWTON'S FIRST LAW

We know from experience that a stationary object remains at rest unless some outside force acts on it. A can of oil will stay on a workbench until someone tips it over. A suspended weight will hang until it is released. We know that forces are necessary to cause anything to move if it is originally at rest.

Less obvious is the fact that an object in motion will continue in motion until an outside force changes the motion. For example, a steel bar that slides on the shop floor soon comes to rest because of its interaction with the floor. The same bar would slide much farther on ice before stopping. This is because the horizontal interaction, called *friction,* between the floor and the bar is much greater than the friction between the ice and the bar. This leads to the idea that a sliding bar on a perfectly frictionless horizontal plane would stay in motion forever. These ideas are a part of Newton's first law of motion.

> **Newton's first law:** *A body at rest remains at rest, and a body in motion remains in uniform motion in a straight line unless acted on by an external unbalanced force.*

Due to the existence of friction, no actual body is ever completely free from external forces. But there are situations in which it is possible to make the resultant force zero or approximately zero. In such cases, the body will behave in accordance with the first law of motion. Since we recognize that friction can never be eliminated completely, we also recognize that Newton's first law is an expression of an *ideal* situation. A flywheel rotating on lubricated ball bearings tends to keep on spinning, but even the slightest friction will eventually bring it to rest.

Newton called the property of a particle that allows it to maintain a constant state of motion or rest *inertia*. His first law is sometimes called the *law of inertia*. When an automobile is accelerated, the passengers obey this law by tending to remain at rest until the external force of the seat compels them to move. Similarly, when the automobile stops, the passengers continue in motion with constant speed until they are restrained by their seat belts or through their own efforts. All matter has inertia. The concept of *mass* is introduced later as an indication of a body's inertia.

4-2 ■ NEWTON'S THIRD LAW

There can be no force unless two bodies are involved. When a hammer strikes a nail, it exerts an ''action'' force on the nail. But the nail must also ''react'' by pushing back against the hammer. In all cases there must be an *acting force* and a *reacting force*. Whenever two bodies interact, the force exerted by the second body on the first (the reaction force) is equal in magnitude and opposite in direction to the force exerted by the first body on the second (the action force). This principle is stated in Newton's third law:

Newton's third law: *To every action there must be an equal and opposite reaction.*

Therefore, there can never be a single isolated force. Consider the examples of action and reaction forces in Fig. 4-1.

Note that the acting and reacting forces do not cancel each other. They are equal in magnitude and opposite in direction, but they act on *different* objects. For two forces to cancel, they must act on the same object. It might be said that the action forces create the reaction forces.

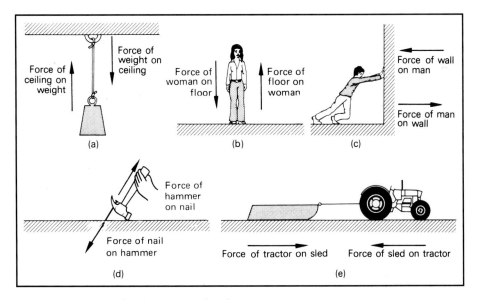

FIG. 4-1 Examples of action and reaction forces.

4-3 ■ FREE-BODY DIAGRAMS

We have seen that forces are easier to understand when they are represented as vectors. The study of objects at rest is particularly assisted by vector diagrams. Consider, for example, the 40-lb weight suspended by ropes shown in Fig. 4-2. There are three forces exerted *on* the knot—those exerted by the ceiling, by the wall, and by the earth (weight). If each of these forces is labeled and represented as a vector, we can draw a vector diagram, as shown in Fig. 4-2*b*. Such a diagram is called a *free-body diagram.*

A **free-body diagram** *is a vector diagram which describes all forces acting on a particular body or object.*

Note that in the case of concurrent forces, all vectors point away from the center of the *x* and *y* axes, which cross at the point of action.

In drawing free-body diagrams, it is important to distinguish between action and reaction forces. In our example, there are forces *on* the knot, but there are also three

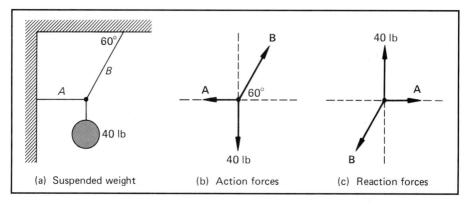

FIG. 4-2 Free-body diagrams showing action and reaction forces.

equal and opposite forces exerted *by* the knot. From Newton's third law, the reaction forces exerted *by* the knot *on* the ceiling, wall, and earth are shown in Fig. 4-2c. To avoid confusion, it is imperative that we pick a point at which all forces are acting and draw those forces which act *on* the body at that point.

4-4 ■ FIRST CONDITION FOR EQUILIBRIUM

Objects at rest or in motion with constant speed are said to be in equilibrium. We know from Newton's first law that all forces must be balanced in such cases. Otherwise, there would be a change in the state of rest or of motion. At least one condition for equilibrium, therefore, must be that the resultant force acting on an object be equal to zero.

> **First condition for equilibrium:** *A body is in translational equilibrium if and only if the vector sum of all forces acting on it is zero.*

The term *translational equilibrium* is used to distinguish this *first* condition for equilibrium from the second condition, which involves rotational motion. Rotational equilibrium will be discussed in Chap. 5.

A graphical understanding of the first condition for equilibrium may be gained by considering four forces acting on a single hook. The free-body diagram in Fig. 4-3a (page 91) shows these forces as vectors **A, B, C,** and **D.** If the vectors are added by the polygon method, the vector polygon closes on itself. The tip of the last vector drawn will always land on the tail of the first vector, indicating a resultant of zero (see Fig. 4-3b and c). This will be true regardless of the order in which the vectors are added.

A system of forces not in equilibrium can be placed in equilibrium by replacing their resultant force with an equal and opposite force called the *equilibrant*.

> *The* **equilibrant** *is a force equal in magnitude to the resultant force but opposite in direction.*

The two forces **A** and **B** in Fig. 4-4a (page 91) have a resultant force **R** in a direction 30° above the horizontal. If we add **E,** which is equal in magnitude to **R** but which has an angle 180° greater, the system will be in equilibrium, as shown in Fig. 4-4b.

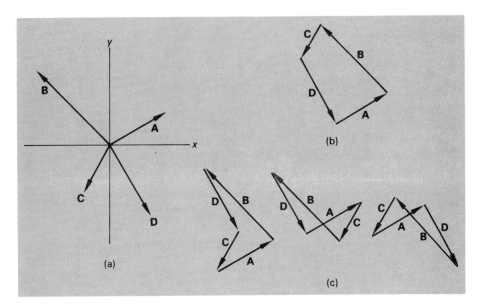

FIG. 4-3 Forces in equilibrium.

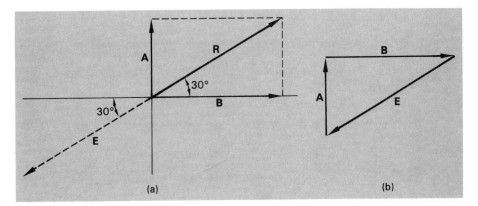

FIG. 4-4 Equilibrant.

4-5 ■ GRAPHICAL SOLUTION OF EQUILIBRIUM PROBLEMS

It is often possible to solve for unknown forces by applying the first condition for equilibrium. A free-body diagram is drawn and labeled from the conditions of a problem. Then a closed vector polygon is constructed with vectors drawn to scale. The unknown force is determined by direct measurement and conversion based on the chosen scale.

EXAMPLE 4-1   Determine the tension in ropes *A* and *B* for the suspended weight in Fig. 4-2.

Solution   The free-body diagram constructed earlier is repeated as Fig. 4-5a. Next we add the vectors by the polygon method, producing a closed triangle as in Fig. 4-5b. The 40-lb vector is drawn first, since its magnitude is known. An appropriate choice of scales might be 1 cm = 10 lb. Next the vectors **A** and **B** are drawn so that the triangle closes with an angle of 60° between **B** and the line of **A**. Careful measurement of lengths indicates that $A = 2.31$ cm and $B = 4.62$ cm. Therefore, the rope tensions are

$$A = 2.31 \ \cancel{cm} \times \frac{10 \ lb}{1 \ \cancel{cm}} = 23.1 \ lb$$

$$B = 4.62 \ \cancel{cm} \times \frac{10 \ lb}{1 \ \cancel{cm}} = 46.2 \ lb$$

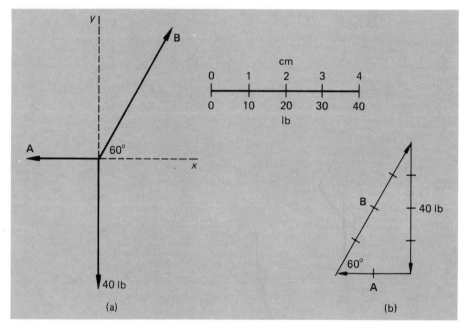

FIG. 4-5 Graphical solution for forces in equilibrium.

In summary, the procedure for graphical solution of equilibrium problems is as follows:

1. Read the problem and draw a rough sketch to help visualize the problem.
2. Draw and label a free-body diagram, indicating all forces acting on the body at some point.
3. Choose an appropriate scale, and construct a closed vector polygon.
4. Measure the unknown lengths or angles, and convert according to the chosen scale.

Probably the most difficult steps in a graphical solution are constructing the free-body diagram and constructing the vector polygon. In drawing the free-body diagram, it is helpful to imagine that the forces in question are acting on *you*. Become the knot in a rope or the block on a table, and try to imagine the forces you would experience. Two additional examples are given in Fig. 4-6*a* and *b*. Note that the force exerted by the light boom in Fig. 4-6*a* is outward and not toward the wall. This is because we are interested in forces exerted *on* the end of the boom and not those exerted *by* the end of the boom. We pick a point at the end of the boom where the two ropes are attached. The 60-N weight and the tension **T** are action forces exerted by the ropes at this point. If the end of the boom is not to move, these forces must be balanced by a third force—the force exerted *by* the wall (through the boom). This third force **B,** acting on the end of the boom, must not be confused with the inward reaction force which acts *on* the wall.

The second example (Fig. 4-6*b*) also shows three action forces exerted *on* the block. The weight **W** is exerted by the earth on the block, the normal force **N** is exerted by the plane on the block, and the friction force acts up the inclined plane. Note that the *x* axis is chosen along the plane so that the normal force and the friction force are each along an axis. This choice makes it easier to determine angles, and it will be especially useful if trigonometric approaches are used.

As an additional example, you might show graphically that the magnitudes of the forces in Fig. 4-6*a* and *b* are $T = 93.3$ N, $B = 71.5$ N, $\mathcal{N} = 173$ lb, and $\mathcal{F} = 100$ lb.

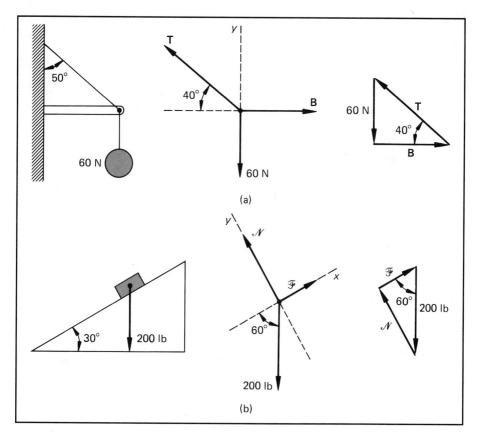

FIG. 4-6 (*a*) Forces on the end of a boom due to a 60-N weight. (*b*) Forces on a block resting on an inclined plane.

*4-6 ■ SOLVING EQUILIBRIUM PROBLEMS WITH TRIGONOMETRY

Problems such as those discussed graphically in the previous section may also be solved with trigonometry. When you are working with just three vectors, the procedure is the same except that the forces are determined with trigonometry instead of with scaled drawings.

EXAMPLE 4-2 Determine the tension in cable *A* and in cable *B* for the arrangement shown in Fig. 4-7.

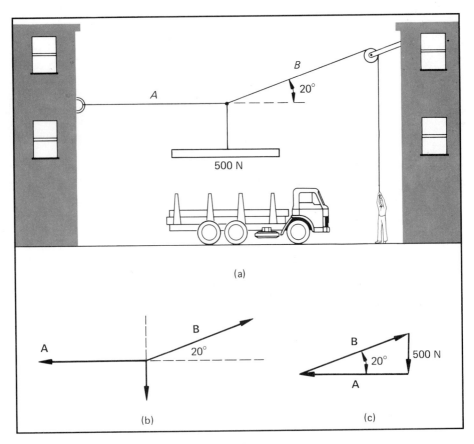

FIG. 4-7 Solving equilibrium problems with trigonometry.

Solution The free-body diagram and closed vector polygon are shown in Fig. 4-7b and c. It is not necessary to draw the diagrams to scale. Notice that **B** is the hypotenuse and 500 N is the side opposite the 20° angle. Thus,

$$\sin 20° = \frac{500 \text{ N}}{B}$$

Multiplying both sides by B, we obtain

$$B \sin 20° = 500 \text{ N}$$

and

$$B = \frac{500 \text{ N}}{\sin 20°} = 1462 \text{ N}$$

Now, A is determined from the tangent function:

$$\tan 20° = \frac{500 \text{ N}}{A}$$

Solving for A, we have

$$A = \frac{500 \text{ N}}{\tan 20°} = 1374 \text{ N}$$

This approach is especially useful when two of the three vectors are perpendicular. The more general cases can best be handled by the component method discussed in the next section.

*4-7 ■ EQUILIBRIUM AND VECTOR COMPONENTS

In Chap. 3 we discussed the component method for adding vectors to find their resultant or vector sum. The x component of the resultant vector was found by adding the x components of each vector. The y component was the sum of all the components along the y axis. A similar procedure may be used to add forces which are in equilibrium. In this case, the first condition for equilibrium tells us that the resultant is zero, or

$$R_x = A_x + B_x + C_x + \cdots = 0$$
$$R_y = A_y + B_y + C_y + \cdots = 0$$

In other words, if the resultant is zero, then the x and y components of the resultant must also be zero.

The first condition for equilibrium, therefore, may be written as two equations:

$$\boxed{\Sigma F_x = 0 \qquad \Sigma F_y = 0} \tag{4-1}$$

The sum of all forces along the x axis must equal zero, and the sum of all the forces along the y axis must be zero.

The following steps might be followed in solving for unknown forces by the component method:

1. Draw a sketch labeling the conditions of the problem.
2. Construct a free-body diagram, and label the x and y components of each vector opposite and adjacent to the reference angle.
3. Calculate the x and y components of each vector:

$$A_x = \pm|A \cos \phi| \qquad A_y = \pm|A \sin \phi|$$

4. Use the first condition for equilibrium to write two equations in terms of the unknown factors:

$$A_x + B_x + C_x + \cdots = 0$$
$$A_y + B_y + C_y + \cdots = 0$$

5. Solve algebraically for the unknown forces.

When you are adding vector components, it is essential to affix the proper sign to each component based on convention.

EXAMPLE 4-3   A 200-N ball is hung from the end of boom B of negligible weight, as shown in Fig. 4-8. Find the tension in rope A and the compression in the boom, using the component method.

Solution   We solve by following the above procedure.
1. Draw and label a sketch (Fig. 4-8a).

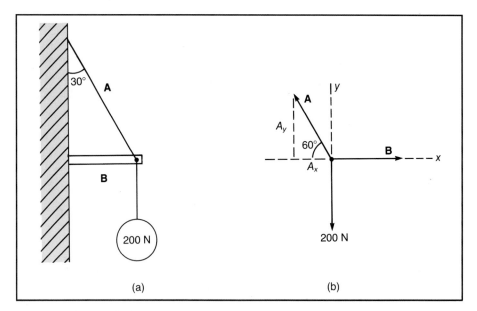

(a)   (b)

FIG. 4-8 When equilibrium problems are solved by the component method, the components should be labeled on the free-body diagram.

2. Draw a free-body diagram (Fig. 4-8b). Note that the force B is exerted by the wall, through the boom, and on the rope at the end.
3. Find the components of each force. Refer to Table 4-1.

TABLE 4-1

Force	ϕ	x Component	y Component
A	60°	$A_x = -A \cos 60°$	$A_y = A \sin 60°$
B	0°	$B_x = +B$	$B_y = 0$
W	-90°	$W_x = 0$	$W_y = -200$ N
	Sum = 0	$\Sigma F_x = -A \cos 60° + B$	$\Sigma F_y = A \sin 60° - 200$ N

4. Apply the first condition for equilibrium by setting the sum of components equal to zero:

$$\Sigma F_x = B - A \cos 60° = 0$$

from which we obtain

$$B = A \cos 60°$$

Since $\cos 60° = 0.5$, this simplifies to

$$B = 0.5 \, A$$

We obtain a second equation from the y components:

$$\Sigma F_y = A \sin 60° - 200 \text{ N} = 0$$
$$A \sin 60° = 200 \text{ N}$$

or since $\sin 60° = 0.866$, we may write

$$0.866 \, A = 200 \text{ N}$$

5. Finally, we solve these equations for the unknown forces. Since the second equation is simplest, we solve first for A:

$$A = \frac{200 \text{ N}}{0.866} = 231 \text{ N}$$

This value of A can now be substituted into the first equation to find B:

$$B = 0.5 \, A = (0.5)(231 \text{ N})$$
$$= 115 \text{ N}$$

All problems involving translational equilibrium may be approached in this manner. Two equations are obtained from summing along each axis. Then unknown forces are found by solving the equations by substitution.

4-8 ■ *FRICTION*

Whenever a body moves while it is in contact with another object, friction forces oppose the relative motion. These forces are caused by the adhesion of one surface to the other and by the interlocking of irregularities in the rubbing surfaces. It is friction that holds a nail in a board, allows us to walk, and makes automobile brakes work. In all these cases friction has a desirable effect.

In many other instances, however, friction must be minimized. For example, it increases the work necessary to operate machinery, it causes wear, and it generates heat, which often causes additional damage. Automobiles and airplanes are streamlined in order to decrease air friction, which is large at high speeds.

Whenever one surface moves past another, the frictional force exerted by each body on the other is parallel or tangent to the two surfaces and acts in such a manner as to oppose relative motion. It is important to note that these forces exist not only when there is relative motion but even when one object only *tends* to slide past another.

Suppose a force is exerted on a block which rests on a horizontal surface, as shown in Fig. 4-9. At first the block will not be moved because of the action of a force called the *force of static friction* $\mathscr{F}_s$. But as the applied force is increased, motion eventually occurs, and the frictional force exerted by the horizontal surface while the block is moving is called the *force of kinetic friction* $\mathscr{F}_k$.

The laws governing friction forces can be determined experimentally in the laboratory by using an apparatus similar to the one shown in Fig. 4-10a. A box of weight W

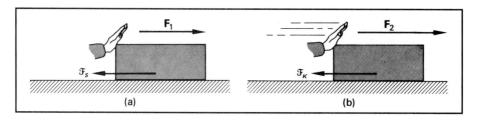

FIG. 4-9 (a) In static friction, motion is impending; (b) in kinetic friction, the two surfaces are in relative motion.

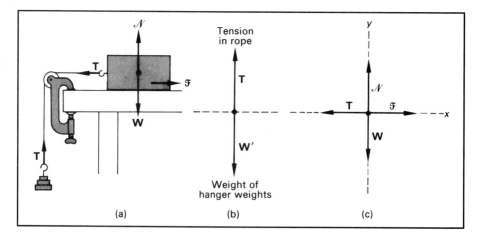

FIG. 4-10 Experiment to determine the force of friction.

is placed on a horizontal table, and a string attached to the box is passed over a light frictionless pulley and attached to a weight hanger. All forces acting on the box and hanger are shown in their corresponding free-body diagrams (Fig. 4-10b and c).

Let us consider that the system is in equilibrium, which requires the box to be stationary or moving with a constant velocity. In either case, we may apply the first condition for equilibrium. Consider the force diagram of the box as shown in Fig. 4-10c.

The first condition for equilibrium tells us that the sums of the forces along the vertical and horizontal axes are zero. From Fig. 4-10b, we note that

$$\Sigma F_x = 0 \qquad \mathcal{F} - T = 0 \qquad \mathcal{F} = T$$
$$\Sigma F_y = 0 \qquad \mathcal{N} - W = 0 \qquad \mathcal{N} = W$$

Thus the force of friction is equal in magnitude to the tension in the string, and the normal (perpendicular) force exerted by the table on the box is equal to the weight of the box. Note that the tension in the string is determined by the weight of the hanger plus that of the added weights.

We begin the experiment by slowly adding weights to the hanger, thus gradually increasing the tension in the string. As the tension is increased, the equal but oppositely directed force of static friction is also increased. If T is increased sufficiently, the box will start to move, indicating that T has overcome the *maximum* force of static friction $\mathcal{F}_{s,max}$. Usually, this maximum value is used in the solution of friction problems. Therefore, in this text $\mathcal{F}_s$ is understood to represent $\mathcal{F}_{s,max}$.

To continue the experiment, suppose we add weights to the box, thereby increasing the normal pressure between the box and the table. Our normal force will now be

$$\mathcal{N} = W + \text{added weights}$$

Repeating the above procedure will show that a *proportionately* larger value of T will be necessary to overcome $\mathcal{F}_s$. In other words, if we double the normal force between two surfaces, the maximum force of static friction which must be overcome is also doubled. If $\mathcal{N}$ is tripled, $\mathcal{F}_s$ is tripled, and so it will be for other factors. Therefore, it can be said that the maximum force of static friction is determined directly from the normal force between the two surfaces. An equation which states this fact is

$$\mathcal{F}_s = \mu_s \mathcal{N} \qquad\qquad (4\text{-}2)$$

where μ_s is a proportionality constant called the *coefficient of static friction*. Equation (4-2) may be rewritten to express this constant in terms of the normal force and the friction force:

$$\mu_s = \frac{\mathcal{F}_s}{\mathcal{N}} \qquad\qquad (4\text{-}3)$$

Note that μ_s is the constant ratio of two forces. It is a dimensionless quantity and has no units.

EXAMPLE 4-4   A horizontal force of 30 lb will just start an empty 600-lb sled moving across packed snow. (*a*) What is the coefficient of static friction? (*b*) What force is required to start the sled if 200 lb of supplies are placed in the sled?

Solution (a)   Figure 4-11 shows a free-body diagram for this situation. A condition of equilibrium exists at the instant before motion occurs. Thus the sum of the horizontal forces is zero.

$$
\begin{array}{lll}
F_x = 0 & \mathscr{F}_s - 30 \text{ lb} = 0 & \mathscr{F}_s = 30 \text{ lb} \\
F_y = 0 & \mathscr{N} - W = 0 & \mathscr{N} = 600 \text{ lb}
\end{array}
$$

The coefficient of static friction is given by Eq. (4-3):

$$
\mu_s = \frac{\mathscr{F}_s}{\mathscr{N}} = \frac{30 \text{ lb}}{600 \text{ lb}} = 0.5
$$

Notice that there are no units for μ_s.

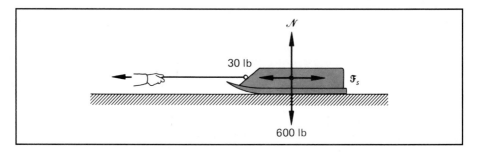

FIG. 4-11

Solution (b)   Now that we know μ_s, we may predict the force required to overcome static friction for any normal force $\mathscr{N}$. If 200 lb of supplies is added to the 600-lb sled, the new normal force is

$$
\mathscr{N} = 600 \text{ lb} + 200 \text{ lb} = 800 \text{ lb}
$$

The maximum friction force to be overcome is, therefore,

$$
\mathscr{F}_s = \mu_s \mathscr{N} = (0.05)(800 \text{ lb}) = 40 \text{ lb}
$$

Now let us return to the laboratory experiment with a box, pulley, and suspended weights. It will be noticed that after $\mathscr{F}_s$ has been overcome, the box will increase in speed or accelerate. This indicates that a lesser value of **T** would be necessary to keep the box moving with a constant speed. Thus, the force of kinetic friction $\mathscr{F}_k$ will be smaller than $\mathscr{F}_s$ for the same surfaces. In other words, it requires more force to start an object moving than it does to keep it moving at constant speed.

When an object slides across a surface with constant speed, it is still in equilibrium.

The same reasoning which led to Eq. (4-2) for static friction will give a similar equation for kinetic friction:

$$\mathscr{F}_k = \mu_k \mathscr{N} \qquad \mu_k = \frac{\mathscr{F}_k}{\mathscr{N}} \qquad\qquad (4\text{-}4)$$

In this case, μ_k is the *coefficient of kinetic friction*.

Problems involving the motion of objects across surfaces with constant speed are also solved in the manner illustrated in Example 4-4. The key words to look for in such problems are *just start moving* and *move with constant speed*. The former implies static friction, and the latter implies kinetic friction. Table 4-2 gives some approximate values for the coefficients of static and kinetic friction for different types of surfaces. It is important to realize that these values are *approximate* and depend on the condition of the surfaces. In practice, it is probably safer to determine μ_s and μ_k experimentally for each application. This can be done by using known loads $\mathscr{N}$ and observing the force required to overcome friction $\mathscr{F}$.

The coefficients μ_s and μ_k can be shown to depend on the roughness of the surfaces but not on the area of contact between the two surfaces. It can be seen from the equations that μ depends only on the frictional force $\mathscr{F}$ and the normal force $\mathscr{N}$. Of course, it must be realized that these equations are only useful approximations. Many factors may interfere with the general application of these formulas. No one who has experience in automobile racing, for instance, will believe that the friction force is *completely* independent of the contact area. Nevertheless, the equations are useful tools for estimating resistive forces in specific cases.

TABLE 4-2 APPROXIMATE COEFFICIENTS OF FRICTION

Material	μ_s	μ_k
Wood on wood	0.7	0.4
Steel on steel	0.15	0.09
Metal on leather	0.6	0.5
Wood on leather	0.5	0.4
Rubber on concrete, dry	0.9	0.7
wet	0.7	0.57

The following points should be considered when you are solving problems involving friction:

1. Friction forces are parallel to the surfaces and directly oppose motion.
2. The force of static friction is larger than the force of kinetic friction for the same materials.
3. In drawing free-body diagrams, it is usually better to choose the x axis along the plane of motion and the y axis perpendicular to the plane of motion.
4. If motion is constant or impending, the first condition for equilibrium will give two equations involving forces along the plane of motion and perpendicular to it.
5. The relations $\mathscr{F}_s = \mu_s \mathscr{N}$ or $\mathscr{F}_k = \mu_k \mathscr{N}$ can be applied to solve for the desired quantity.

It is important to remember that the normal force $\mathcal{N}$ is the *net* force exerted by each surface on the other surface. It is not always equal to the *weight* of an object. For example, if we push downward on an object as we attempt to move it horizontally, we increase the normal force $\mathcal{N}$. In Fig. 4-12*a*, the normal force is

$$\mathcal{N} = W + F_y$$

This action will result in greater friction.

Similarly, dragging an object with an inclined rope, as in Fig. 4-12*b*, will reduce the normal force. The first condition for equilibrium applied for vertical forces gives

$$\mathcal{N} = W - F_y$$

The smaller normal force results in less friction, since $\mathcal{F} = \mu\mathcal{N}$.

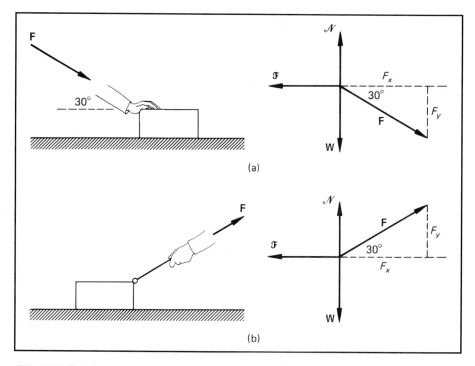

FIG. 4-12 Pushing downward at an angle increases friction, whereas pulling at an upward angle decreases friction.

*4-9 ■ LIMITING ANGLE FOR FRICTION

The normal force also changes as the angle of inclination changes. For example, the normal force is much less when an object rests on the side of a hill than it is when the object is on a horizontal surface. The steeper the incline, the smaller the force pressing the surfaces together. If we make the angle of inclination large enough, the downward component of the weight will be large enough to overcome the maximum force of static friction. When this happens, the object will slide. The minimum angle of inclination for which motion will occur is called the *limiting angle*.

Consider a block of weight **W** resting on a plane surface inclined at an angle, as shown in Fig. 4-13a. It can be shown that the angle of repose is determined by the coefficient of static friction between the surfaces. Applying the first condition of equilibrium to Fig. 4-13b, we note that

$$\Sigma F_x = 0 \qquad \mathcal{F}_s - W_x = 0 \qquad \mathcal{F}_s = W_x$$
$$\Sigma F_y = 0 \qquad \mathcal{N} - W_y = 0 \qquad \mathcal{N} = W_y$$

Knowledge of trigonometry will show that the ratio W_x/W_y is equal to the tangent of θ:

$$\tan \theta = \frac{W_x}{W_y} = \frac{\mathcal{F}_s}{\mathcal{N}}$$

Since $\mu_s = \mathcal{F}_s/\mathcal{N}$, we may finally write

$$\boxed{\tan \theta = \mu_s} \qquad (4\text{-}5)$$

Thus a block, regardless of its weight, will remain at rest on an inclined plane unless $\tan \theta$ exceeds μ_s. The angle in this case will be the limiting angle.

Similar calculations will show that an object will slide down an incline at *constant speed* when $\tan \theta = \mu_k$. In this case, the angle θ is referred to as the *angle of uniform slip*.

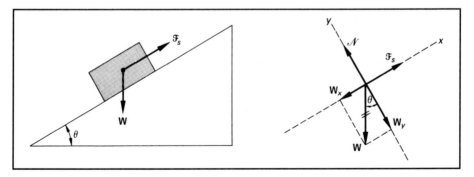

FIG. 4-13 The limiting angle for static friction.

4-10 ■ ROLLING FRICTION

That friction can be reduced by allowing one surface to roll on another has been known since ancient times. Wheels are used universally to assist in the movement of materials from one place to another. Modern industry also uses ball or roller bearings in components of machines to reduce wear and to increase efficiency.

Although many factors determine the value of rolling friction in specific cases, an approximation can be found from

$$\mathscr{F}_R = \mu_R \mathscr{N}$$

(4-6)

where μ_R is the coefficient of rolling friction. Application of this relation is so similar to previous applications that no additional examples will be necessary.

Summary

In this chapter, we have defined objects which are at rest or in motion with constant speed to be in equilibrium. Through the use of vector diagrams and Newton's laws, we have found it possible to determine unknown forces for systems which are known to be in equilibrium. The following items will summarize the more important concepts to be remembered:

- *Newton's first law of motion* states that an object at rest will remain at rest and an object in motion with constant speed will remain at constant speed unless acted on by an external unbalanced force.
- *Newton's third law of motion* states that every action must produce an equal but oppositely directed reaction. The action and reaction forces are on different bodies.
- *Free-body diagrams:* From the conditions of the problem a neat sketch is drawn, and all known quantities are labeled. Then, a force diagram is constructed, indicating all forces acting on a body. Such a diagram is shown in Fig. 4-14*a*, with all angles and components labeled.

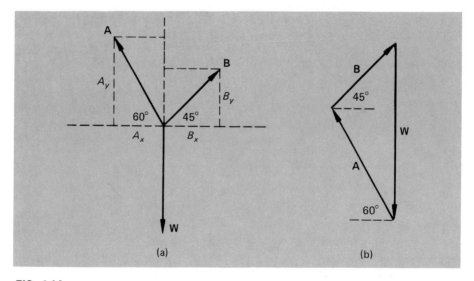

(a)   (b)

FIG. 4-14

- Unknown forces can be determined if a condition of equilibrium exists. All force vectors may be added graphically to form a closed polygon, such as the one shown in Fig. 4-14b. When known vectors have been drawn to scale and other vectors at given angles, the unknown forces may be measured.
- *The first condition for equilibrium:* A body in translational equilibrium has no resultant force acting on it. In such cases, the sum of all the x components is zero, and the sum of all the y components is zero. A mathematical statement is

$$R_x = \Sigma F_x = 0 \qquad R_y = \Sigma F_y = 0$$

- Applying these equations to Fig. 4-14a, we obtain two equations in two unknowns:

$$B \cos 45° - A \cos 60° = 0$$
$$B \sin 45° + A \sin 60° - 200 \text{ N} = 0$$

These equations may be solved by substitution to find A and B.
- *Static friction* exists between two surfaces when motion is impending; *kinetic friction* occurs when the two surfaces are in relative motion; *rolling friction* opposes motion when one surface rolls on another. They are related to the *normal force* and are found from

$$\mathcal{F}_s = \mu_s \mathcal{N} \qquad \mathcal{F}_k = \mu_k \mathcal{N} \qquad \mathcal{F}_R = \mu_R \mathcal{N}$$

These forces must be considered in many problems.

Questions

4-1. Define the following terms:
a. inertia
b. equilibrium
c. Newton's first law
d. Newton's third law
e. normal force
f. friction force
g. equilibrant
h. reaction force
i. coefficient of friction
j. static friction
k. kinetic friction
l. free-body diagram

4-2. A popular magic trick consists of placing a coin on a card and the card on top of a glass. The edge of the card is flipped briskly with the forefinger, causing the card to fly off the top of the glass as the coin drops into the glass. Explain. What law does this illustrate?

4-3. When the head of a hammer becomes loose, you can reseat it by holding the hammer vertically and tapping the base of the handle against the floor. Explain. What law does this illustrate?

4-4. Explain the part played by Newton's third law of motion in the following activities: (a) walking, (b) rowing, (c) rockets, and (d) parachuting.

4-5. Can a moving body be in equilibrium? Give several examples.

4-6. According to Newton's third law of motion, every force has an equal and oppositely directed reaction force. Therefore, the concept of a resultant unbalanced force must be an illusion that really does not hold under close examination. Do you agree with this statement? Give the reasons for your answer.

4-7. A brick is suspended from the ceiling by a light string. A second identical string is attached to the bottom of the brick and hangs within the reach of a student. When the student pulls the lower string slowly, the upper string breaks, but when the student jerks the lower string, it breaks. Explain.

Problems

4-1. Consider the suspended weight in Fig. 4-15. Visualize the forces acting on the knot and draw a free-body diagram. Label all forces and indicate known angles. Choose a scale and draw the weight vector as a vertical arrow pointing downward. Represent the tension in ropes A and B by vectors, and construct a closed vector triangle. Determine the tensions in ropes A and B by measuring the scaled drawings.

Answer A = 67.1 N; B = 104 N.

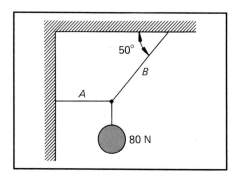

FIG. 4-15

*4-2. Use trigonometry and the first condition for equilibrium to solve for A and B in Prob. 4-1. Draw and label the x and y components of A and B on the free-body diagram. Set up a table listing components of each vector in terms of A and B. Write two equations by setting the sums equal to zero. Solve for A and B by substitution.

4-3. A mechanic pulls with a force of 80 lb along rope C, as shown in Fig. 4-16. The pulley serves only to change the direction of this force, so that the tension in rope A is also 80 lb. Using the polygon method, draw vector A to the tip of vector C, then close the triangle with vector B. Determine vector B by measurement.

Answer 139 lb, upward.

*4-4. Use trigonometry and the first condition for equilibrium to solve for the upward force B exerted by the ceiling on the pulley in Prob. 4-3.

4-5. Determine graphically the tension in rope A and the compression B in the strut for Fig. 4-17. Make sure that each force in your free-body diagram is exerted *on* the rope at the end of the strut.

Answer A = 231 N; B = 462 N.

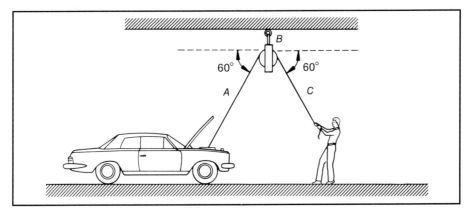

FIG. 4-16

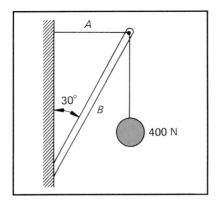

FIG. 4-17

*4-6. Solve Prob. 4-5 by using trigonometry and the first condition for equilibrium. Verify that the same answer can be obtained by applying trigonometry to a closed vector polygon.

4-7. Rope *B* in Fig. 4-15 has a breaking strength of 200 N. What maximum weight can be suspended in place of the 80-N weight?

Answer 153 N.

4-8. If the breaking strength of cable *A* in Fig. 4-18 is 300 lb, what is the maximum weight *W* which can be supported by this apparatus? *Hint:* Set $A = 300$ lb, then determine *W* for equilibrium.

4-9. Determine the compression in the center strut *B* and the tension in rope *A* for the situation described by Fig. 4-19.

Answer $A = 643$ N; $B = 940$ N.

4-10. An 80-lb engine block is suspended from the ceiling of a garage by a chain hoist. What is the tension in the chain when the engine block is at rest? What is the tension in the chain if the engine is lifted at constant speed?

4-11. Two 20-N weights are suspended at opposite ends of a rope which passes over a light frictionless pulley. The pulley is attached to the ceiling with a short chain. (*a*) What is the tension in the rope? (*b*) What is the tension in the chain?

Answer (*a*) 20 N; (*b*) 40 N.

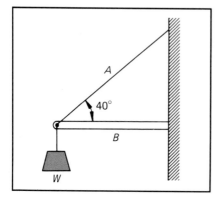

FIG. 4-18

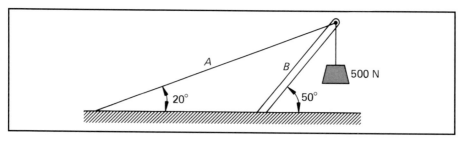

FIG. 4-19

4-12. Find the tension in ropes *A* and *B* for the arrangements shown in Fig. 4-20.

4-13. Find the tension in the cable and the force *F* exerted by the boom on its end in the arrangements of Fig. 4-21.

> *Answer* (*a*) $F = 34.6$ lb, $T = 43.2$ lb;
> (*b*) $F = 91.5$ lb, $T = 61.2$ lb.

4-14. A rope is stretched between two trees which are 60 ft apart. A 70-lb weight is suspended from the midpoint, causing the rope to sag a distance of 5 ft. What is the tension in the rope?

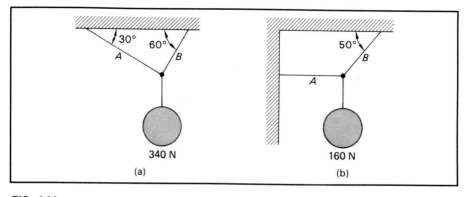

FIG. 4-20

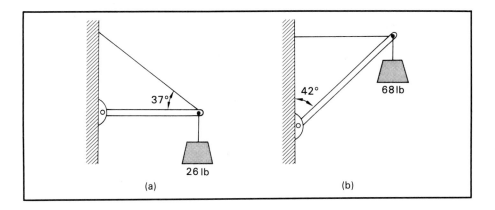

(a)   (b)

FIG. 4-21

4-15. A cable is stretched horizontally across the top of two vertical poles. From the midpoint of the cable a 400-N traffic light is suspended so that the cable sags downward. If the cable makes an angle of 75° with each vertical pole, what is the tension in the cable? *Hint:* Since the weight is at the midpoint, the tension will be the same on each side.

Answer 773 N.

4-16. Consider the truss illustrated in Fig. 4-22. Draw a free-body diagram showing all forces acting at the top of the truss. What is the maximum force that the diagonal braces must be able to support if the truss is not to collapse?

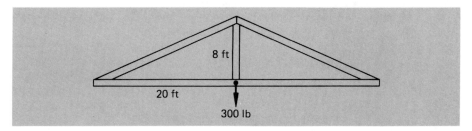

FIG. 4-22

*4-17. Find the tension in the ropes and the compression in the boom for those arrangements shown in Fig. 4-23.

Answer (a) A = 1405 N, *B* = 1146 N;
(b) T = 33.7 lb, *F* = 23.8 lb.

4-18. A 5-lb weight is suspended from the ceiling by rope *A;* a 10-lb weight is attached to the bottom of the first weight by rope *B;* and a third rope *C* is used to attach a 15-lb weight to the bottom of the second weight. What are the tensions in each rope?

4-19. A 200-lb refrigerator is placed on a blanket and dragged across a tile floor. If $\mu_s = 0.4$ and $\mu_k = 0.2$, what horizontal force is required to just start motion, and what force will move the refrigerator with constant speed?

Answer 80 lb; 40 lb.

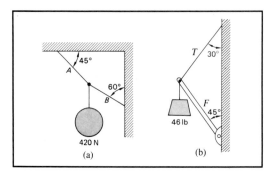

FIG. 4-23

4-20. A 90-lb anvil rests on a metal table. A horizontal pull of 40 lb is required to just start the anvil moving. After motion is started, a pull of only 25 lb will cause the anvil to move with constant speed. (*a*) What is the normal force? (*b*) What is the coefficient of static friction? (*c*) What is the coefficient of kinetic friction?

4-21. A cake of ice slides with constant speed across a wooden floor when a horizontal force of 8 lb is applied. What is the weight of the cake of ice if $\mu_k = 0.15$?

Answer 53.3 lb.

4-22. Suppose that the block in Fig. 4-12*a* has a weight of 80 N. When the *x* component of the downward force is 40 N and the *y* component is 30 N, motion just begins to occur. (*a*) Use the first condition for equilibrium to write two equations. (*b*) What is the maximum force of static friction? (*c*) What is the normal force? (*d*) What is the coefficient of static friction?

4-23. In Fig. 4-12*b*, suppose that the block weighs 300 lb. When the upward component is 35 lb and the horizontal component is 42 lb, the block will move at constant speed. What is the coefficient of kinetic friction?

Answer 0.158.

4-24. A 400-lb box rests on a horizontal floor. It is determined that a force of 50 lb is necessary to just start it moving across the floor, and a force of 30 lb will keep it moving at constant speed. What are μ_s and μ_k?

4-25. Suppose that 200 lb of supplies is placed inside the box in Prob. 4-24. What horizontal force would be needed to move the box and its contents with constant speed?

Answer 45 lb.

4-26. If the coefficient of rolling friction between the tires and the road is 0.03 and a car weighs 8000 N, what horizontal force is required to move it with constant speed?

4-27. A wagon is pulled along a horizontal floor at constant speed by a rope which exerts an upward force of 20 lb and a horizontal force of 30 lb. If the wagon has a weight of 400 lb, what is the coefficient of rolling friction?

Answer 0.079.

*4-28. In Fig. 4-13, assume that the weight of the block is 200 N and the angle θ is 40°. Determine the friction force and the normal force. If the 40° angle is the limiting angle, what is the coefficient of static friction? Would a much heavier object have the same limiting angle if it were constructed of the same material?

*4-29. Assume that $\theta = 60°$, $W = 80$ lb, and $\mu_k = 0.2$ for the situation shown in Fig. 4-13. What push P directed up the plane will cause the block to move *up* the plane with constant speed? Note that in this case the friction force will be *down* the plane.

Answer 77.3 lb.

4-30. A mobile conveyer belt (Fig. 4-24) is used to load crates into a cargo plane. What is the maximum angle of elevation if the crates are not to slip on the surface of the belt? (Assume that $\mu_s = 0.3$.)

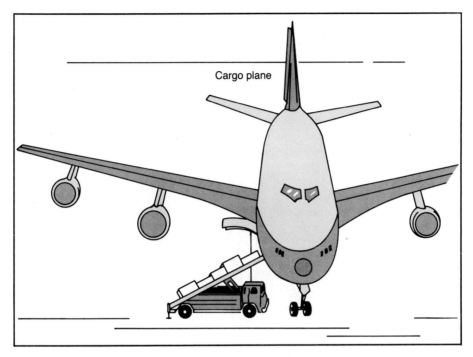

FIG. 4-24 A mobile conveyer belt for loading a cargo plane.

*4-31. In the arrangement of Fig. 4-25, assume that the coefficient of static friction is 0.3. Find the maximum weight which can be hung at point *O* in order to maintain equilibrium.

Answer 21.9 lb.

4-32. A wooden roof is sloped at a 40° angle. What is the minimum coefficient of static friction between the sole of a roofer's shoe and the roof if the roofer is not to slip?

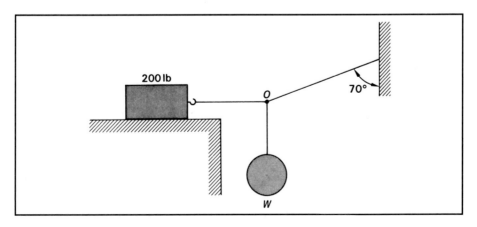

FIG. 4-25

5

TORQUE AND ROTATIONAL EQUILIBRIUM

OBJECTIVES

As a result of completing this chapter, you should be able to

1. Illustrate by example and definition your understanding of the terms *moment arm* and *torque*.
2. Calculate the *resultant torque* about any axis when given the magnitude and position of forces applied to an extended object.
3. Solve for unknown forces or distances by applying the *first and second conditions for equilibrium*.
4. Define and illustrate by example what is meant by the *center of gravity*.

In previous chapters, we have discussed forces which act at a single point. Translational equilibrium exists when the vector sum of the forces is zero. However, there are many cases in which the forces acting on an object do not have a common point of application. For example, a mechanic exerts a force on the handle of a wrench to tighten a bolt. A carpenter uses a long lever to pry the lid from a wooden box. The steering wheel of an automobile is turned by forces which do not have a common point of application. In such cases, there may be a *tendency to rotate* which we will define as *torque*. If we learn to measure or predict the torques produced by certain forces, we can obtain desired rotational effects. If no rotation is desired, there must be no resultant torque. This leads us naturally to a condition for *rotational equilibrium* which is very important for industrial and engineering applications.

5-1 ■ CONDITIONS FOR EQUILIBRIUM

When a body is in equilibrium, it is either at rest or in uniform motion. According to Newton's first law, only the application of a resultant force can change this condition. We have seen that if all forces acting on such a body intersect at a single point and their vector sum is zero, the system must be in equilibrium. When a body is acted on by forces which do not have a common *line of action,* the body may be in translational equilibrium but not in rotational equilibrium. In other words, it may not move to the right or left or up or down, but it may still rotate. In studying equilibrium, we must consider the point of application of each force as well as its magnitude.

Consider the forces exerted on the lug wrench in Fig. 5-1a. Two equal opposing forces **F** are applied to the right and to the left. The first condition for equilibrium tells us that the vertical and horizontal forces are balanced. Hence, the system is said to be in equilibrium. However, if the same two forces are applied as shown in Fig. 5-1b, the wrench has a definite tendency to rotate. This is true even though the vector sum of the forces is still zero. Clearly, we need a second condition for equilibrium to cover rotational motion. A formal statement of this condition will be given later.

In Fig. 5-1b, the forces **F** do not have the same *line of action*.

> The **line of action** of a force is an imaginary line extended indefinitely along the vector in both directions.

When the lines of action of forces do not intersect at a common point, rotation may occur about a point called the *axis of rotation*. In our example, the axis of rotation is an imaginary line passing through the stud perpendicular to the page.

5-2 ■ MOMENT ARM

The perpendicular distance from the axis of rotation to the line of action of a force is called the *moment arm* of that force. It is the moment arm that determines the effectiveness of a given force in causing rotational motion. For example, if we exert a force **F** at increasing distances from the center of a large wheel, it becomes easier and easier to rotate the wheel about its center. (See Fig. 5-2.)

If the line of action of a force passes through the axis of rotation (point A of Fig. 5-2), the moment arm is zero. No rotational effect is observed regardless of the magnitude of the force. In this simple example, the moment arms at points B and C are simply the distance from the axis of rotation to the point of application of the force. Note, however, that the line of action of a force is a mere geometric construction. The moment arm is drawn perpendicular to this line. It may be equal to the distance from the axis to the point of application of a force, but this is true only when the applied force is directed perpendicular to this distance. In the examples of Fig. 5-3, r repre-

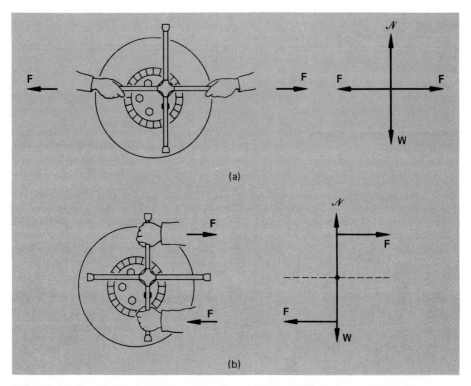

FIG. 5-1 (*a*) Equilibrium exists; the forces have the same line of action. (*b*) Equilibrium does not exist because opposing forces do not have the same line of action.

sents the moment arm and O represents the axis of rotation. Study each example, observing how the moment arms are drawn and reasoning whether the rotation is clockwise or counterclockwise about O.

5-3 ■ *TORQUE*

Force has been defined as a push or pull that tends to cause motion. *Torque* can be defined as the tendency to produce a change in rotational motion. It is also called the *moment of force* in some textbooks. As we have seen, rotational motion is affected by both the magnitude of a force and its moment arm. Thus, we will define torque τ as the product of a force and its moment arm:

$$\text{Torque} = \text{force} \times \text{moment arm}$$

$$\tau = Fr$$

$$(5\text{-}1)$$

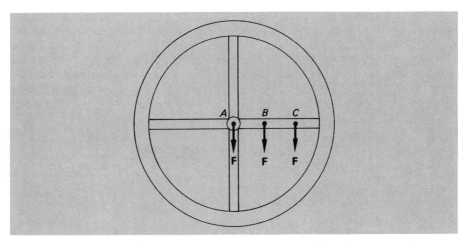

FIG. 5-2 The unbalanced force F has no rotational effect at point A but becomes increasingly effective as the moment arm gets longer.

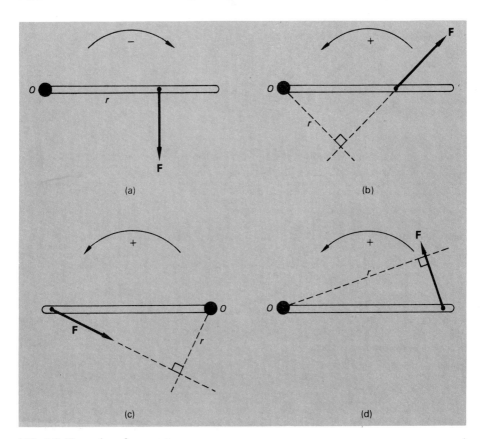

FIG. 5-3 Examples of moment arms.

It must be understood that r in Eq. (5-1) is measured perpendicular to the line of action of the force **F.** The units of torque are the units of force times distance, for example, newton-meters (N·m) and pound-feet (lb·ft).

Earlier, we established a sign convention to indicate the direction of forces. The direction of torque depends on whether it tends to produce clockwise (cw) or counterclockwise (ccw) rotation. We shall follow the same convention we used for measuring angles. If the force **F** tends to produce counterclockwise rotation about an axis, the torque will be considered positive. Clockwise torques will be considered negative. In Fig. 5-3, all the torques are positive (ccw) except for that in Fig. 5-3a.

EXAMPLE 5-1   A force of 20 N is exerted on a cable wrapped around a drum which has a diameter of 0.12 m. What is the torque produced about the center of the drum? (Refer to Fig. 5-4.)

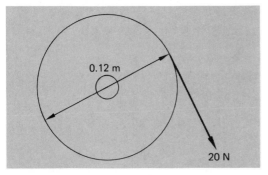

FIG. 5-4 The tangential force exerted by a cable wrapped around a drum.

Solution   Notice that the line of action of the 20-N force is perpendicular to the diameter of the drum. The moment arm is, therefore, equal to the radius of the drum (0.12 m/2). The torque is given by

$$\tau = Fr = -(20 \text{ N})(0.06 \text{ m}) = -1.2 \text{ N·m}$$

The torque is negative because it tends to cause clockwise rotation.

EXAMPLE 5-2   A mechanic exerts a 20-lb force at the end of a 10-in. wrench, as shown in Fig. 5-5. If this pull makes an angle of 60° with the handle, what is the torque produced on the nut?

Solution   First, we draw a neat sketch to scale, making sure that the handle is 10 in. long and that the 20-lb vector is at 60°. Then we extend the line of action of the force. The moment arm r is drawn perpendicular to this line and from the axis of rotation. Using a ruler, we measure the length of r as 8.66 in. Thus, the torque is

$$\tau = Fr = (20 \text{ lb})(8.66 \text{ in.}) = 173 \text{ lb·in.}$$

It is positive because the force tends to rotate the nut in a counterclockwise direction. Notice that the moment arm of the 20-lb force is *not* equal to the length of the wrench. How could the mechanic obtain more torque with the same force?

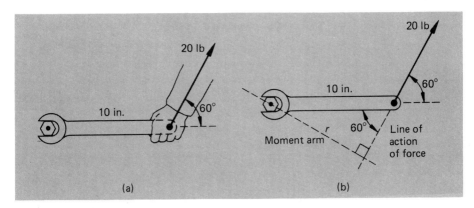

FIG. 5-5 Calculating the torque.

Problems involving angles, such as the preceding one, may also be solved by using trigonometry if you have sufficient mathematical skills. The moment arm r would be calculated from $\sin 60° = r/10$ in., or

$$r = (10 \text{ in.})(\sin 60°) = 8.66 \text{ in.}$$

The use of trigonometry is, of course, more accurate than graphical approaches.

In some applications, it is more useful to work with the *components* of a force to obtain the resultant torque. In the previous example, for instance, we could have used graphical or trigonometric approaches to find the components of the 20-lb vector. Instead of finding the torque of a single 20-lb force, we would then need to find the torque of the two component forces. As shown in Fig. 5-6, the 20-lb vector has a 17.3-lb component perpendicular to the wrench and a 10-lb component along the wrench. We can obtain these values graphically by drawing the 20-lb force to scale at 60° and measuring the component vectors. The components may also be found with trigonometry.

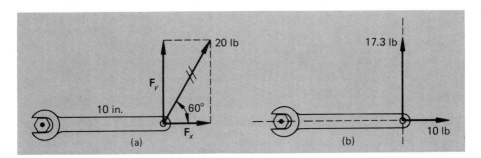

FIG. 5-6 Component method of calculating torque.

Notice from Fig. 5-6b that the line of action of the 10-lb force passes through the axis of rotation. It does not produce any torque because the moment arm is zero. The entire torque is, therefore, due to the 17.3-lb component perpendicular to the wrench. The moment arm of this force is the length of the wrench, since this distance is perpendicular. The torque is found from

$$\tau = F_y \cdot r$$
$$= (17.3 \text{ lb})(10 \text{ in.}) = 173 \text{ lb·in.}$$

Note that the same result is obtained by using this method. No more calculations are required because the horizontal component has a zero moment arm. If we choose the components of a force along and perpendicular to a known distance, we need only concern ourselves with the torque of the perpendicular component.

5-4 ■ RESULTANT TORQUE

In Chap. 3, we demonstrated that the resultant of a number of forces could be obtained by adding the x and y components of each force to find the components of the resultant:

$$R_x = A_x + B_x + C_x + \cdots \qquad R_y = A_y + B_y + C_y + \cdots$$

This procedure applies to forces which have a common point of intersection. Forces which do not have a common line of action may produce a resultant torque in addition to a resultant translational force. When the applied forces act in the same plane, the resultant torque is the algebraic sum of the positive and negative torques due to each force:

$$\boxed{\tau_R = \Sigma\tau = \tau_1 + \tau_2 + \tau_3 + \cdots} \qquad (5\text{-}2)$$

Remember that counterclockwise torques are positive and clockwise torques are negative.

EXAMPLE 5-3 What is the resultant torque about the center of the pulley in Fig. 5-7 if the diameter of the outer pulley is 4 ft and the diameter of the inner pulley is 2 ft?

Solution The sum of the torques about O, following the adopted sign convention, is

$$\tau_R = \tau_1 + \tau_2 = (16 \text{ lb})(1 \text{ ft}) - (12 \text{ lb})(2 \text{ ft})$$
$$= 16 \text{ lb·ft} - 24 \text{ lb·ft} = -8 \text{ lb·ft}$$

The minus sign on the resultant torque indicates that the resultant torque is clockwise.

One essential element of good problem-solving techniques is organization. The following procedure is useful for calculating resultant torque.

CALCULATING RESULTANT TORQUE
1. Read the problem, then draw and label a figure.
2. Construct a free-body diagram showing all forces, distances, and the axis of rotation.
3. Extend lines of action of each force with dotted lines.
4. Draw and label the moment arms for each force.

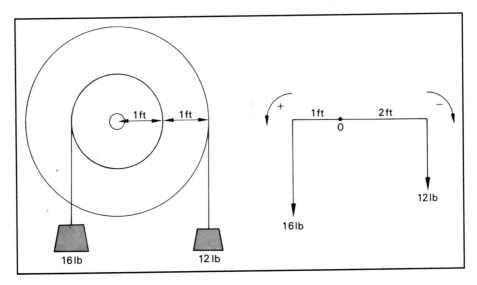

FIG. 5-7 Computing the resultant torque about the center of the pulley.

5. Calculate the moment arms if necessary.
6. Calculate the torques due to each force independent of others, and affix appropriate signs.
7. The resultant torque is the algebraic sum of the torques due to each force. See Eq. (5-2).

EXAMPLE 5-4 Calculate the resultant torque about axis A due to the forces shown in Fig. 5-8.

Solution Draw and label the free-body diagram as in Fig. 5-8b, showing the moment arms for each force. Next calculate the moment arms. This can be done graphically, but trigonometric methods are more accurate:

$$r_1 = (2 \text{ m})(\sin 60°) = 1.73 \text{ m}$$
$$r_2 = (3 \text{ m})(\sin 30°) = 1.5 \text{ m}$$
$$r_3 = 0 \text{ m} \quad \text{(line of action is through the axis)}$$

Observe that the 70-N force F_1 tends to produce clockwise $(-)$ torque about A; whereas the 60-N force F_2 tends to cause counterclockwise $(+)$ rotation. The torques due to each force are

$$\tau_1 = -(70 \text{ N})(1.73 \text{ m}) = -121 \text{ N·m}$$
$$\tau_2 = +(60 \text{ N})(1.5 \text{ m}) = +90 \text{ N·m}$$
$$\tau_3 = (20 \text{ N})(0 \text{ m}) = 0$$

The resultant torque is then found from Eq. (5-2):

$$\tau_R = \tau_1 + \tau_2 + \tau_3$$
$$= -121 \text{ N·m} + 90 \text{ N·m} + 0$$
$$= -31 \text{ N·m}$$

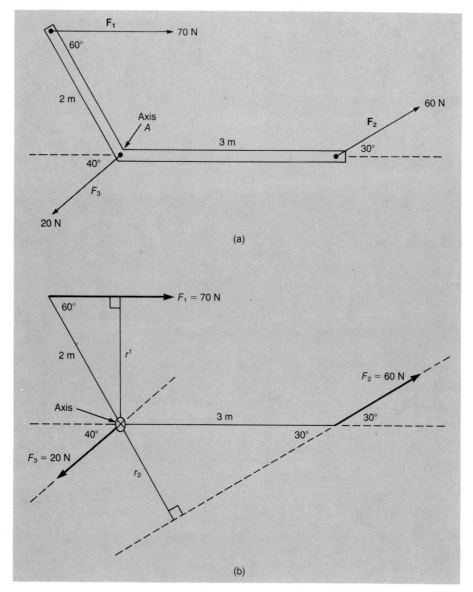

FIG. 5-8

The significance of the negative sign is that the resultant tendency is for clockwise rotation about axis A.

5-5 ■ EQUILIBRIUM

We are now ready to discuss the necessary conditions for rotational equilibrium. In Chap. 4, we stated the condition for translational equilibrium as two equations:

$$\boxed{\Sigma F_x = 0} \qquad \boxed{\Sigma F_y = 0} \qquad\qquad \text{(5-3)}$$

If we are to ensure that rotational effects are also balanced, we must require that there be no resultant torque. The *second condition for equilibrium* is

The algebraic sum of all the torques about any axis must be zero.

$$\boxed{\Sigma \tau_R = \tau_1 + \tau_2 + \tau_3 + \cdots = 0} \qquad\qquad \text{(5-4)}$$

The second condition for equilibrium simply tells us that the clockwise torques are exactly equal in magnitude to the counterclockwise torques. Moreover, since rotation is not occurring about any point, we may choose the axis of rotation at whatever point we wish. As long as the moment arms are measured from the line of action of each force to a common axis, the resultant torque will be zero. By choosing the axis at the point of application of a known force, that force will have a zero moment arm. Such forces affect translational equilibrium, but they do not contribute to resultant torque.

The solution to most equilibrium problems can be simplified by following the procedure outlined below:

EQUILIBRIUM PROBLEMS
1. Read the problem and draw a rough sketch with all given information.
2. Construct a free-body diagram showing the distances between each force and all other given information.
3. Choose and label an axis of rotation at the point of application of an unknown force.
4. Apply the first and second conditions for equilibrium to obtain three equations:

$$\Sigma F_x = 0 \qquad \Sigma F_y = 0 \qquad \Sigma \tau_A = 0$$

5. Solve for the unknowns by substitution.

EXAMPLE 5-5 Consider the arrangement shown in Fig. 5-9. A uniform beam weighing 200 N is held up by two supports A and B. Given the distances and forces listed in the figure, what are the forces exerted by the supports?

Solution A free-body diagram is drawn to show clearly all forces and distances between forces. Note that all the weight of the beam is considered to act at the center of the beam. The justification of this assumption is explained in the next section. The first condition for equilibrium requires that the sum of the horizontal and vertical forces be equal to zero. Only vertical forces are present in this case, so we have

$$\Sigma F_y = A + B - 300 \text{ N} - 200 \text{ N} - 400 \text{ N} = 0$$

from which

$$A + B = 900 \text{ N}$$

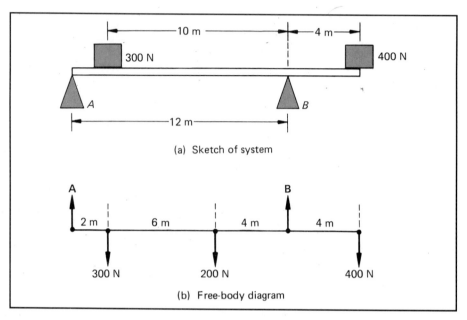

(a) Sketch of system

(b) Free-body diagram

FIG. 5-9 The solution of torque problems is aided by first drawing a rough sketch of the system. Then a free-body diagram can be drawn to indicate forces and distances.

Since this equation has two unknowns, we must have more information. Since the torques are also balanced, we can apply the second condition for equilibrium.

First, we must select an axis about which we can measure moment arms. The logical choice is at the point of application of an unknown force. Choosing the axis at B gives this force a zero moment arm. The sum of the torques about B results in the following equation:

$$\Sigma \tau_B = -A(12 \text{ m}) + (300 \text{ N})(10 \text{ m}) + (200 \text{ N})(4 \text{ m}) - (400 \text{ N})(4 \text{ m}) = 0$$

Note that the 400-N force and force A tend to produce clockwise rotation about B. (Their torques were negative.) Simplifying gives

$$-(12 \text{ m})A + 3000 \text{ N·m} + 800 \text{ N·m} - 1600 \text{ N·m} = 0$$

Adding $(12 \text{ m})A$ to both sides and simplifying, we obtain

$$2200 \text{ N·m} = (12 \text{ m})A$$

from which

$$A = 183 \text{ N}$$

Now to find the force exerted by support B, we can return to the equation found from the first condition for equilibrium:

$$A + B = 900 \text{ N}$$

Solving for B, we have

$$B = 900 \text{ N} - A = 900 \text{ N} - 183 \text{ N}$$
$$= 717 \text{ N}$$

As a check on this solution, we could choose the axis of rotation at A, then apply the second condition for equilibrium to find B.

5-6 ■ CENTER OF GRAVITY

Every particle on the earth has at least one force in common with every other particle—its *weight*. In the case of a body made up of many particles, these forces are essentially parallel and directed toward the center of the earth. Regardless of the shape and size of the body, there exists a point at which the entire weight of the body may be considered to be concentrated. This point is called the *center of gravity* of the body. Of course, not all the weight, in fact, acts at this point. But we would calculate the same torque about a given axis if we considered the weight to act at that point.

> **Center of gravity:** *The center of gravity of a body is that point at which the entire weight of a body may be considered as acting and produce the same resultant torque.*

If the weight of a body is uniformly distributed and if the body has a regular shape, such as a uniform sphere, cube, rod, or beam, then its center of gravity will be located at its geometric center. This fact was used in the examples of the previous section, where we considered the weight of an entire beam as acting at its center. Although the center of gravity is a fixed point, it does not necessarily lie within the body. For example, a hollow sphere, a circular hoop, and a carpenter's square all have centers of gravity outside the material of the body.

From the definition of the center of gravity, it is recognized that any body which is suspended at this point will be in equilibrium. This is true because the weight vector which represents the sum of forces acting on each portion of the body will have a zero moment arm. Thus, we can compute the center of gravity of an object by finding the point at which a single upward force will produce rotational equilibrium.

EXAMPLE 5-6 Compute the center of gravity of the two spheres in Fig. 5-10 if they are connected by a 30-in. rod of negligible weight.

Solution We first draw an upward vector indicating the force at the center of gravity which would balance the system. Suppose we choose to locate this vector at a distance from the center of the 16-lb sphere. The distance x would be drawn and labeled on the figure. Since the upward force must equal the sum of the downward forces, the first condition for equilibrium tells us that

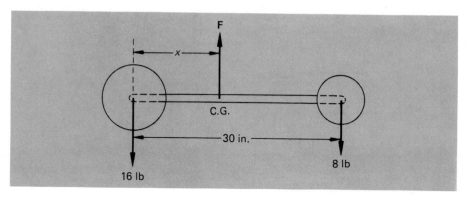

FIG. 5-10

$$F = 16 \text{ lb} + 8 \text{ lb} = 24 \text{ lb}$$

Now we will choose our axis of rotation at the center of the 16-lb sphere. This is a good choice because the distance x is measured from this point. The second condition for equilibrium can now be applied:

$$\Sigma \tau = (24 \text{ } lb)x - (8 \text{ lb})(30 \text{ in.}) = 0$$
$$(24 \text{ lb})x = 240 \text{ lb·in}$$
$$x = 10 \text{ in.}$$

If the rod joining the two spheres is suspended from the ceiling at a point 10 in. from the center of the 16-lb sphere, the system will be in equilibrium. This point is the center of gravity. You should show that the same conclusion follows if you sum torques about the right end or about any other point.

Summary

When forces acting on a body do not have the same line of action or do not intersect at a common point, rotation may occur. In this chapter, we introduced the concept of torque as a measure of the tendency to rotate. The major concepts are summarized below:

- The **moment arm** of a force is the perpendicular distance from the line of action of the force to the axis of rotation.
- The **torque** about a given axis is defined as the product of the magnitude of a force and its moment arm:

Torque = force × moment arm

$$\tau = Fr$$

It is positive if it tends to produce counterclockwise motion and negative if the motion produced is clockwise.

⊕ ccw

⊖ cw

- The **resultant torque** τ_R about a particular axis A is the algebraic sum of the torques produced by each force. The signs are determined by the above convention.

$$\tau_R = \Sigma\tau_A = F_1 r_1 + F_2 r_2 + F_3 r_3 + \cdots$$

- **Rotational equilibrium:** A body in rotational equilibrium has no resultant torque acting on it. In such cases, the sum of all the torques about *any* axis must equal zero. The axis may be chosen anywhere because the system is not tending to rotate about any point. This is called the second condition for equilibrium and may be written

$$\Sigma\tau = \tau_1 + \tau_2 + \tau_3 + \cdots = 0$$

- **Total equilibrium** exists when the first and second conditions are each satisfied. In such cases, three independent equations can be written:

$$\Sigma F_x = 0 \qquad \Sigma F_y = 0 \qquad \Sigma\tau = 0$$

By using these equations, unknown forces, distances, or torques may be determined for a given situation.
- The **center of gravity** of a body is that point at which all the weight of the body may be considered as acting and still produce the same resultant torque. In applications involving torque, it is useful to assume that the entire weight of an object acts at its center of gravity.

Questions

5-1. Define the following terms:
   a. line of action   **d.** torque
   b. axis of rotation   **e.** rotational equilibrium
   c. moment arm   **f.** center of gravity

5-2. You lift a heavy suitcase with your right hand. Describe and explain the position of your body.

5-3. You are asked to stand against a wall with your feet together so that the side of your right foot rests against the wall. You are then told to raise your left foot off the floor. Why can't you do this without falling?

5-4. Consider several vehicles such as vans, trucks, sports cars, and sedans. Discuss the factors which determine their stability as they negotiate sharp turns.

5-5. Pick up a meter stick and support it at random points by your left and right index fingers. Slowly move your fingers together, allowing the stick to slide on top of the fingers. Try again, choosing different points of initial support. Explain your observations based on your understanding of the principles of friction and torque.

5-6. Sometimes the first and second conditions for equilibrium are written in terms of the equality of opposites without regard to sign:

$$\Sigma F(\text{right}) = \Sigma F(\text{left}) \qquad \Sigma F(\text{up}) = \Sigma F(\text{down})$$
$$\Sigma\tau(\text{ccw}) = \Sigma\tau(\text{cw})$$

Discuss the advantages of this approach and the problems which might arise from its application.

5-7. If you know the weight of a brick to be 6 lb, describe how you could use a meter stick and a pivot to measure the weight of a can of nails.

5-8. Describe and explain the arm and leg motions used by a tightrope walker to maintain balance.

5-9. Discuss the following items and their use of the principle of torque: (*a*) screwdriver, (*b*) wrench, (*c*) pliers, (*d*) wheelbarrow, (*e*) nutcracker, and (*f*) crowbar.

Problems

Note: *Problems with angles may be worked graphically or by using trigo-nometry. Check with your instructor.*

5-1. (*a*) Draw and label the moment arms for the forces shown in Fig. 5-11. (*b*) Calculate the moment arms and state whether the torque will be positive or negative. (*c*) Determine the torque due to each force.
> *Answer (b)* 2 m, 3 m, 3.21 m; *(c)* −40 N·m, −150 N·m, 96.4 N·m.

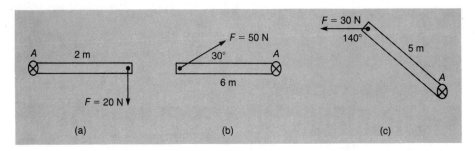

(a) (b) (c)

FIG. 5-11

5-2. Verify the torques found in Prob. 5-1 by working with the components of each force along and perpendicular to the bar.

5-3. An 8-in. wrench is used to *tighten* a nut onto a bolt. Is this negative or positive torque? Find the moment arms when the force on the end of the wrench makes the following angles with the handle: (*a*) 90°, (*b*) 60°, (*c*) 30°, and (*d*) 0°.
> *Answer (a)* 8 in.; *(b)* 6.93 in.; *(c)* 4 in.; *(d)* 0.

5-4. Assume that the force is 80 lb for each of the situations described in Prob. 5-3. What are the torques?

5-5. Find the resultant torque for each of the arrangements shown in Fig. 5-12 when the axis of rotation is at point *A*.
> *Answer (a)* 240 N·m; *(b)* −112 N·m; *(c)* 66.8 N·m.

5-6. Find the resultant torque if the axis of rotation is at point *B* for each of the arrangements in Fig. 5-12.

5-7. A leather belt is wrapped around a pulley 12 in. in diameter. A force of 6 lb is applied to the belt. What is the torque at the center of the shaft?
> *Answer 3 lb·ft.*

5-8. In Fig. 5-13, what is the resultant torque (*a*) about point *A* and (*b*) about the left end of the wooden beam? Neglect the weight of the beam.

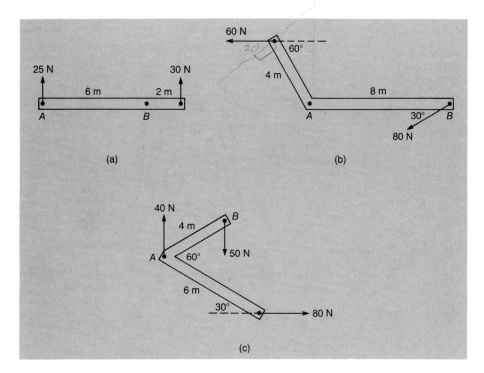

FIG. 5-12

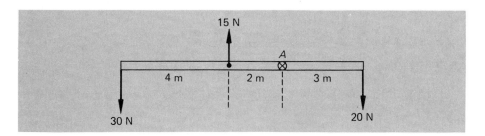

FIG. 5-13

5-9. Determine the resultant torque about point *A* and about point *C* in Fig. 5-14.

Answer (a) 74.3 lb in; *(b)* −80 lb in.

5-10. A V belt is wrapped around a pulley 16 in. in diameter. If a resultant torque of 4 lb·ft is required, what force must be applied along the belt?

5-11. Determine the unknown forces for the arrangements in Fig. 5-15. Assume that equilibrium exists and that the weight of the bar is negligible in each case.

Answer (a) A = 26.7 lb, F = 107 lb; *(b)* F_1 = 198 lb, F_2 = 87.5 lb;
(c) A = 50.9 N, B = 49.1 N.

5-12. A bridge 20 ft long is supported at each end. Consider the weight of the bridge to be 1200 lb, and assume that the entire weight acts at the center. What are the forces exerted at each end when a 600-lb mower is located 8 ft from the left end of the bridge?

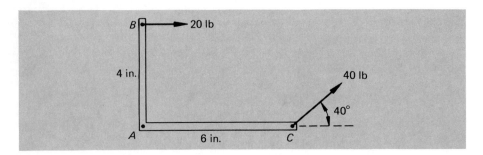

FIG. 5-14

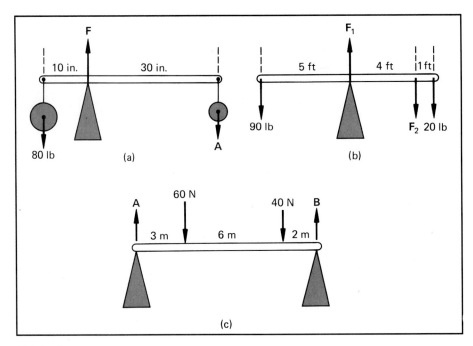

FIG. 5-15

5-13. The wheel and axle (see Fig. 5-16) is a simple machine used to lift heavy loads with less effort. If the radius R of the outer wheel is 16 in. and the radius r of the inner wheel is 4 in., what input force F_i is required to lift an 80-lb weight?

Answer 20 lb.

5-14. Torque is sometimes transmitted by a belt drive, such as the one shown in Fig. 5-17. A motor attached to the smaller pulley delivers a torque of 600 lb·ft. If the radius of the small pulley is 4 in. and the radius of the large pulley is 8 in., what is the output torque at the larger pulley? *Hint:* The force exerted on the belt by the smaller pulley is transmitted through the belt to the edge of the large pulley.

5-15. What is the resultant torque about the hinge in Fig. 5-18?

Answer −3.42 Nm.

5-16. What is the resultant torque about point B in Fig. 5-14?

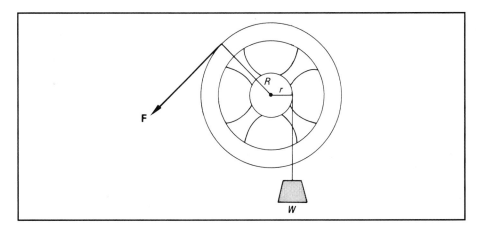

FIG. 5-16 Wheel and axle.

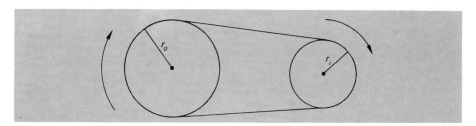

FIG. 5-17 Belt drive.

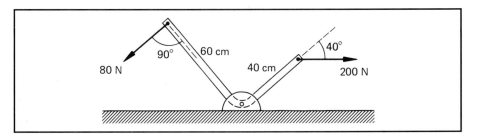

FIG. 5-18

5-17. A 10-ft platform is placed across two stepladders, one at each end. The platform weighs 40 lb, and a painter located 4 ft from the right end weighs 180 lb. What are the forces exerted by each of the ladders?

Answer A = 92 lb; B = 128 lb.

5-18. For the wheel and axle described in Fig. 5-16, assume that $R = 60$ cm and $r = 20$ cm. What is the resultant torque about the axle if the weight W is 60 N and the input force is 80 N?

5-19. A 200-lb vertical force is applied to the end of a lever which is attached to a shaft, as in Fig. 5-19. (*a*) What is the torque of the 200-lb force about O? (*b*) If the 200-lb

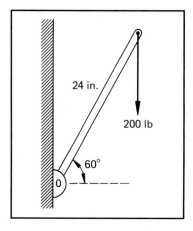

24 in.

200 lb

60°

0

FIG. 5-19

force were removed, what horizontal force acting at the same end would produce the same torque? Disregard the weight of the lever.

Answer (a) −200 lb·ft; *(b)* 115 lb.

5-20. A horizontal metal rod 80 cm in length is supported 20 cm from the right end. What downward force at the left end is necessary to lift a 60-N weight at the right end? Neglect the weight of the rod.

5-21. Weights of 2, 5, 8, and 10 N are hung from a 10-m rod at distances of 2, 4, 6, and 8 m, respectively, from the left end. (*a*) What total upward force is needed to support the rod and weights? (Neglect the weight of the rod.) (*b*) At what distance from the left end should a single support be placed in order to balance the system?

Answer (a) 25 N; *(b)* 6.08 m.

5-22. Weights of 100, 200, and 500 lb are placed on a board resting on two supports, as shown in Fig. 5-20. Neglecting the weight of the board, what are the forces exerted by the supports?

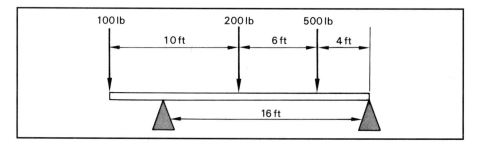

100 lb   200 lb   500 lb

10 ft   6 ft   4 ft

16 ft

FIG. 5-20

5-23. A uniform bar 24 ft long and weighing 400 lb is supported by a fulcrum 8 ft from the right end. If an 800-lb weight is attached to the right end, what downward force must be exerted at the left end to balance the system? What is the force exerted by the support?

Answer 300 lb; 1500 lb.

5-24. What upward force *F* is needed to support the wheelbarrow in Fig. 5-21?

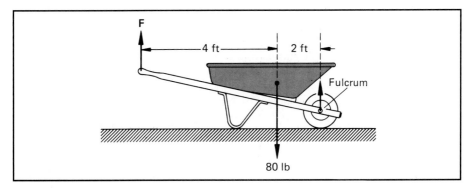

FIG. 5-21 Wheelbarrow.

5-25. Consider the boom arrangement of Fig. 5-22. The 12-ft boom has a weight of 100 lb, and the weight *W* is 400 lb. What is the tension in the cable if it is attached 2 ft from the right end? *Hint:* If you choose your axis at *A*, you do not need to know the force *F* exerted by the wall at the hinge.

Answer 897 lb.

5-26. In Fig. 5-22, the beam is 2 m long, and the cable is attached 20 cm from the right end. The breaking strength of the cable is 800 N, and the beam weighs 200 N. What is the largest weight that can be hung from the right end?

5-27. The handle of a screwdriver is 1 in. in diameter, and the metal shaft is only ¼ in. in diameter. If a tangential force of 40 lb is applied to the handle, what tangential force is transmitted to the shaft for the purpose of driving screws?

Answer 160 lb.

5-28. A 30-lb box and a 50-lb box are positioned at opposite ends of a 16-ft board. The board is supported only at its midpoint. Where should a third box weighing 40 lb be placed in order to balance the system?

5-29. Compute the center of gravity of a sledgehammer if the metal head weighs 12 lb and the 32-in. supporting handle weighs 2 lb. Assume that the handle is of uniform weight and construction.

Answer 2.29 in. from head.

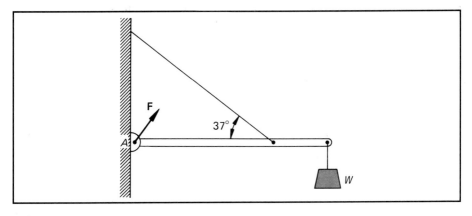

FIG. 5-22 Horizontal boom.

5-30. Find the center of gravity for two spheres connected by a rod, as in Fig. 5-10. In this example, assume that the rod is 20 cm long and of negligible weight. The left sphere weighs 40 N, and the right sphere weighs 10 N.

5-31. A beam 6 m long has weights of 400, 600, and 300 N placed at distances of 1, 2, and 5 m from the left end, respectively. Consider that the beam weighs 1200 N and that this entire weight acts at the center of the beam. Draw a diagram showing all forces and the distances between the forces. How far from the left end should a single support be placed in order to balance the system? *Hint:* Indicate the upward force F on your figure at some arbitrary point. Label it as being a distance x from the left end. The first condition for equilibrium should allow you to find the value of F. Finally, you can find x by summing the torques about the left end of the beam.

Answer 2.68 m from left.

UNIFORMLY ACCELERATED MOTION

After completing this chapter, you should be able to

1. Define and give formulas for *average speed* and *average acceleration*.
2. Solve problems involving *time, distance, average speed,* and *average acceleration*.
3. Apply one of the four general equations for uniformly accelerated motion to solve for one of the five parameters: *initial speed, final speed, acceleration, time,* and *distance*.
4. Solve acceleration problems involving free-falling bodies in a gravitational field.

Everything in the physical world is in motion, from the largest galaxies in the universe to the elementary particles within atoms. We must study the motions of objects if we are to understand their behavior and learn to control them. Uncontrolled or erratic motion, as in falling buildings, destructive vibrations, or a runaway car, can create dangerous situations, but controlled motion often serves our convenience. It is important to be able to analyze motion and to represent it in terms of fundamental formulas.

6-1 ■ SPEED AND VELOCITY

The simplest kind of motion an object can experience is uniform motion in a straight line. If an object covers the same distance in each successive unit of time, it is said to move with *constant speed.* For example, if a train covers a distance of 26 ft of track every second that it moves, we say that it has a constant speed of 26 ft/s. Whether the speed is constant or not, the average speed of a moving object is defined by

$$\text{Average speed} = \frac{\text{distance traveled}}{\text{time elapsed}}$$

$$\bar{v} = \frac{s}{t} \tag{6-1}$$

The bar over the v means that the speed represents an average value for the time interval t.

Remember that the dimension of speed is the ratio of a length to a time interval. Common units are *kilometers per hour* (km/h), *miles per hour,* (mi/h), *meters per second* (m/s), and *feet per second* (ft/s).

EXAMPLE 6-1 A golfer sinks a putt 3 s after the ball leaves the face of the club. If the ball traveled with an average speed of 0.8 m/s, how long was the putt?

Solution We solve Eq. (6-1) for s and then substitute known values for $\bar{v}$ and t:

$$s = \bar{v}t = (0.8 \text{ m/s})(3 \text{ s}) = 2.4 \text{ m}$$

Thus, the distance of the putt is 2.4 m.

It is important to recognize that speed is a scalar quantity which is completely independent of direction. In Example 6-1, we did not need to know either the speed of the ball at any point or the nature of its path. Similarly, the average speed of a car traveling from Atlanta to Chicago is a function of only the distance registered on the car's odometer and the time required to make the trip. It makes no difference as far as computation is concerned whether the driver of the car took the direct or the scenic route, or even if he or she stopped for meals.

Sometimes it is necessary to distinguish between the scalar quantity *speed* and its directional counterpart *velocity.* This is best done by recalling the difference between *distance* and *displacement,* as discussed in Chap. 3. Suppose that an object moves along the broken path from A to B in Fig. 6-1, covering a distance $s = 500$ mi. The actual displacement might be $\mathbf{D} = (400 \text{ mi}, 45°)$. The *speed* is dependent on only the distance traveled and the time elapsed. If the trip took 8 h, we have

$$\bar{v} = \frac{s}{t} = \frac{500 \text{ mi}}{8 \text{ h}} = 62.5 \text{ mi/h}$$

The average *velocity*, however, must consider the *displacement* magnitude and direction:

$$\bar{\mathbf{v}} = \frac{\mathbf{D}}{t} = \frac{400 \text{ mi}, 45°}{8 \text{ h}} = 50 \text{ mi/h}, 45°$$

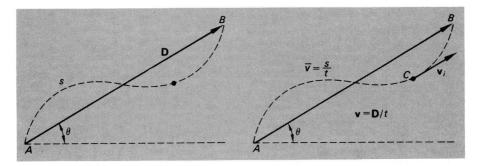

FIG. 6-1 Displacement and velocity are vector quantities, whereas distance and speed are independent of direction: s, distance; $\mathbf{D}$, displacement; $\mathbf{v}$, velocity; t, time.

We can see from this example that if an object moves along a curved path, the difference between speed and velocity is one of both magnitude and direction.

Automobiles cannot always travel at constant speeds for long periods. In traveling from point A to point B, we may be required to slow down or speed up because of road conditions. For this reason, it is sometimes useful to talk of *instantaneous speed* or *instantaneous velocity*.

The **instantaneous speed** *is a scalar quantity representing the speed at the instant the car is at an arbitrary point* C. *It is, therefore, the time rate of change in distance.*

The **instantaneous velocity** *is a vector quantity representing the velocity at any point* C. *It is the time rate of change in displacement.*

In this chapter we shall be concerned primarily with average speed and average velocity. Note that if the motion is along a straight path (no change in direction), the terms *speed* and *velocity* might be used interchangeably. However, it is a good practice to reserve the term *velocity* for the more complete description of motion.

6-2 ■ *ACCELERATED MOTION*

In most cases, the velocity of a moving object changes as motion continues. This type of motion is called *accelerated motion*. The time rate at which velocity changes is called the *acceleration*. For example, suppose we observe the motion of a body for a time t. The initial velocity v_0 of the body will be defined as its velocity at the beginning of the time interval, i.e., when $t = 0$. The final velocity is defined as the velocity v_f the body has at the end of the time interval, when $t = t$. Thus, if we are able to measure the initial and final velocities of a moving object, we can say that its average acceleration is given by

$$\text{Average acceleration} = \frac{\text{change in velocity}}{\text{time interval}}$$

$$\bar{\mathbf{a}} = \frac{\mathbf{v}_f - \mathbf{v}_0}{t} \qquad (6\text{-}2)$$

Acceleration written in the above manner is a vector quantity and thus depends upon changes in direction as well as changes in magnitude. If the direction of motion is a straight line, only the speed of the object is changing. If the object follows a curved path, both magnitude and directional changes occur and the acceleration is no longer in the same direction as the motion. In fact, if the curved path follows a perfect circle, the acceleration can be at right angles to the motion. In this case, only the direction of motion is changing while the speed at any point on the circle is constant. The latter type of motion will be treated in a later chapter.

6-3 ■ *UNIFORMLY ACCELERATED MOTION*

The simplest kind of acceleration is motion in a straight line in which the speed changes at a constant rate. This special kind of motion is generally referred to as *uniformly accelerated motion,* or *constant acceleration*. Since there is no change in direction, the vector difference in Eq. (6-2) becomes simply the algebraic difference between the magnitude of the final velocity v_f and the magnitude of the initial velocity v_0. Also, for uniform acceleration, the average value $\bar{a}$ is the same as the constant instantaneous value a:

$$a = \frac{v_f - v_0}{t} \qquad (6\text{-}3)$$

For example, consider a car which moves with constant acceleration from point A to point B, as in Fig. 6-2. The car's speed at A is 40 ft/s, and its speed at B is 60 ft/s. If the increase in speed requires 5 s, the car's acceleration can be determined from Eq. (6-3). Hence

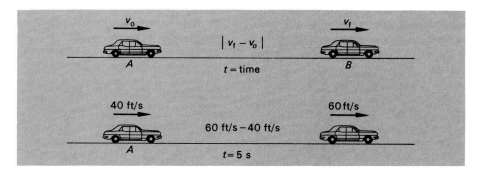

FIG. 6-2 Uniformly accelerated motion.

$$a = \frac{v_f - v_0}{t} = \frac{60 \text{ ft/s} - 40 \text{ ft/s}}{5 \text{ s}}$$

$$= \frac{20 \text{ ft/s}}{5 \text{ s}} = 4 \text{ ft/s}^2$$

The answer is read as *4 feet per second per second* or *4 feet per second squared*. It means that every second the car increases its speed by 4 ft/s. Since the car had already reached a speed of 40 ft/s when we started our time ($t = 0$), 1,2, and 3 s later it would have speeds of 44, 48, and 52 ft/s, respectively.

EXAMPLE 6-2 A train reduces its speed from 60 to 20 mi/h in 8 s. Find the acceleration.

Solution We first note that there is an inconsistency of units between velocity in miles per hour and time in seconds. The velocity is converted to feet per second as follows:

$$60 \frac{\text{mi}}{\text{h}} \times \frac{5280 \text{ ft}}{1 \text{ mi}} \times \frac{1 \text{ h}}{3600 \text{ s}} = 88 \text{ ft/s}$$

Similarly, it is determined that 20 mi/h is equal to 29.3 ft/s. Substitution into Eq. (6-3) gives

$$a = \frac{v_f - v_0}{t} = \frac{29.3 \text{ ft/s} - 88 \text{ ft/s}}{8 \text{ s}}$$

$$= -7.33 \text{ ft/s}^2$$

The minus sign tells us that the speed is *reduced* by 7.33 ft/s every second. This type of acceleration is sometimes referred to as *deceleration*.

Many times the same equation is used to solve for different quantities. You should, therefore, solve each equation literally for each symbol in the equation. A very convenient form of Eq. (6-3) arises when it is solved explicitly for the final speed. Thus

$$v_f = v_0 + at \tag{6-4}$$

Final speed = initial speed + change in speed

EXAMPLE 6-3   An automobile maintains a constant acceleration of 8 m/s². If its initial speed was 20 m/s, what will its speed be after 6 s?

Solution   The final speed is obtained from Eq. (6-4):

$$v_f = v_0 + at = 20 \text{ m/s} + (8 \text{ m/s}^2)(6 \text{ s})$$

or

$$v_f = 20 \text{ m/s} + 48 \text{ m/s}$$

Hence the final speed is

$$v_f = 68 \text{ m/s}$$

Now that the concept of initial and final velocities is understood, let us return to the equation for average speed and express it in terms of initial and final states. The average velocity of an object moving with uniform acceleration is found just as the arithmetic mean of two numbers. Given an initial speed and a final speed, the average speed is simply

$$\overline{v} = \frac{v_f + v_0}{2} \tag{6-5}$$

Utilizing this relation in Eq. (6-1) gives us a more useful expression for computing distance:

$$\boxed{s = \overline{v}t = \frac{v_f + v_0}{2} t} \tag{6-6}$$

EXAMPLE 6-4   A moving object increases its speed uniformly from 200 to 400 cm/s in 2 min. (*a*) What is its average speed, and (*b*) how far did it travel in 2 min?

Solution   The average speed is found by direct substitution into Eq. (6-5):

$$\overline{v} = \frac{v_f + v_0}{2} = \frac{400 \text{ cm/s} + 200 \text{ cm/s}}{2}$$

or

$$= \frac{600 \text{ cm/s}}{2} = 300 \text{ cm/s}$$

The distance traveled is then found from Eq. (6-1):

$$s = \bar{v}t = (300 \text{ cm/s})(2 \text{ min})$$

The units of time are inconsistent, but since 2 min = 120 s, we have

$$s = (300 \text{ cm/s})(120 \text{ s}) = 36,000 \text{ cm}$$

6-4 ■ OTHER USEFUL RELATIONS

Thus far we have presented two fundamental relations. One arises from the definition of velocity and the other from the definition of acceleration. They are

$$s = \bar{v}t = \frac{v_f + v_0}{2}t \tag{6-6}$$

and

$$v_f = v_0 + at \tag{6-4}$$

Although these are the only formulas necessary to attack the many problems presented in this chapter, two other very useful relationships can be obtained from them. The first is derived by eliminating the final velocity from Eqs. (6-6) and (6-4). Substituting the latter into the former yields

$$s = \frac{(v_0 + at) + v_0}{2}t$$

Simplifying gives

$$s = \frac{(2v_0 + at)t}{2} = \frac{2v_0t + at^2}{2}$$

or

$$s = v_0t + \tfrac{1}{2}at^2 \tag{6-7}$$

The second relation is derived by eliminating t from the basic equations. Solving Eq. (6-4) for t yields

$$t = \frac{v_f - v_0}{a}$$

which on substitution into Eq. (6-6) gives

$$s = \left(\frac{v_f + v_0}{2}\right)\left(\frac{v_f - v_0}{a}\right)$$

from which

$$2as = v_f^2 - v_0^2 \qquad \qquad (6\text{-}8)$$

Although these two equations add no new information, they are very useful in solving problems in which either the final velocity or the time is not given and you need to find one of the other parameters.

6-5 ■ SOLUTION OF ACCELERATION PROBLEMS

Although the solution to problems involving constant acceleration depends primarily upon choosing the correct formula and substituting known values, there are several suggestions that can help the beginning student. Physical problems of this kind frequently involve motion which either started from rest or is brought to a stop from some initial speed. In either case, the formulas discussed can be simplified by the substitution of either $v_0 = 0$ or $v_f = 0$, as the case may be. Table 6-1 summarizes the general formulas.

TABLE 6-1 SUMMARY OF ACCELERATION FORMULAS

(1)	$s = \dfrac{v_f + v_0}{2}t$
(2)	$v_f = v_0 + at$
(3)	$s = v_0 t + \frac{1}{2}at^2$
(4)	$2as = v_f^2 - v_0^2$

A close look at the four general equations will reveal a total of five parameters: s, v_0, v_f, a, and t. Given any three of these quantities, the remaining two can be found from the general equations. Therefore, a good starting point in solving any problem is to read it thoroughly with a view to establishing the three quantities required for solution. It is also important to choose a direction to call positive and apply it consistently to speed, distance, and acceleration when you insert the values into equations.

If you have difficulty in deciding which equation should be used, it may help to recall the conditions such an equation must satisfy. First, it must contain the unknown parameter. Second, all other parameters that appear in the equation must be known. For example, if a problem gives you values for v_f, v_0, and t, you may solve for a in Eq. (2) of Table 6-1.

The following procedure illustrates a technique which might be followed for solving acceleration problems:

1. Read the problem; then draw and label a sketch.
2. Indicate the consistent positive direction.
3. Establish three given parameters and two that are unknown. Make sure the signs and units are consistent.

Given: _____   Find: _____

    _____   _____

    _____

4. Select an equation which contains one of the unknown parameters but not the other one.
5. Substitute known quantities and solve the equation.

The examples which follow are abbreviated and do not include figures, but they do illustrate the approach.

EXAMPLE 6-5   A motorboat starting from rest attains a speed of 30 mi/h in 15 s. What was its acceleration, and how far did it travel?

Given: $v_0 = 0$ Find: $a = ?$
      $v_f = 30$ mi/h $= 44$ ft/s $s = ?$
      $t = 15$ s

Solution   In solving for acceleration, we must choose an equation which contains a but not s. Equation (2) can be used where $v_0 = 0$. Hence

$$v_f = at$$

from which

$$a = \frac{v_f}{t} = \frac{44 \text{ ft/s}}{15 \text{ s}}$$
$$= 2.93 \text{ ft/s}^2$$

The distance can be found from Eq. (1), as follows:

$$s = \left(\frac{v_f + v_0}{2}\right)t = \left(\frac{44 \text{ ft/s} + 0}{2}\right)t = \left(\frac{44 \text{ ft/s}}{2}\right)(15 \text{ s})$$
$$= 330 \text{ ft}$$

EXAMPLE 6-6   An airplane lands on a carrier deck at 200 mi/h and is brought to a stop in 600 ft. Find the acceleration and the time required to stop.

Given: $v_0 = 200$ mi/h $= 293$ ft/s   Find: $a = ?$
      $v_f = 0$ $t = ?$
      $s = 600$ ft

Solution   Choosing Eq. (4), we solve for a as follows:

$$2as = v_f^2 - v_0^2$$
$$(2a)(600 \text{ ft}) = 0 - (293 \text{ ft/s})^2$$

$$a = \frac{-(293 \text{ ft/s})^2}{(2)(600 \text{ ft})} = \frac{-86,000 \text{ ft}^2/\text{s}^2}{1200 \text{ ft}}$$

$$= -71.7 \text{ ft/s}^2$$

Then, using Eq. (2) to solve for the time yields

$$t = \frac{v_f - v_0}{a} = \frac{-v_0}{a} = \frac{-293 \text{ ft/s}}{-71.7 \text{ ft/s}^2}$$

$$= 4.09 \text{ ft/s}^2$$

EXAMPLE 6-7   A train traveling initially at 16 m/s is under a constant accelera-tion of 2 m/s². How far will it travel in 20 s? What will its final velocity be?

Given: $v_0 = 16$ m/s   Find: $s = ?$
$a = 2$ m/s² $v_f = ?$
$t = 20$ s

Solution   From Eq. (3) we have

$$s = v_0 t + \tfrac{1}{2} a t^2$$
$$= (16 \text{ m/s})(20 \text{ s}) + \tfrac{1}{2}(2 \text{ m/s}^2)(20 \text{ s})^2$$
$$= 320 \text{ m} + 400 \text{ m} = 720 \text{ m}$$

The final velocity is found from Eq. (2):

$$v_f = v_0 + at = 16 \text{ m/s} + (2 \text{ m/s}^2)(20 \text{ s})$$
$$= 56 \text{ m/s}$$

6-6 ■ SIGN CONVENTION IN ACCELERATION PROBLEMS

The signs of acceleration, displacement, and velocity are mutually independent and are each determined by different criteria. Probably no other point is more confusing for beginning students. Whenever the direction of motion changes, such as with an object tossed into the air or with an object attached to an oscillating spring, the signs of displacement and acceleration are particularly difficult to visualize. Only the sign of velocity is determined by the direction of motion. The sign of displacement depends on the location or position of the object, and the sign of acceleration is determined by the force which produces the acceleration.

Once a choice is made for the positive direction, the following convention will determine the signs of velocity, displacement, and acceleration:

Velocity *is positive or negative depending on whether the direction of motion is with or against the chosen positive direction.*

Acceleration *is positive or negative depending on whether the resultant force is with or against the chosen positive direction.*

Displacement *is positive or negative depending on the location or position of the object relative to its zero position.*

You will recall from Newton's first law that an object will not experience a change in velocity unless it is acted on by a resultant force. It should be apparent, then, that the resultant force is what produces the acceleration. The direction and, therefore, the sign of acceleration are determined by this force. An object can be moving to the right and have an acceleration to the left. An object may be increasing its speed and still have a negative acceleration. These events are confusing only if we confuse the motion of the object with the resultant force on the object.

Whenever the direction of motion changes during acceleration, the parameter *s* refers to the *displacement* of the object rather than to the total distance traveled. In the example of a rock thrown vertically upward, the value of *s* may be +8 m on the way up as well as later on the way down. The position becomes negative only when the rock drops below the point of release ($s = 0$).

6-7 ■ GRAVITY AND FREELY FALLING BODIES

Much of our knowledge about the physics of falling bodies originated with the Italian scientist Galileo Galilei (1564–1642). He was the first to demonstrate that in the absence of friction all bodies, large or small, heavy or light, fall to the earth with the same acceleration. This was a revolutionary idea, for it contradicted what a person might expect. Until the time of Galileo, people followed the teachings of Aristotle that heavy objects fall proportionally faster than lighter objects. The classic explanation for the paradox rests with the fact that heavier bodies are proportionately more difficult to accelerate. This resistance to motion is a property of a body called its *inertia*. Thus, in a vacuum, a feather will fall at the same rate as a steel ball because the larger inertial effect of the steel ball compensates exactly for its larger weight. (See Fig. 6-3.)

In the treatment of falling bodies given in this chapter, the effects of air friction are neglected entirely. Under these circumstances, gravitational acceleration is uniformly accelerated motion. At sea level and 45° latitude, this acceleration has been measured to be 32.17 ft/s², or 9.806 m/s², and is denoted by *g*. For our purposes, the following values will be sufficiently accurate:

$$\boxed{\begin{aligned} g &= 32 \text{ ft/s}^2 \\ g &= 9.8 \text{ m/s}^2 \end{aligned}} \tag{6-9}$$

Since gravitational acceleration is constant acceleration, the same general equations of motion apply. However, one of the parameters is known in advance and need not be stated in the problem. If the constant *g* is inserted into the general equations (Table 6-1), the following modified forms will result:

FIG. 6-3 All bodies fall with the same acceleration in a vacuum.

$$(1a) \qquad s = \frac{v_f + v_0}{2}t \qquad s = \bar{v}t$$

$$(2a) \qquad v_f = v_0 + gt$$
$$(3a) \qquad s = v_0t + \tfrac{1}{2}gt^2$$
$$(4a) \qquad 2gs = v_f^2 - v_0^2$$

Before we use these equations, a few general comments are in order. In problems dealing with free-falling bodies, it is extremely important to choose a direction to call positive and to follow through consistently in the substitution of known values. The sign of the answer is necessary to determine the location of a point or the direction of the velocity at specific times. For example, the distance s in the above equations represents the distance above or below the origin. If the upward direction is chosen as positive, a positive value for s indicates a distance above the starting point; if s is negative, it represents a distance below the starting point. Similarly, the signs of v_0, v_f, and g indicate their directions.

EXAMPLE 6-8   A rubber ball is dropped from rest, as shown in Fig. 6-4. Find its velocity and position after 1, 2, 3, and 4 s.

Solution   Since all parameters will be measured downward, it will be more convenient to choose the downward direction as positive. Organizing the data, we have

Given: $v_0 = 0$ Find: $v_f = ?$
$g = +32 \text{ ft/s}^2$ $s = ?$
$t = 1, 2, 3, \text{ and } 4 \text{ s}$

The velocity as a function of time is given by Eq. (2a), where $v_0 = 0$:

$$v_f = v_0 + gt = gt$$
$$= (32 \text{ ft/s}^2)t$$

After 1 s we have

$$v_f = (32 \text{ ft/s}^2)(1 \text{ s}) = 32 \text{ ft/s} \qquad \text{downward}$$

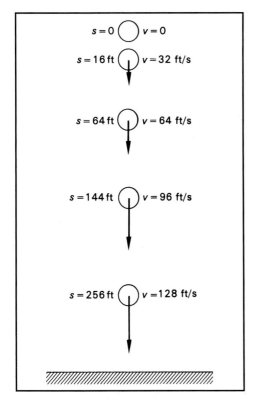

FIG. 6-4 A freely falling body has a constant downward acceleration of 32 ft/s².

Similar substitution of $t = 2$, 3, and 4 s will yield final velocities of 64, 96, and 128 ft/s, respectively. All these velocities are directed downward, since that direction was chosen as positive.

The position as a function of time is calculated from Eq. (3a). Since the initial velocity is zero, we write

$$s = v_0t + \tfrac{1}{2}gt^2 = \tfrac{1}{2}gt^2$$

from which

$$s = \tfrac{1}{2}(32 \text{ ft/s}^2)t^2 = (16 \text{ ft/s}^2)t^2$$

After 1 s the body will have fallen a distance

$$s = (16 \text{ ft/s}^2)(1 \text{ s})^2 = (16 \text{ ft/s}^2)(1 \text{ s}^2)$$
$$= 16 \text{ ft}$$

After 2 s,

$$s = (16 \text{ ft/s}^2)(2 \text{ s})^2 = (16 \text{ ft/s}^2)(4 \text{ s}^2)$$
$$= 64 \text{ ft}$$

Similarly, calculations give 144 and 256 ft for the positions after 3 and 4 s, respectively. The above results are summarized in Table 6-2.

TABLE 6-2

Time t, s	Speed at the end of time t, ft/s	Position at the end of time t, ft
0	0	0
1	32	16
2	64	64
3	96	144
4	128	256

EXAMPLE 6-9   Assuming that a ball is projected upward with an initial velocity of 96 ft/s, explain without using equations how its upward motion is just the reverse of its downward motion.

Solution   We shall assume the upward direction to be positive, making the acceleration due to gravity equal to -32 ft/s^2. The negative sign indicates that the speed of an object projected vertically will have its speed reduced by 32 ft/s every second it rises. (Refer to Fig. 6-5.) If its initial speed is 96 ft/s, its speed after 1 s will be reduced to 64 ft/s. After 2 s its speed will be 32 ft/s, and after 3 s its speed will be reduced to zero. When the speed becomes zero, the ball has reached its maximum height and

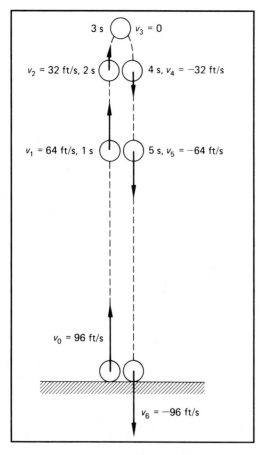

FIG. 6-5 A ball that is thrown vertically upward returns to the ground with the same speed.

begins to fall freely from rest. However, now the speed of the ball will be *increasing* by 32 ft/s every second, since both the direction of motion and the acceleration due to gravity are in the negative direction. Its speed after 4, 5, and 6 s will be -32, -64, and -96 ft/s, respectively. Except for the sign, which indicates the direction of motion, the speeds are the same at equal heights above the ground.

EXAMPLE 6-10   A baseball thrown vertically upward from the roof of a tall building has an initial velocity of 20 m/s. (*a*) Calculate the time required to reach its maximum height. (*b*) Find the maximum height. (*c*) Determine its position and velocity after 1.5 s. (*d*) What are its position and velocity after 5 s? (See Fig. 6-6.)

Solution (a)   Let us choose the upward direction as positive, since the initial speed is directed upward. At the highest point, the final velocity of the ball will be zero. Organizing the known data, we have

$$\text{Given: } v_0 = 20 \text{ m/s} \qquad \text{Find: } t = ?$$
$$v_f = 0 \qquad\qquad\quad s = ?$$
$$g = -9.8 \text{ m/s}^2$$

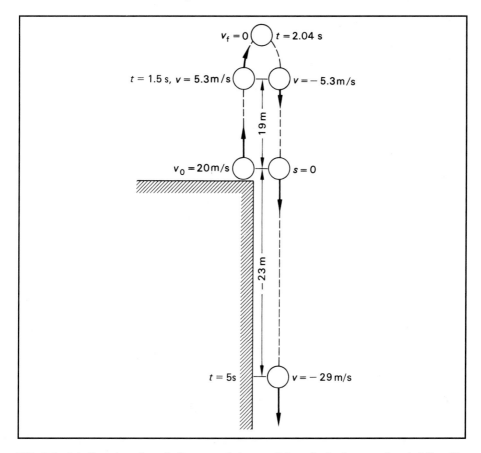

FIG. 6-6 A ball projected vertically upward rises until its velocity is zero, then it falls with increasing downward velocity.

The time required to reach the maximum height can be determined from Eq. (2a):

$$t = \frac{v_f - v_0}{g} = \frac{v_0}{g}$$

$$= \frac{-20 \text{ m/s}}{-9.8 \text{ m/s}^2} = 2.04 \text{ s}$$

Solution (b)   The maximum height is found by setting $v_f = 0$ in Eq. (1a):

$$s = \frac{v_f + v_0}{2} t = \frac{v_0}{2} t$$

$$= \frac{20 \text{ m/s}}{2} (2.04 \text{ s}) = 20.4 \text{ m}$$

Solution (c)   To find the position and velocity after 1.5 s, we must establish new conditions:

$$\text{Given: } v_0 = 20 \text{ m/s} \qquad \text{Find: } s = ?$$
$$g = -9.8 \text{ m/s}^2 \qquad\qquad v_f = ?$$
$$t = 1.5 \text{ s}$$

We can now calculate the position as follows:

$$s = v_0 t + \tfrac{1}{2} g t^2$$
$$= (20 \text{ m/s})(1.5 \text{ s}) + \tfrac{1}{2}(-9.8 \text{ m/s}^2)(1.5 \text{ s})^2$$
$$= 30 \text{ m} - 11 \text{ m} = 19 \text{ m}$$

The velocity after 1.5 s is given by

$$v_f = v_0 + gt$$
$$= 20 \text{ m/s} + (-9.8 \text{ m/s}^2)(1.5 \text{ s})$$
$$= 20 \text{ m/s} - 14.7 \text{ m/s} = 5.3 \text{ m/s}$$

Solution (d)   The same equations apply to find the position and velocity after 5 s. Thus

$$s = v_0 t + \tfrac{1}{2} g t^2$$
$$= (20 \text{ m/s})(5 \text{ s}) + \tfrac{1}{2}(-9.8 \text{ m/s}^2)(5 \text{ s})^2$$
$$= 100 \text{ m} - 123 \text{ m} = -23 \text{ m}$$

The negative sign indicates that the ball is located 23 m below the point of release. The speed after 5 s is given by

$$v_f = v_0 + gt$$
$$= 20 \text{ m/s} + (-9.8 \text{ m/s}^2)(5 \text{ s})$$
$$= 20 \text{ m/s} - 49 \text{ m/s} = -29 \text{ m/s}$$

In this case, the negative sign indicates that the ball is traveling downward.

Summary

A convenient way of describing objects in motion is to discuss their *velocity* or their acceleration. In this chapter, a number of applications have been presented which involve these physical quantities.

- **Average velocity** is the distance traveled per unit of time, and **average acceleration** is the change in velocity per unit of time:

$$\bar{v} = \frac{s}{t} \qquad a = \frac{v_f - v_0}{t}$$

- The definitions of velocity and acceleration result in the establishment of four basic equations involving uniformly accelerated motion:

$$s = \left(\frac{v_0 + v_f}{2}\right) t$$

$$v_f = v_0 + at$$
$$s = v_0 t + \tfrac{1}{2}at^2$$
$$2as = v_f^2 - v_0^2$$

Given any three of the five parameters (v_0, v_f, a, s, and t), the other two can be determined from one of these equations.

- To solve acceleration problems, read the problem with a view to establishing the three given parameters and the two which are unknown. You might set up columns like

Given: $a = 4$ m/s²   Find: v_f = ?
$s = 500$ m v_0 = ?
$t = 20$ s

This procedure helps you choose the appropriate equation. Remember to choose a direction as positive and to apply it consistently throughout the problem.

- **Gravitational acceleration:** Problems involving gravitational acceleration can be solved like other acceleration problems. In this case, one of the parameters is known in advance to be

$$a = g = 9.8 \text{ m/s}^2 \qquad \text{or} \qquad 32 \text{ ft/s}^2$$

The sign of the gravitational acceleration is plus or minus depending on whether you choose up or down as the positive direction.

Questions

6-1. Define the following terms:

  a. constant speed
  b. average speed
  c. velocity
  d. acceleration

  e. instantaneous velocity
  f. instantaneous speed
  g. uniformly accelerated motion
  h. acceleration due to gravity

6-2. Distinguish clearly between the terms *speed* and *velocity*. A stock car racer drives 500 laps around a 1-mi track in a time of 5 h. What was the average speed? What was the average velocity?

6-3. A bus driver travels a distance of 180 mi in 4 h. At the same time, a tourist travels the 180 mi by car but stops twice for a 30-min rest along the route. Nevertheless, the tourist arrives at the destination at the same instant as the bus driver. Compare the average speed of the bus driver with the average speed of the tourist.

6-4. Give some examples of motion in which the speed is constant but the velocity is not.

6-5. Two evenly spaced bowling balls are rolling along a level return trough at constant speed. Describe their later speed and separation in view of the fact that the first ball starts uphill to the rack before the second ball.

6-6. A sports announcer states, "The newly designed race car negotiated the 500-mi obstacle course in a record-breaking endurance run, reaching average speeds of 150 mi/h along the way." What is wrong with this statement?

6-7. A long strip of pavement is marked off in 100-ft intervals. Students stationed on a nearby ridge use their stopwatches to measure the time required for a car to cover the distance. The following data are obtained:

Distance, ft	0	100	200	300	400	500
Time, s	0	2.1	4.3	6.4	8.4	10.5

Plot a graph with distance as the ordinate and time as the abscissa. What is the significance of the slope of the curve? What is the speed of the car?

6-8. The acceleration due to gravity on a distant planet is one-fourth of the acceleration experienced on the earth. Does this mean that a rock dropped from a height of 4 ft will hit the ground in one-fourth the time required on the earth? Explain.

6-9. A spring-loaded gun fires a Ping-Pong ball vertically upward. On the moon the ball is observed to rise to a height of 6 times that observed on the earth. What can we say about the acceleration due to gravity on the surface of the moon?

6-10. The symbol g is sometimes referred to as *gravity* or the *acceleration of gravity*. Comment on the appropriateness of these expressions.

6-11. A rock is thrown vertically upward from the edge of a roof. It moves to its highest point, returns to its starting point, and then drops to the street below. Imagine that you take a snapshot at various times during the flight of the rock. Discuss the signs of velocity, acceleration, and distance at each instant.

6-12. Assume that the upward direction is positive in Question 6-11. What is the acceleration at the instant the rock leaves the hand? When it reaches its highest point? Just before it strikes the street below?

6-13. In the absence of air resistance, the same time is required for a projectile to reach its maximum height as is required for it to return to its starting position. Will this statement still be true if air resistance is not neglected? Draw free-body force diagrams for each situation.

Problems

6-1. A car travels 86 km. If the average speed was 8 m/s, how many hours were required for the trip?

Answer 2.99 h.

6-2. Sound travels through the air at an average speed of 340 m/s. A boy drops a rock from a bridge to the water 20 m below. How much time will be required for the sound of the splash to reach the boy's ear?

6-3. A transfer truck traveled 640 mi on a run from Atlanta to New York (state). The entire trip took 14 h, but the driver made two 30-min stops for meals. What was the average speed for the trip?

Answer 45.7 mi/h.

6-4. A dump truck gets about 9 mi on a gallon of fuel, which is priced at a dollar a gallon. What will be the cost of driving this truck for 2 h if its average speed is 30 mi/h?

6-5. An arrow leaves the bow 0.5 s after being released from a cocked position. If it has reached a speed of 40 m/s in this time, what was the average acceleration?

Answer 80 m/s^2.

6-6. An automobile traveling at a constant speed of 50 km/h accelerates at a rate of 4 m/s^2 for 3 s. What is its speed at the end of the 3-s interval?

6-7. A truck traveling at a speed of 60 mi/h suddenly brakes to a stop. The skid marks are observed to be 180 ft long. What was the average acceleration, and how much time was required to stop the truck after the brakes were applied?

Answer −21.5 ft/s^2; 4.09 s.

6-8. An arresting device is used to land airplanes on a carrier deck. The average acceleration produced by this device is −150 ft/s^2, and the airplanes are generally stopped in a time of 3 s. Under these conditions, what would be the approach velocity and the stopping distance?

6-9. A car travels at a constant speed of 55 mi/h. If the driver's mind wanders for a couple of seconds, how far will the car have traveled?

Answer 161 ft.

6-10. A truck travels for 2 h at an average speed of 60 km/h. Then it travels for 3 h at an average speed of 40 km/h. What were the total distance traveled and the average speed for the entire trip?

6-11. An elevator is lifted at a constant speed of 40 ft/s. How much time is required for the elevator to be lifted 200 ft?

Answer 5 s.

6-12. Two cities are 2000 km apart. What must be the average speed of a light plane if it is to make the trip in 10 h?

6-13. A bullet leaves a 28-in. rifle barrel with a muzzle velocity of 2700 ft/s. What was its average acceleration in the barrel, assuming that it started from rest? How long did the bullet remain in the barrel after the rifle was fired?

Answer 1.56 × 10^6 ft/s^2; 0.00173 s.

6-14. A monorail train traveling at 80 km/h must be brought to a stop in a distance of 40 m. What average acceleration is required, and what is the stopping time?

6-15. In a braking test, a car is observed to come to rest in 3 s. What were the acceleration and the stopping distance if the initial speed of the car was 60 km/h?

Answer −5.56 m/s^2, 25.0 m.

6-16. A rocket traveling at 400 ft/s in space is given a sudden acceleration. If it reaches a speed of 600 ft/s in 4 s, what was its average acceleration? How far did it travel in this time?

6-17. A railroad car loaded with coal starts from rest and coasts freely down a gentle slope. If the average acceleration is 4 ft/s^2, what will be the speed of the car in 5 s? What distance will be covered in this time?

Answer 20 ft/s; 50 ft.

6-18. A crane is used to lift a steel I beam to the top of a 100-ft building. For the first 2 s, the beam is lifted from rest with an upward acceleration of 8 ft/s^2. If the speed remains constant for the remainder of the trip, how much total time was required to raise the beam from the ground to the roof?

6-19. A rocket starting from rest on the launching pad acquires a vertical velocity of 140 m/s in 9 s. Find (a) the acceleration, (b) the distance above the ground at the end of 9 s, and (c) the velocity at the end of 3 s.

Answer (a) 15.6 m/s^2; (b) 630 m; (c) 46.7 m/s.

6-20. A train accelerates from rest at 4 ft/s^2. After covering a distance of 200 ft, the train then travels at a constant velocity for 4 s. At that instant, the train is braked to a stop in 6 s. What is the total distance traveled, and how much total time was required?

6-21. A brick is dropped from the top of the bridge 80 m above the water. (a) How long is the brick in the air? (b) With what velocity does the brick strike the water?

Answer (a) 4.04 s; (b) 39.6 m/s.

6-22. A bolt is accidentally dropped from the top of a building. Five seconds later it strikes the street below. (a) How high is the building? (b) What is the final speed?

6-23. An arrow is shot vertically upward with an initial velocity of 80 ft/s. (a) For how long will it rise? (b) How high will it rise? (c) What are its position and velocity after 2 s? (d) What are its position and velocity after 6 s?

Answer (a) 2.5 s; (b) 100 ft; (c) 96 ft, 16 ft/s; (d) −96 ft, −112 ft/s.

6-24. A hammer is thrown vertically upward to the top of a roof 50 ft high. (a) What was the minimum initial velocity required? (b) How much time was required?

*6-25. A ball is dropped from the window of a skyscraper, and 2 s later a second ball is thrown vertically downward. What must the initial velocity of the second ball be if it is to overtake the first ball just as the second ball strikes the ground 400 m below?

Answer 22.3 m/s.

6-26. A stone is thrown vertically downward from the top of a bridge. Four seconds later it strikes the water below with a final velocity of 60 m/s. (a) What was the initial velocity of the stone? (b) How high is the bridge above the water?

6-27. An object is tossed vertically upward and rises to a maximum height of 16 m. What was the initial speed, and how much time was required?

Answer 17.7 m/s; 1.81 s.

6-28. A ball is thrown vertically upward with an initial velocity of 23 m/s. What are its position and velocity after 2, 4, and 8 s?

6-29. A ball is rolled up an inclined plane so that it just reaches the top of the incline in 3 s. The speed of the ball changes by 4 m/s every second that it travels. What is the length of the incline, and what was the initial speed?

Answer 12 m/s; 18 m.

7

FORCE AND ACCELERATION

After completing this chapter, you should be able to

1. **Describe the relationship among force, mass, and acceleration and give the consistent units for each in SI and USCS units.**
2. **Define the units *newton* and *slug* and explain why they are derived units rather than fundamental units.**
3. **Demonstrate by definition and example your understanding of the distinction between mass and weight.**
4. **Determine mass from weight and weight from mass at a point where the acceleration due to gravity is known.**
5. **Draw a free-body diagram for objects in motion with constant acceleration, set the resultant force equal to the total mass times the acceleration, and solve for unknown parameters.**

According to Newton's first law of motion, an object will undergo a change in its state of rest or motion *only* when acted on by a resultant, unbalanced force. We now know that a change in motion, i.e., a change in speed, results in *acceleration*. In many industrial applications, we need to be able to predict the acceleration that will be produced by a given force. For example, the forward force required to accelerate a car from rest to a speed of 60 km/h in 8 s is of interest to the automotive industry. In this chapter, you will study the relationships among force, mass, and acceleration.

7-1 ■ NEWTON'S SECOND LAW OF MOTION

Before studying the relationship between a resultant force and the acceleration it produces in a formal way, let us consider a simple experiment.

A linear air track is an apparatus for studying the motion of objects under conditions that approximate zero friction. Hundreds of tiny air jets create an upward force which balances the weight of the glider in Fig. 7-1. A string is attached to the front of the glider, and a spring scale of negligible weight is used to measure the applied horizontal force as shown. The acceleration the glider receives can be measured by determining the change in speed for a known time. The first applied force F_1 in Fig. 7-1a causes an acceleration a_1. For example, when we apply a force of 4 lb, let's assume that we measure an acceleration of 2 ft/s^2. Next we double the applied force, as indicated in Fig. 7-1b, and again observe the acceleration.

Our observations will show that twice the force $2F_1$ will produce twice the acceleration $2a_1$. In other words, an 8-lb force would cause an acceleration of 4 ft/s^2 in our example. Similarly, tripling the force will increase the acceleration by a factor of 3. A force of 12 lb will produce an acceleration of 6 ft/s^2.

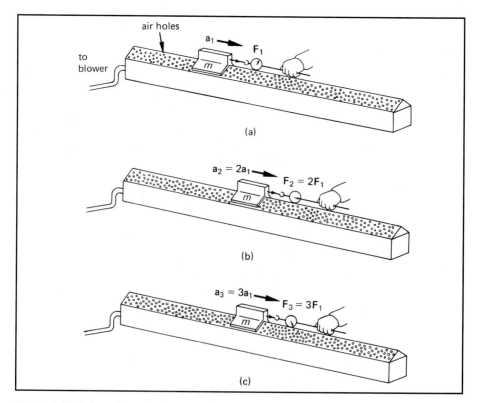

FIG. 7-1 Variation of acceleration with force.

Thus it is found that the acceleration of a body is directly proportional to the applied force and in the direction of that force. This means that the ratio of force to acceleration is always a constant:

$$\frac{F_1}{a_1} = \frac{F_2}{a_2} = \frac{F_3}{a_3} = \text{constant}$$

In our example, for forces of 4, 8, and 12 lb, the constant is equal to 2 lb·s²/ft.
We shall later see that this constant ratio is a property of the body called its mass m, where

$$m = \frac{F}{a}$$

Newton's second law is a statement of how the acceleration of a body varies with the applied force and the *mass* of the body. For the moment, the mass of an object may be thought of as the amount of matter that makes up the object. To understand the variation of acceleration with mass, let us return to our experiment. This time we shall keep the applied force F constant. The mass will be changed by adding in succession more gliders of equal size and weight. Note from Fig. 7-2 that if the force is unchanged, an increase in mass will result in a proportionate *decrease* in the acceleration. If the constant applied force is equal to 12 lb, we shall observe accelerations of 6, 3, and 2 ft/s² for the cases shown in Fig. 7-2a, b, and c, respectively.
From the above observations we are now prepared to state *Newton's second law of motion:*

Whenever an unbalanced force acts on a body, it produces in the direction of the force an acceleration that is directly proportional to the force and inversely proportional to the mass of the body.

Accordingly, we can write the proportion

$$a \propto \frac{F}{m}$$

Whenever mass is held constant, an increase in the applied force will result in a similar increase in acceleration. However, if the force is unchanged, increasing the mass results in a proportionate decrease of acceleration. By choosing appropriate units, we can write this proportion as an equation:

Resultant force = mass × acceleration

$$F = ma$$

Newton's Second Law (7-1)

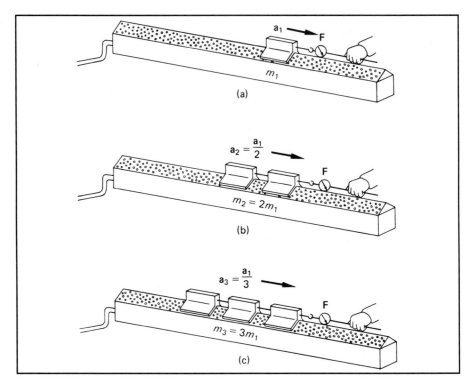

FIG. 7-2 Variation of acceleration with mass.

In order for a unit of force to be consistent with combined units of mass and acceleration, it is necessary to define one of the three parameters *F, m,* and *a* in terms of the other two. For example, we can choose the unit *kilogram* for mass *m* and the unit *meters per second squared* for acceleration *a*. From Eq. (7-1), the only consistent unit for force would be the kilogram-meter per second squared. This awkward combination is redefined as a *newton*. Because of the use of derived units in Newton's second law, it is important to know the consistent units for each quantity.

1. The fundamental SI unit for mass is the *kilogram,* and the acceleration unit is the *meter per second per second* (m/s²). The SI force unit derived from these units is called the *newton.* One **newton** is that resultant force required to give a one-kilogram mass an acceleration of one meter per second per second:

$$\text{Force (N)} = \text{mass (kg)} \times \text{acceleration (m/s}^2)$$

2. In USCS units, the mass unit is derived from the chosen units of *pound* for force and *feet per second per second* for acceleration. This new unit for mass is called the *slug* (from the sluggish or inertial property of mass). One **slug** (slug) is that mass to which a resultant force of one pound will give an acceleration of one foot per second per second:

$$\text{Force (lb)} = \text{mass (slugs)} \times \text{acceleration (ft/s}^2)$$

The SI unit of force is less than the USCS unit, and a mass of 1 slug is much greater than a mass of 1 kg. The following conversion factors might be helpful:

$$1 \text{ lb} = 4.448 \text{ N} \qquad 1 \text{ slug} = 14.59 \text{ kg}$$

A 1-lb bag of apples might contain four or five apples—each weighing about a newton. A person weighing 160 lb on earth would have a mass of 5 slugs, or 73 kg.

It is important to recognize that the F in Newton's second law represents a *resultant* or unbalanced force. If more than one force acts on an object, it will be necessary to determine the resultant force *along the direction of motion*. The resultant force will always be along the direction of motion, since it is the *cause* of the acceleration. All components of forces which are perpendicular to the acceleration will be balanced. If the x axis is chosen along the direction of motion, we can determine the x component of each force and write

$$\Sigma F_x = ma_x \qquad (7\text{-}2)$$

A similar equation could be written for the y components if the y axis were chosen along the direction of motion.

7-2 ■ RELATIONSHIP BETWEEN WEIGHT AND MASS

Before we consider examples of Newton's second law, we must have a clear understanding of the difference between the *weight* of a body and its *mass*. Perhaps no other point is more confusing to the beginning student. The pound, which is a unit of force, is often used as a mass unit, the pound-mass (lb_m). The kilogram, which is a unit of mass, is often used in industry as a unit of force, the kilogram-force (kgf). These seemingly inconsistent units result from the fact that there are many different systems of units in use. In this text, there should be less cause for confusion because we use only the SI and USCS units. Therefore, in this book the pound will always refer to *weight,* which is a force, and the unit *kilogram* will always refer to the *mass* of a body.

The weight of any body is the force with which the body is pulled vertically downward by gravity. When a body falls freely to the earth, the only force acting on it is its weight **W.** This net force produces an acceleration **g,** which is the same for all falling bodies. Thus, from Newton's second law we can write the relationship between a body's weight and its mass:

$$W = mg \qquad \text{or} \qquad m = \frac{W}{g} \qquad (7\text{-}3)$$

In either system of units, (1) the mass of a particle is equal to its weight divided by the acceleration of gravity, (2) weight has the same units as the unit of force, and (3) the acceleration of gravity has the same units as acceleration.

Therefore, we can summarize as follows:

SI: W (N) $= m$ (kg) $\times g$ (9.8 m/s^2)
USCS: W (lb) $= m$ (slug) $\times g$ (32 ft/s^2)

The values for g, and hence the weights, in the above relations apply only at points on the earth near sea level, where g has these values.

Two things must be remembered in order to understand fully the difference between mass and weight:

> **Mass** *is a universal constant equal to the ratio of a body's weight to the gravitational acceleration due to its weight.*

> **Weight** *is the* force *of gravitational attraction and is very much dependent upon the acceleration of gravity.*

Therefore, the mass of a body is only a measure of its inertia and is not in any way dependent upon gravity. In outer space, a hammer has negligible weight, but it serves to drive nails just the same, since its mass is unchanged.

In USCS units a body is usually described by stating its weight **W** in pounds. The mass, if desired, is computed from this weight and has the unit of slugs. In SI units a body is usually described in terms of its mass in kilograms. The weight, if desired, is computed from the given mass and has the units of newtons. In the following examples, all parameters are measured at points where $g = 32$ ft/s^2, or 9.8 m/s^2.

EXAMPLE 7-1   Find the mass of a person who weighs 150 lb.

Solution

$$m = \frac{W}{g} = \frac{150 \text{ lb}}{32 \text{ ft/s}^2} = 4.69 \text{ slugs}$$

EXAMPLE 7-2   Find the weight of an 18-kg block.

Solution

$$W = mg = 18 \text{ kg}(9.8 \text{ m/s}^2) = 176 \text{ N}$$

EXAMPLE 7-3   Find the mass of a body whose weight is 100 N.

Solution

$$m = \frac{W}{g} = \frac{100 \text{ N}}{9.8 \text{ m/s}^2} = 10.2 \text{ kg}$$

7-3 ■ APPLICATION OF NEWTON'S SECOND LAW TO SINGLE-BODY PROBLEMS

The primary difference between the problems discussed in this chapter and earlier problems is that a net, unbalanced force is acting to produce an acceleration. Thus, after we construct the free-body diagrams which describe the situation, our first step is to determine the unbalanced force and set it equal to the product of mass and acceleration. The desired quantity can then be determined from the relation

$$\text{Resultant force} = \text{mass} \times \text{acceleration}$$

$$F(\text{resultant}) = ma \tag{7-1}$$

The following examples will serve to demonstrate the relationship among force, mass, and acceleration.

EXAMPLE 7-4   What acceleration will a force of 20 N impart to a 10-kg body?

Solution   There is only one force acting, and so

$$F = ma \qquad \text{or} \qquad a = \frac{F}{m}$$

and

$$a = \frac{20 \text{ N}}{10 \text{ kg}} = 2 \text{ m/s}^2$$

EXAMPLE 7-5   What resultant force will give a 32-lb body an acceleration of 5 ft/s²?

Solution   In order to find the resultant force, we must first determine the mass of the body from its given weight:

$$m = \frac{W}{g} = \frac{32 \text{ lb}}{32 \text{ ft/s}^2} = 1 \text{ slug}$$

then

$$F = ma = (1 \text{ slug})(5 \text{ ft/s}^2) = 5 \text{ lb}$$

EXAMPLE 7-6   What mass has a body if a force of 60 N will give it an acceleration of 4 m/s?

Solution   Solving for m in Newton's law, we have

$$m = \frac{F}{a} = \frac{60 \text{ N}}{4 \text{ m/s}^2} = 15 \text{ kg}$$

In the preceding examples the unbalanced forces were easily determined. However, as the number of forces acting on a body increases, the problem of determining the resultant force is less obvious. In these cases, perhaps it is wise to point out a few considerations.

According to Newton's second law, a resultant force always produces an acceleration *in the direction of the resultant force*. This means that the net force and the acceleration it causes have the same algebraic sign, and they each have the same line of action. Therefore, if the direction of motion (acceleration) is considered positive, fewer negative factors will be introduced into the equation $F = ma$. For example, in Fig. 7-3b, it is preferable to choose the direction of motion (left) as positive, since the equation

$$P - \mathscr{F}_k = ma$$

is preferable to the equation

$$\mathscr{F}_k - P = -ma$$

which would result if we chose the right direction as positive.

Another consideration which results from the above discussion is that the forces acting perpendicular to the line of motion will be in equilibrium if the resultant force is constant. Thus, in problems involving friction, normal forces can be determined from the first condition for equilibrium.

In summary, we apply the following equations to acceleration problems:

$$\Sigma F_x = ma_x \qquad \Sigma F_y = ma_y = 0 \qquad\qquad (7\text{-}4)$$

where ΣF_x and a_x are taken as positive and along the line of motion and ΣF_y and a_y are taken normal to the line of motion.

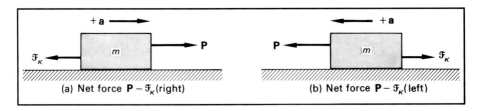

(a) Net force $\mathbf{P} - \mathscr{F}_k$ (right) (b) Net force $\mathbf{P} - \mathscr{F}_k$ (left)

FIG. 7-3 The direction of acceleration should be chosen as positive.

EXAMPLE 7-7   A force of 100 lb pulls a 64-lb block horizontally across the floor. If $\mu_k = 0.1$, find the acceleration of the block.

Solution   A free-body diagram is superimposed on the sketch in Fig. 7-4. We will choose right as positive. To avoid confusing the weight of the block with its mass, it is often desirable to calculate each in advance. The weight (64 lb) is given, and the mass is found from $m = W/g$:

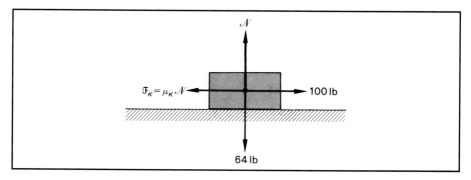

FIG. 7-4

$$W = 64 \text{ lb} \qquad m = \frac{64 \text{ lb}}{32 \text{ ft/s}^2} = 2 \text{ slugs}$$

The resultant force along the floor is 100 lb less the force of friction $\mathcal{F}_k$. Applying Newton's second law, we obtain

$$\text{Resultant force} = \text{mass} \times \text{acceleration}$$
$$100 \text{ lb} - \mathcal{F}_k = ma$$

Recalling that $\mathcal{F}_k = \mu_k \mathcal{N}$, we may write

$$100 \text{ lb} - \mu_k \mathcal{N} = ma$$

Since no motion occurs in the vertical direction, we note from the figure that

$$\mathcal{N} = W = 64 \text{ lb}$$

Substituting $\mathcal{N} = 64$, $\mu_k = 0.1$, and $m = 2$ slugs, we have

$$100 \text{ lb} - (0.1)(64 \text{ lb}) = (2 \text{ slugs})a$$

Simplifying and solving for a yield

$$100 \text{ lb} - 6.4 \text{ lb} = (2 \text{ slugs})a$$
$$a = \frac{93.6 \text{ lb}}{2 \text{ slugs}} = 46.8 \text{ ft/s}^2$$

You should verify that pounds per slug is equivalent to feet per second squared.

7-4 ■ PROBLEM-SOLVING TECHNIQUES

Solving all physics problems requires an ability to organize the given data and to apply formulas in a consistent manner. Often a procedure is helpful to the beginning student, and this is particularly true for the problems in this chapter. A logical sequence of operations for problems involving Newton's second law is outlined below.

1. Read the problem carefully for general understanding.
2. Draw a sketch, labeling given information.
3. Draw a free-body diagram with an axis along the direction of motion.
4. Indicate the positive direction of acceleration.
5. Determine the mass and weight of each object:

$$W = mg \qquad m = \frac{W}{g}$$

6. From the free-body diagram, determine the resultant force along the direction of motion ΣF.
7. Determine total mass $m_t = m_1 + m_2 + \cdots$.
8. Set the resultant force ΣF equal to total mass m_t times acceleration a:

$$\boxed{\Sigma F = m_t a}$$

9. Solve for the unknown quantity.

EXAMPLE 7-8 A 2000-lb elevator is lifted upward with an acceleration of 4 ft/s^2. What is the tension in the supporting cable?

Solution Read the problem, then draw a sketch from which a free-body diagram may be drawn (see Fig. 7-5). Notice that the positive direction of acceleration (upward) is indicated on the free-body diagram.

Now we determine the mass and the weight of the 2000-lb elevator. The weight, of course, is 2000 lb. The mass must be calculated from $m = W/g$:

$$W = 2000 \text{ lb} \qquad m = \frac{2000 \text{ lb}}{32 \text{ ft/s}^2} = 62.5 \text{ slugs}$$

Since the elevator is the only object moving, the 62.5 slugs represents the *total* mass m_t.

The resultant force from the free-body diagram is

$$\Sigma F = T - 2000 \text{ lb}$$

From Newton's second law, we now write

$$\text{Resultant force} = \text{total mass} \times \text{acceleration}$$
$$T - 2000 \text{ lb} = (62.5 \text{ slugs})(4 \text{ ft/s}^2)$$
$$= 250 \text{ lb}$$

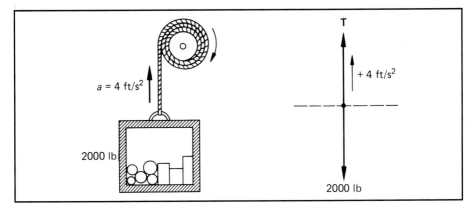

FIG. 7-5 Upward acceleration.

Finally, we solve for the unknown T by adding 2000 lb to both sides of the equation:

$$T = 250 \text{ lb} + 2000 \text{ lb}$$
$$= 2250 \text{ lb}$$

EXAMPLE 7-9 A 100-kg ball is lowered by means of a cable with a downward acceleration of 5 m/s^2. What is the tension in the cable?

Solution As before, we construct a sketch and a free-body diagram (Fig. 7-6). Note that the downward direction is chosen as positive since that is the direction of motion.

This time the mass is given and the weight must be calculated from $W = mg$:

$$m = 100 \text{ kg} \qquad W = (100 \text{ kg})(9.8 \text{ m/s}^2) = 980 \text{ N}$$

The resultant force is the *net* downward force, or

$$\Sigma F = W - T \qquad \text{(remember, down is positive)}$$

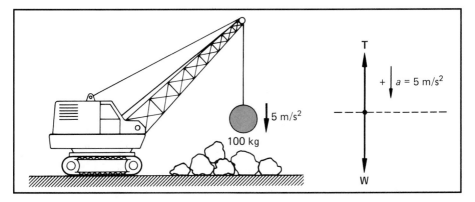

FIG. 7-6 Downward acceleration.

Now, from Newton's second law, we write

Net downward force = total mass × downward acceleration

$$W - T = ma$$

Substituting known quantities, we obtain

$$980 \text{ N} - T = (100 \text{ kg})(5 \text{ m/s}^2)$$
$$= 500 \text{ N}$$

from which we solve for T by adding T to both sides and subtracting 500 N from both sides:

$$980 \text{ N} - 500 \text{ N} = T$$
$$T = 480 \text{ N}$$

*EXAMPLE 7-10   An Atwood machine consists of a single pulley with masses suspended on each side. It is a simplified version of many industrial systems in which counterweights are used for balance. Assume that the mass on the right side is 10 kg and that the mass on the left side is 2 kg. (*a*) What is the acceleration of the system? (*b*) What is the tension in the cord?

Solution (a)   We first draw a sketch and a free-body diagram for each mass (Fig. 7-7). The weight and mass of each object are determined:

$m_1 = 2 \text{ kg}$ $W_1 = m_1g = (2 \text{ kg})(9.8 \text{ m/s}^2)$ or $W_1 = 19.6 \text{ N}$
$m_2 = 10 \text{ kg}$ $W_2 = m_2g = (10 \text{ kg})(9.8 \text{ m/s}^2)$ or $W_2 = 98 \text{ N}$

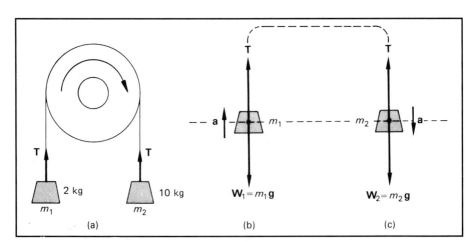

FIG. 7-7 Two masses suspended from a single pulley. Free-body diagrams are drawn; the positive direction of acceleration is chosen to be upward on the left and downward on the right.

Now the problem is to determine the net unbalanced force on the entire system. Note that the pulley merely changes the direction of the forces, and for all practical purposes the diagram might look like Fig. 7-8.

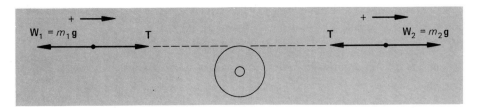

FIG. 7-8

The resultant force on the entire system is

$$\Sigma F = W_2 - T + T - W_1$$
$$= W_2 - W_1$$

The unbalanced force *on the entire system,* therefore, is just the difference in the weights. This is just what we would expect from experience. Notice that the tension **T** is the same on either side, since there is only one rope. Thus, the tension cancels and does not figure in the resultant force.

The total mass of the system is simply the sum of all the masses in motion:

$$m_t = m_1 + m_2 = 2 \text{ kg} + 10 \text{ kg}$$
$$= 12 \text{ kg} \qquad \text{(total mass)}$$

From Newton's second law of motion, we have

$$\text{Resultant force} = \text{total mass} \times \text{acceleration}$$
$$W_2 - W_1 = (m_1 + m_2)a$$

Substituting for W_2, W_1, m_1, and m_2, we have

$$98 \text{ N} - 19.6 \text{ N} = (2 \text{ kg} + 10 \text{ kg})a$$

From which we may solve for a as follows:

$$78.4 \text{ N} = (12 \text{ kg})a$$
$$a = \frac{78.4 \text{ N}}{12 \text{ kg}} = 6.53 \text{ m/s}^2$$

Solution (b) In order to solve for the tension in the cord, we must consider either of the masses by itself, since considering the system as a whole would not involve cord tension. Suppose we consider the forces acting on m_1:

$$\text{Resultant force} = \text{mass} \times \text{acceleration}$$
$$T - W_1 = m_1 a$$

But $a = 6.53$ m/s^2 and the mass and weight are known, so that we have

$$T - 19.6 \text{ N} = (2 \text{ kg})(6.53 \text{ m/s}^2)$$
$$= 13.06 \text{ N}$$
$$T = 32.7 \text{ N}$$

We would obtain the same value for the tension if we applied Newton's law to the second mass. You should demonstrate this fact as an additional exercise.

Summary

In this chapter, we have considered the fact that a resultant force will always produce an acceleration in the direction of the force. The magnitude of the acceleration is directly proportional to the force and inversely proportional to the mass, according to Newton's second law of motion. The following concepts are essential to applications of this fundamental law:

- The mathematical formula which expresses Newton's second law of motion may be written in three different forms:

$$\text{Force} = \text{mass} \times \text{acceleration}$$

$$F = ma \qquad m = \frac{F}{g} \qquad a = \frac{F}{m}$$

In SI units: $\qquad\qquad$ 1 N = (1 kg)(1 m/s^2)
In USCS units: $\qquad$ 1 lb = (1 slug)(1 ft/s^2)

- Weight is the force due to a particular acceleration g. Thus weight W is related to mass m by Newton's second law:

$$W = mg \qquad m = \frac{F}{g} \qquad g = 9.8 \text{ m/s}^2 \text{ or } 32 \text{ ft/s}^2$$

For example, a mass of 1 kg has a weight of 9.8 N. A weight of 1 lb has a mass of $\frac{1}{32}$ slug. In a given problem you must determine whether weight or mass is given. Then you must determine what is needed in an equation. Conversions of mass to weight and weight to mass are common.

- Application of Newton's second law:
  a. *Construct a free-body diagram for each body undergoing an acceleration. Indicate on this diagram the direction of positive acceleration.*
  b. *Determine an expression for the net resultant force on a body or a system of bodies.*
  c. *Set the resultant force equal to the total mass of the system multiplied by the acceleration of the system.*
  d. *Solve the resulting equation for the unknown quantity.*

Questions

7-1. Define the following terms:

  a. Newton's second law   **d.** slug

  b. mass   **e.** newton

  c. weight

7-2. In a laboratory experiment the acceleration of a small car is measured by the separation of spots burned at regular intervals in a paraffin-coated tape. Larger and larger weights are transferred from the car to a hanger at the end of a tape which passes over a light, frictionless pulley. In this manner, the mass of the entire system is kept constant. Since the car moves on a horizontal air track with negligible friction, the resultant force is equal to the weights at the end of the tape. The following data are recorded:

Weight, N	2	4	6	8	10	12
Acceleration, N/s^2	1.4	2.9	4.1	5.6	7.1	8.4

Plot a graph of weight vs. acceleration. What is the significance of the slope of this curve? What is the mass of the system?

7-3. In the experiment described in Question 7-2, the student places a constant weight of 4 N at the free end of the tape. Several runs are made, increasing the mass of the car each time by adding weights. What happens to the acceleration as the mass of the system is increased? What should be the value of the product of the mass of the system and the acceleration for each run? Is it necessary to include the mass of the constant 4-N weight in these experiments?

7-4. Distinguish clearly between the mass of an object and its weight, and give the appropriate units for each in SI and USCS units.

7-5. What exactly do we mean when we describe an athlete as a 160-lb person? What would be the mass of this person on the moon?

7-6. A round piece of brass found in the laboratory is labeled 500 g. Is this its weight or its mass? How can you be sure?

7-7. A state of equilibrium is maintained on a force table by hanging masses from pulleys mounted at various locations on the circular edge. In calculating the equilibrant, we sometimes use grams instead of newtons. Are we justified in doing this?

7-8. When drawing free-body diagrams, why is it usually to our advantage to choose either the x or the y axis along the direction of motion even if it means rotated axes? Use the example of motion along an inclined plane as an illustration.

7-9. In the example of an Atwood machine (Example 7-10), we neglected the mass of the cord which connects the two masses. Discuss how this problem is altered if the mass of the cord is large enough to affect the motion.

7-10. In industry we often hear of a kilogram-force unit which is defined as a force equivalent to the weight of a 1-kg mass near the earth's surface. In the United States, we also speak often of the pound-mass unit which is the mass of an object which has a weight of 1 lb near the surface of the earth. Find the value of these quantities in appropriate SI units, and discuss the problems caused by their use.

Problems

7-1. A 3-kg mass is acted on by a resultant force of 12 N. What is the acceleration? What force will give this mass an acceleration of 8 m/s^2?

Answer 4 m/s^2; 24 N.

7-2. A resultant force of 20 N will increase the speed of an object from 5 to 11 m/s in 2 s. What is the mass?

7-3. In outer space a force of 40 lb will give a ball an acceleration of 2 ft/s^2. What is the mass of this object in space? What are the mass and weight of the ball at the surface of the earth?

Answer 20 slugs; 20 slugs, 640 lb.

7-4. Express the mass and weight of a 120-lb woman in SI units. If the acceleration due to gravity on the moon is 1.6 m/s^2, what are the mass and weight of this woman on the surface of the moon?

7-5. A 2-kg mass is acted on by a resultant force of (*a*) 8 N and (*b*) 4 lb. What is the resulting acceleration for each case (in metric units)?

Answer (*a*) 4 m/s^2; (*b*) 8.9 m/s^2.

7-6. A resultant force of 200 lb produces an acceleration of 5 ft/s^2. What is the mass of the object being accelerated? What is its weight?

7-7. Find the mass and the weight of a body if a resultant force of 16 N will give it an acceleration of 5 m/s^2.

Answer 3.2 kg; 31.4 N.

7-8. What resultant force is necessary to give a 4-kg hammer an acceleration of 6 m/s^2?

7-9. Find the weight of a body whose mass is (*a*) 5 slugs, (*b*) 8 kg, and (*c*) 0.25 kg.

Answer (*a*) 160 lb; (*b*) 78.4 N; (*c*) 2.45 N.

7-10. Find the mass of a body whose weight is (*a*) 600 lb, (*b*) 40 N, and (*c*) 0.80 lb.

7-11. A 2500-lb automobile is speeding at 55 mi/h. What retarding force is required to stop the car in 200 ft on a level road?

Answer −1273 lb.

7-12. What horizontal push is required to drag a 60-kg sled with an acceleration of 4 m/s^2? Assume that a horizontal friction force of 20 N opposes the motion.

7-13. A horizontal force of 100 N pulls an 8-kg block horizontally across the floor. If the coefficient of kinetic friction between the block and the floor is 0.2, find the acceleration of the block.

Answer 10.5 m/s^2.

7-14. A car's speed increases from 30 to 60 mi/h in 5 s under the action of a resultant force of 1150 lb. What is the mass of the car? What is its weight?

7-15. A 5-ton tractor pulls a 10-ton trailer on a level road and gives it an acceleration of 6 ft/s^2. If this tractor exerts the same force on an 8-ton trailer, what acceleration will result?

Answer 6.92 ft/s^2.

7-16. A net force of 200 N gives a wagon an acceleration of 6 m/s^2. What are the mass and weight of the wagon? Assume that a 10-kg load is added to the wagon. What resultant force is required to give the same acceleration?

7-17. A 3-kg mass is being moved by two perpendicular forces. A 50-N force is directed northward while a 90-N force is directed to the west. What are the magnitude and direction of the resulting acceleration?

Answer 34.3 m/s^2, 29.1°N of W.

7-18. A 60-lb block is being moved horizontally by a 40-lb force which produces an acceleration of 8 m/s². What are the friction force and the coefficient of kinetic friction?

7-19. A 64-lb load hangs at the end of a rope. Find the acceleration of the load if the tension in the cable is (a) 64, (b) 40, and (c) 96 lb.

Answer (a) 0; (b) 12 ft/s² down; (c) 16 ft/s² up.

7-20. A 10-kg mass is lifted upward by a light cable. What is the tension in the cable if the acceleration is (a) zero, (b) 6 m/s² upward, and (c) 6 m/s² downward?

7-21. An 800-kg elevator is lifted vertically by a strong rope. Find the acceleration of the elevator if the tension in the rope is (a) 9000, (b) 7840, and (c) 2000 N.

Answer (a) 1.45 m/s² up; (b) 0; (c) −7.3 m/s² down.

7-22. An 18-lb block is accelerated upward with a cord whose breaking strength is 20 lb. Find the maximum acceleration which can be given to the block without breaking the cord.

7-23. A 2000-lb elevator is lifted vertically with an acceleration of 8 ft/s². Find the minimum breaking strength of the cable. A 200-lb man standing on a scale will read his weight differently in the moving elevator. What is the weight recorded by the spring scale?

Answer 2500 lb, 250 lb.

7-24. A 9-kg load of cement is accelerated upward with a cord whose breaking strength is 200 N. What is the maximum upward acceleration such that the cord does not break?

7-25. If the coefficient of friction between the tire and the road is 0.7, what is the minimum distance in which a 1600-kg car can be stopped if it is traveling at 60 km/h?

Answer 20.2 m.

7-26. What horizontal force is required to drag a 6-kg block with an acceleration of 5 m/s² if the coefficient of friction is 0.3?

7-27. A light cord passing over a light frictionless pulley, as in Fig. 7-7, has masses m_1 and m_2 attached to its ends. What will be the acceleration of the system and the tension in the cord if (a) $m_1 = 12$ kg and $m_2 = 10$ kg; (b) $m_1 = 20$ g and $m_2 = 50$ g?

Answer (a) 0.891 m/s², 107 N; (b) 4.2 m/s², 0.28 N.

7-28. Suppose the masses in Prob. 7-27 are replaced with the weights $W_1 = 24$ lb and $W_2 = 16$ lb. What will be the resultant acceleration and tension in the cord?

*7-29. A 30-kg mass rests on a 37° inclined plane, as in Fig. 7-9. The coefficient of kinetic friction is 0.3. A push **P** applied parallel to the plane and directed up the plane causes the mass to accelerate at 3 m/s². (a) What is the normal force exerted by the plane on the mass? (b) What is the friction force? (c) What is the resultant force up the plane? (d) What is the magnitude of the push P?

Answer (a) 235 N; (b) 70.5 N; (c) 90 N; (d) 338 N.

*7-30. A 400-lb sled slides down a hill sloped at 60° with the horizontal. The coefficient of sliding friction is 0.2. (a) What is the resultant force down the hill? (b) What is the acceleration?

7-31. Consider the system shown in Fig. 7-10. If $m_1 = 16$ kg and $m_2 = 10$ kg, what is the acceleration of the system? What is the tension in the cord? Neglect friction.

Answer 3.77 m/s²; 60.3 N.

7-32. Answer the questions in Prob. 7-31 if the only change is that the coefficient of kinetic friction between m_1 and the table is 0.3.

7-33. If the *weight* of m_1 in Fig. 7-10 is 200 lb and if $\mu_k = 0.3$, what must be the weight of m_2 if the system is to have an acceleration of 6 ft/s²?

Answer $W_2 = 120$ lb.

7-34. If $m_1 = 10$ kg, $m_2 = 8$ kg, and $m_3 = 6$ kg in Fig. 7-11, what is the acceleration of the system, neglecting friction?

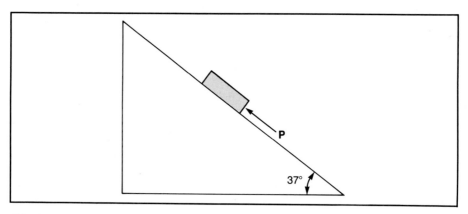

FIG. 7-9

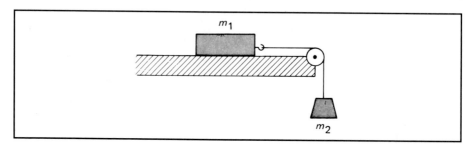

FIG. 7-10

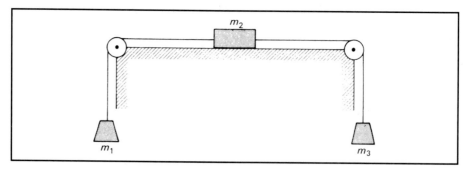

FIG. 7-11

ENERGY AND MOMENTUM

As a result of completing this chapter, you should be able to

1. Define and write formulas for work, gravitational potential energy, kinetic energy, power, impulse, and momentum.
2. Apply the concepts of work, energy, power, and momentum to the solution of problems similar to those given in the text.
3. Discuss and apply your knowledge of the relationship between the performance of work and the corresponding change in kinetic energy.
4. Discuss and apply your knowledge of the principles of conservation of energy and linear momentum.
5. Determine the power of a system and understand its relationship to time, force, distance, and velocity.

The principal reason for the application of a resultant force is to cause a displacement. For example, a large crane lifts a steel beam to the top of a building, the compressor in an air conditioner forces a fluid through its cooling cycle, and electromagnetic forces move electrons across a television screen. Whenever a force acts through a distance, we will learn, *work* is done in a way that can be measured or predicted. The capacity for doing work is called *energy,* and the rate at which work is accomplished will be defined as *power.*

Conservation laws are frequently encountered in the description of physical events. We will study two such laws in this chapter. The first has to do with the conservation of energy, and the second has to do with the conservation of linear *momentum*. For example, should two objects collide, we might ask how the energy and velocities are distributed after collision.

8-1 ■ WORK

When we attempt to drag a block with a rope, as shown in Fig. 8-1a, nothing happens at first. We are applying a force, but the block has not moved. If we continue to increase our pull, eventually the block will be displaced. In this case, we have actually accomplished something in return for our effort. This accomplishment is defined in physics as *work*. The term *work* has a very specific meaning. For work to be done, three things are necessary.

1. There must be an applied force **F**.
2. The force must act through a certain distance s, called the displacement.
3. The force must have a component F_x along the displacement.

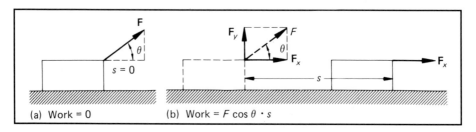

(a) Work = 0   (b) Work = $F \cos \theta \cdot s$

FIG. 8-1 Work done by a force **F** undergoing a displacement s.

Assuming that we are given these conditions, a formal definition of work may be stated:

Work *is a scalar quantity equal to the product of the magnitudes of the displacement and the component of the force in the direction of the displacement.*

Work = force component × displacement

$$\text{Work} = F_x\, s \tag{8-1}$$

In this equation F_x is the component of **F** along the displacement s. In Fig. 8-1, only F_x contributes to work. Its magnitude can be found graphically by drawing **F** and s to scale, as discussed in Chap. 2. If trigonometry is used, Eq. (8-1) may be written in terms of the angle θ between **F** and s:

$$\text{Work} = (F \cos \theta)s \tag{8-2}$$

Quite often the force causing work is directed entirely along the displacement. This happens when a weight is lifted vertically upward or when a horizontal force drags an object along the floor. In these simple cases, $F_x = F$, and Eq. (8-1) becomes

$$\text{Work} = Fs \qquad\qquad (8\text{-}3)$$

Another special case occurs when the applied force is perpendicular to the displacement. In this instance, the work will be zero, since $F_x = 0$. An example is motion parallel to the earth's surface in which gravity acts vertically downward and is perpendicular to all horizontal displacements. Then the force of gravity does no work.

EXAMPLE 8-1   What work is done by a 60-N force in dragging the block of Fig. 8-1 a distance of 50 m when the force is transmitted by a rope making an angle of 30° with the horizontal?

Solution   We must first determine the component F_x of the 60-N force **F**. Only this component contributes to work. Graphically, this is accomplished by drawing the 60-N vector to scale at a 30° angle. Measuring F_x and converting to newtons gives

$$F_x = 52.0 \text{ N}$$

With trigonometry, we could accomplish the same calculation by using the cosine function:

$$F_x = (60 \text{ N})(\cos 30°) = 52.0 \text{ N}$$

Now, applying Eq. (8-1), we obtain the work:

$$\begin{aligned}\text{Work} = F_x s &= (52.0 \text{ N})(50 \text{ m}) \\ &= 2600 \text{ N·m}\end{aligned}$$

Note that the units of work are the units of force times distance. Thus, in SI units, work is measured in newton-meters. By agreement, this combination unit is renamed the *joule,* which is denoted by J.

> One **joule** *(1 J) is equal to the work done by a force of one newton in moving an object through a parallel distance of one meter.*

In Example 8-1, the work done in dragging the block would be written as 2600 J.
    In the United States, work is sometimes also given in USCS units. When the force is given in pounds and the displacement is given in feet, the corresponding work unit is called the *foot-pound* (ft·lb):

> One **foot-pound** *(1 ft·lb) is equal to the work done by a force of one pound in moving an object through a parallel distance of one foot.*

No special name is given to this unit.

The following conversion factors will be useful when you are comparing work units in the two systems:

$$1 \text{ J} = 0.7376 \text{ ft·lb} \qquad 1 \text{ ft·lb} = 1.356 \text{ J}$$

8-2 ■ RESULTANT WORK

When we consider the work of several forces acting on the same object, it is often useful to distinguish between positive and negative work. In this text, we will follow the convention that the work of a particular force is positive if the force component is in the same direction as the displacement. Negative work is done by a force component which opposes the actual displacement. Hence, work done by a crane in lifting a load is positive, but the gravitational force exerted by the earth on the load is doing negative work. Similarly, when we stretch a spring, the work on the spring is positive; the work on the spring is negative when the spring contracts, pulling us back. Another important example of negative work is that performed by a friction force which is opposite to the direction of displacement.

If several forces act on a body in motion, the resultant (total) work is the sum of the work done by each force. Consider the following example.

EXAMPLE 8-2 The tractor in Fig. 8-2 exerts a 500-lb horizontal force in pulling a loaded trailer. The total weight of tractor and trailer is 6000 lb. What is the net work done by the tractor if it moves the load a distance of 400 ft? Assume that the coefficient of rolling friction is 0.03. Neglect air friction.

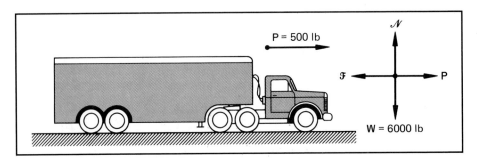

FIG. 8-2 Work done by a tractor.

Solution First, we will draw a free-body diagram showing all forces acting on the tractor-trailer system (see Fig. 8-2b). There are four forces acting on the system: **P**, $\mathcal{F}_R$, $\mathcal{N}$, and **W**. The pull **P** is entirely along the displacement and in the direction of the displacement. Thus,

$$(\text{Work})_P = Ps = (500 \text{ lb})(400 \text{ ft}) = 200{,}000 \text{ ft·lb}$$

The magnitude of the friction force $\mathcal{F}_K$ is computed from the values of μ_R and $\mathcal{N}$:

$$\mathcal{F}_R = \mu_R \mathcal{N} = (0.03)(6000 \text{ lb}) = 180 \text{ lb}$$

Since this friction force is directed opposite to the displacement, it does negative work, given by

$$(\text{Work})_{\mathscr{F}} = (-180 \text{ lb})(400 \text{ ft}) = -72,000 \text{ ft·lb}$$

The normal force $\mathscr{N}$ and the weight $\mathbf{W}$ do no work because they have no components along the displacement:

$$(\text{Work})_{\mathscr{N}} = 0 \qquad (\text{work})_W = 0$$

Now the net work is equal to the sum of the individual works:

$$
\begin{aligned}
\text{Net work} &= (\text{work})_P + (\text{work})_{\mathscr{F}} + (\text{work})_{\mathscr{N}} + (\text{work})_W \\
&= 200,000 \text{ ft·lb} - 72,000 \text{ ft·lb} + 0 + 0 \\
&= 128,000 \text{ ft·lb}
\end{aligned}
$$

When we are interested only in the *net* work and not in the work of some individual force, it is easier to calculate the work of the *resultant force*. Since each force acts through the same distance, it is the vector sum of the individual forces that really determines the total work done.

EXAMPLE 8-3   Show that the work of the resultant force in the previous example is the same as the sum of the individual works.

Solution   Since the forces $\mathscr{N}$ and $\mathbf{W}$ are balanced, the resultant force is the difference between the pull of the tractor and the resistance of friction. The resultant force is

$$F_R = P - \mathscr{F}_R = 500 \text{ lb} - 180 \text{ lb} = 320 \text{ lb}$$

directed parallel to the displacement. The work done by the resultant force, therefore, is

$$(\text{Work})_R = F_R s = (320 \text{ lb})(400 \text{ ft}) = 128,000 \text{ ft·lb}$$

which equals the value obtained by computing the work done by each force separately.

8-3 ■ ENERGY

Energy may be thought of as *anything which can be converted to work*. When we say that an object has energy, we mean that it is capable of exerting a force on another object in order to do work on it. Conversely, if we do work on some object, we have added to it an amount of energy equal to the work done. The units of energy are the same as those for work: the *joule* and the *foot-pound*.

In mechanics we shall be concerned with two kinds of energy:

Kinetic energy E_k: *Energy possessed by a body by virtue of its motion.*

Potential energy E_p: *Energy possessed by a system by virtue of its position or condition.*

One can readily think of many examples of each kind of energy. For instance, a moving car, a moving bullet, and a rotating flywheel all have the ability to do work because of their motion. Similarly, a lifted object, a compressed spring, and a cocked rifle have the potential for doing work because of position. Several examples are provided in Fig. 8-3.

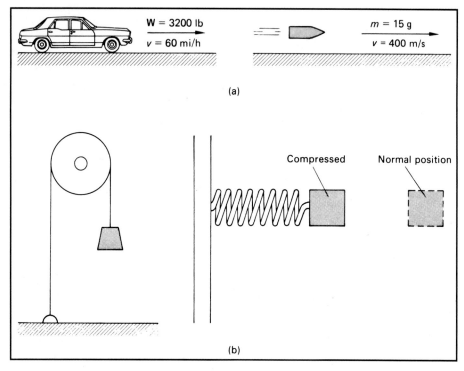

FIG. 8-3 Examples of (*a*) kinetic energy and (*b*) potential energy.

8-4 ■ WORK AND KINETIC ENERGY

When we perform work on a body, what becomes of it? Consider a body with an initial speed v_0 on which a steady force **F** acts through a distance s. If the body has a mass m, Newton's second law of motion tells us that it will gain speed, or accelerate, at a rate given by

$$a = \frac{F}{m} \qquad (8\text{-}4)$$

until it reaches a final speed v_f. From Chap. 5 recall the relation

$$2as = v_f^2 - v_0^2$$

from which

$$a = \frac{v_f^2 - v_0^2}{2s}$$

Substitution of this into Eq. (8-4) yields

$$\frac{F}{m} = \frac{v_f^2 - v_0^2}{2s}$$

which can be solved for the product Fs to obtain

$$Fs = \tfrac{1}{2}mv_f^2 - \tfrac{1}{2}mv_0^2 \qquad (8\text{-}5)$$

The quantity on the left side of Eq. (8-5) represents the work done on the mass m. The quantity on the right side must be the change in kinetic energy as a result of this work. Thus we can define kinetic energy E_k as

$$\boxed{E_k = \tfrac{1}{2}mv^2} \qquad (8\text{-}6)$$

Following this notation, $\tfrac{1}{2}mv_f^2$ and $\tfrac{1}{2}mv_0^2$ would represent final and initial kinetic energies, respectively. The important result can be stated as follows:

The work of a resultant external force on a body is equal to the change in kinetic energy of the body.

A close look at Eq. (8-5) will show that an *increase* in kinetic energy ($v_f > v_0$) will result from *positive* work, whereas a decrease in kinetic energy ($v_f < v_0$) will result from *negative* work. In the special case in which zero work is done, the kinetic energy is a constant, given by Eq. (8-6).

EXAMPLE 8-4   Compute the kinetic energy of a 4-kg sledgehammer at the instant its velocity is 24 m/s.

Solution   Applying Eq. (8-6), we obtain

$$E_k = \tfrac{1}{2}mv^2 = \tfrac{1}{2}(4 \text{ kg})(24 \text{ m/s})^2$$
$$= 1152 \text{ N·m} = 1152 \text{ J}$$

EXAMPLE 8-5   Compute the kinetic energy of a 3200-lb automobile traveling at 60 mi/h (88 ft/s).

Solution   We compute as before except that we must determine the mass from the weight:

$$E_k = \tfrac{1}{2}mv^2 = \frac{1}{2}\left(\frac{W}{g}\right)v^2$$

Substituting given values for W and v, we have

$$E_k = \frac{1}{2}\left(\frac{3200 \text{ lb}}{32 \text{ ft/s}^2}\right)(88 \text{ ft/s})^2$$

$$= 3.87 \times 10^5 \text{ ft·lb}$$

EXAMPLE 8-6 What average force F is necessary to stop a 16-g bullet traveling at 260 m/s as it penetrates into wood a distance of 12 cm?

Solution The total work required to stop the bullet will be equal to the change in kinetic energy. (See Fig. 8-4.) Since the bullet is stopped, $v_f = 0$, and so Eq. (8-5) yields

$$Fs = -\tfrac{1}{2}mv_0^2$$

Substituting gives

$$F(0.12 \text{ m}) = -\tfrac{1}{2}(0.016 \text{ kg})(260 \text{ m/s})^2$$

Dividing by 0.12 m, we have

$$F = \frac{-(0.016 \text{ kg})(260 \text{ m/s})^2}{(2)(0.12 \text{ m})}$$

$$= -4510 \text{ N}$$

The minus sign indicates that the force was opposite to the displacement. It should be noted that this force is about 30,000 times the weight of the bullet.

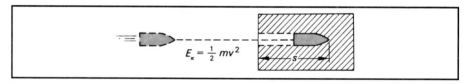

FIG. 8-4 The work done in stopping the bullet is equal to the change in kinetic energy of the bullet.

8-5 ■ POTENTIAL ENERGY

The energy that systems possess by virtue of their positions or conditions is called *potential energy*. Since energy expresses itself in the form of work, potential energy implies that there must be a potential for doing work. For example, suppose the pile driver in Fig. 8-5 is used to lift a body of weight W to a height h above the ground stake. We say that the body-earth system has gravitational potential energy. When such a body is released, it will do work when it strikes the stake. If the body is heavy enough and if it has fallen from a great enough height, the work done will result in driving the stake through a distance s.

The external force F required to lift the body must at least be equal to the weight W. Thus, the work done on the system is given by

$$\text{Work} = Wh = mg \cdot h$$

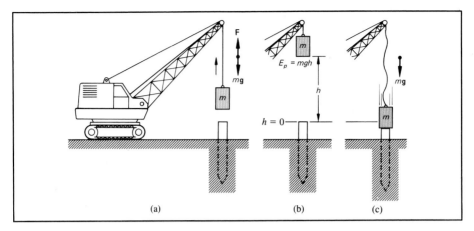

FIG. 8-5 (*a*) Lifting a mass *m* to a height *h* requires the work *mgh*. (*b*) The body-earth system, therefore, has a potential energy $E_p = mgh$. (*c*) When the mass is released, it has the capacity for doing work *mgh* on the stake.

This amount of work can also be done *by* the body after it has dropped a distance *h*. Thus, the body has potential energy equal in magnitude to the external work required to lift it. This energy does not come from the earth-body system, but results from work done on the system by an external agent. Only *external forces,* such as **F** in Fig. 8-5 or friction, can add or remove energy from the system made up of the body and the earth.

Note from the preceding discussion that potential energy E_p can be found from

$$E_p = Wh = mgh \qquad \textit{Potential Energy} \quad (8\text{-}7)$$

where *W* and *m* are the weight and the mass of an object, respectively, located a distance *h* above some reference point.

The potential energy depends on the choice of a particular reference level. The gravitational potential energy for an airplane is quite different when it is measured with respect to a mountain peak, a skyscraper, or sea level. The capacity for doing work is much greater if the aircraft falls to sea level. Potential energy has physical significance only in the event that a reference level is established.

EXAMPLE 8-7   A 250-g carburetor is held 200 mm above a workbench which is 1 m above the floor. Compute the potential energy relative to (*a*) the bench top and (*b*) the floor.

Solution (a)   The height *h* of the carburetor above the bench is 200 mm (or 0.2 m), and the mass is 250 g (or 0.25 kg). Thus, the potential energy relative to the bench is

$$E_p = mgh = (0.25 \text{ kg})(9.8 \text{ m/s}^2)(0.2 \text{ m})$$
$$= 0.49 \text{ J}$$

Notice that kilograms, meters, and seconds, respectively, are the only units of mass, length, and time which are consistent with the definition of a joule.

Solution (b)   The potential energy with reference to the floor is based on a different value of h:

$$E_p = mgh = (0.25 \text{ kg})(9.8 \text{ m/s}^2)(1.2 \text{ m})$$
$$= 2.94 \text{ J}$$

EXAMPLE 8-8   An 800-lb commercial air-conditioning unit is lifted by a chain hoist until it is 22 ft above the floor. What is the potential energy relative to the floor?

Solution   Applying Eq. (8-7), we obtain

$$E_p = Wh = (800 \text{ lb})(22 \text{ ft}) = 17,600 \text{ ft·lb}$$

We have stated that the potential for doing work is a function of only the weight and the height h above some reference point. The potential energy at a particular position is not dependent on the path taken to reach that position. This is because the same work must be done against gravity regardless of the path. In Example 8-8, work of 17,600 ft·lb was required to lift the air conditioner through a vertical distance of 22 ft. If we chose to exert a lesser force by moving it up an incline, a greater distance would be required. In either case, the work done against gravity is 17,600 ft·lb, because the end result is the placement of an 800-lb weight at a height of 22 ft.

8-6 ■ CONSERVATION OF ENERGY

Quite often, at relatively low speeds, an interchange takes place between kinetic and potential energies. For example, suppose a mass m is lifted to a height h and dropped, as shown in Fig. 8-6. An external force has increased the energy of the system, giving it a potential energy $E_p = mgh$ at the highest point. This is the total energy available to the system, and it cannot change unless an external resistive force is encountered. As the mass falls, its potential energy decreases because its height above the ground is reduced. The lost potential energy reappears in the form of kinetic energy of motion. In the absence of air resistance, the total energy $E_p + E_k$ remains the same. Potential energy continues to be converted to kinetic energy until the mass reaches the ground ($h = 0$). At this final position, the kinetic energy is equal to the total energy, and the potential energy is zero. The important point to be made is that the sum of E_p and E_k is the same at any point during the fall (see Fig. 8-6):

$$\text{Total energy} = E_p + E_k = \text{constant}$$

We say that mechanical energy is *conserved*. In our example the total energy at the top is mgh, and the total energy at the bottom is $\frac{1}{2}mv^2$, if we neglect air resistance. We are now prepared to state the principle of *conservation of mechanical energy:*

Conservation of mechanical energy: *In the absence of air resistance or other dissipative forces, the sum of the potential and kinetic energies is a constant provided that no energy is added to the system.*

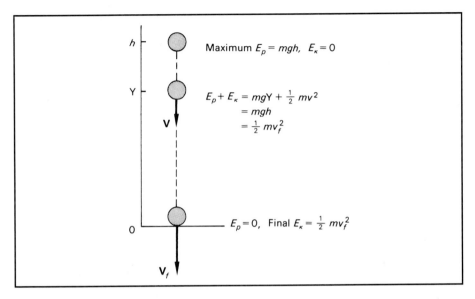

FIG. 8-6 In the absence of friction, the total mechanical energy is constant. At any point, it is equal to either the potential energy at the top or the kinetic energy at the bottom.

Under these conditions, the final kinetic energy of a mass m dropped from a height h is equal to the initial potential energy:

$$\tfrac{1}{2}mv_f^2 = mgh \tag{8-8}$$

Solving this relationship for v_f gives a useful equation for determining the final velocity from energy considerations:

$$v_f = \sqrt{2gh} \tag{8-9}$$

A great advantage of this method is that the final velocity is determined from the initial and final energy states. The actual path taken does not matter in the absence of friction.

EXAMPLE 8-9 In Fig. 8-7, a 40-kg ball is pulled to one side until it is 1.6 m above its lowest position. Neglecting friction, what will its velocity be as it passes through its lowest point?

Solution Conservation of mechanical energy requires that the final kinetic energy be equal to the initial potential energy:

$$\tfrac{1}{2}mv_f^2 = mgh$$

Thus, Eq. (8-9) applies, and we simply solve for v_f:

$$v_f = \sqrt{2gh} = \sqrt{2(9.8 \text{ m/s}^2)(1.6 \text{ m})}$$
$$= 5.60 \text{ m/s}$$

As an additional exercise you should show that the total energy of the system is 627 J.

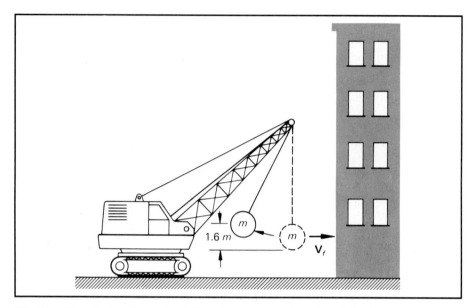

FIG. 8-7 The velocity of a suspended mass as it passes through its lowest point can be found from energy considerations.

Now let us consider the more general case in which some of the mechanical energy is lost due to friction. The change in mechanical energy resulting from such a force will always be equal to the negative work done by the friction force. The final kinetic energy will be reduced because some of the total energy available initially will be lost in doing work against friction. We may write this fact as follows:

$$\left|\begin{matrix}\text{Final kinetic}\\ \text{energy}\end{matrix}\right| = \left|\begin{matrix}\text{initial potential}\\ \text{energy}\end{matrix}\right| - \left|\begin{matrix}\text{work against}\\ \text{friction}\end{matrix}\right|$$

$$\tfrac{1}{2}mv_f^2 = mgh - \mathcal{F}s \qquad (8\text{-}10)$$

Perhaps a better way of writing this statement would be to express it in terms of the total energy available initially:

$$mgh = \tfrac{1}{2}mv_f^2 + \mathcal{F}s \qquad (8\text{-}11)$$

This equation is a mathematical statement of the principle of conservation of energy, which can now be restated as follows:

Conservation of energy: *The total energy of a system is always constant, although energy changes from one form to another may occur within the system.*

EXAMPLE 8-10 A 64-lb block rests initially at the top of a 300-ft plane inclined at an angle of 30°, as shown in Fig. 8-8. If the block encounters a friction force of 8 lb, find (a) the potential energy at the top, (b) the work done against friction, (c) the kinetic energy at the bottom, and (d) the velocity at the bottom.

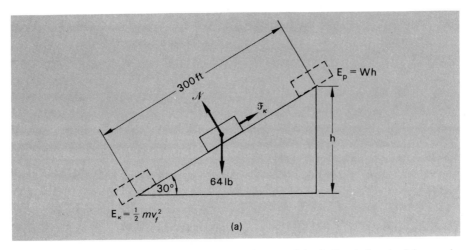

FIG. 8-8 Some of the initial potential energy at the top of the incline is lost in doing work against friction as the block slides down.

Solution (a) The potential energy at the top is a function of the height h and the weight W. If the triangle is drawn to scale, the height h can be determined graphically to be 150 ft. With trigonometry we can obtain this same value from the sine function $h = (300 \text{ ft})(\sin 30°)$. The potential energy is

$$E_p = Wh = (64 \text{ lb})(150 \text{ ft}) = 9600 \text{ ft·lb}$$

Solution (b) The 9600 ft·lb value is the total energy available to the system. Some of this energy is lost to work against friction. Since the 8-lb friction force acts through a distance of 300 ft, the work done is

$$(\text{Work})_{\mathscr{F}} = \mathscr{F}s = (8 \text{ lb})(300 \text{ ft}) = 2400 \text{ ft·lb}$$

Solution (c) According to the principle of conservation of energy, the kinetic energy at the bottom must equal the potential energy at the top less the work of friction. Thus,

$$E_k = 9600 \text{ ft·lb} - 2400 \text{ ft·lb} = 7200 \text{ ft·lb}$$

Solution (d) The velocity is determined from the kinetic energy:

$$E_k = \tfrac{1}{2}mv_f^2 = 7200 \text{ ft·lb}$$

The mass of the block is found from its weight:

$$m = \frac{W}{g} = \frac{64 \text{ lb}}{32 \text{ ft/s}^2} = 2 \text{ slugs}$$

Substituting known values for m and E_k, we obtain

$$\tfrac{1}{2}(2 \text{ slugs})v_f^2 = 7200 \text{ ft·lb}$$

from which

$$v_f = \sqrt{7200 \frac{\text{ft·lb}}{\text{slug}}} = 84.9 \text{ ft/s}$$

You should show that the final velocity would have been 98 ft/s if there had been no friction.

8-7 ■ POWER

In our definition of work, *time* is not involved in any way. The same amount of work is done whether the task takes an hour or a year. Given enough time, even the weakest motor can lift the pyramids of Egypt. However, if we wish to perform a task efficiently, the *rate* at which work is done becomes a very important engineering quantity:

Power *is the rate at which work is performed.*

$$P = \frac{\text{work}}{t} \tag{8-12}$$

In USCS units, we use foot-pounds per second as a unit of power. The corresponding SI unit has a special name, the *watt,* defined as

$$1 \text{ W} = 1 \text{ J/s}$$

The watt and the foot-pound per second are inconveniently small units for most industrial purposes. Therefore, the *kilowatt* (kW) and the *horsepower* (hp) are defined:

$$1 \text{ kW} = 1000 \text{ W}$$
$$1 \text{ hp} = 550 \text{ ft·lb/s}$$

In the United States, the watt and kilowatt are used almost exclusively in connection with electric power; horsepower is reserved for mechanical power. This practice is purely a convention and by no means necessary. It is perfectly proper to speak of an 0.08-hp light bulb or to brag about a 238-kW engine. The conversion factors are

$$1 \text{ hp} = 746 \text{ W} = 0.746 \text{ kW}$$
$$1 \text{ kW} = 1.34 \text{ hp}$$

Since work is frequently done in a continuous fashion, an expression for power which involves velocity is useful. Thus

$$P = \frac{\text{work}}{t} = \frac{Fs}{t} \tag{8-13}$$

from which

$$P = F\frac{s}{t} = Fv \tag{8-14}$$

where v is the velocity of the body on which the parallel force F is applied.

EXAMPLE 8-11   A 40-kg load is raised to a height of 25 m. If the operation requires 1 min, find the power required. What is the power in units of horsepower?

Solution   The work done in lifting the load is

$$\text{Work} = Fs = mgh = (40 \text{ kg})(9.8 \text{ m/s}^2)(25 \text{ m})$$
$$= 9800 \text{ J}$$

The power is then

$$P = \frac{\text{work}}{t} = \frac{9800 \text{ J}}{60 \text{ s}} = 163 \text{ W}$$

Since 1 hp = 746 W, the horsepower developed is

$$P = (163 \text{ W})\left(\frac{1 \text{ hp}}{746 \text{ W}}\right) = 0.219 \text{ hp}$$

EXAMPLE 8-12   A 60-hp motor provides power for the elevator of a hotel. If the weight of the elevator is 2000 lb, how much time is required to lift the elevator 120 ft?

Solution   The work required is given by

$$\text{Work} = Fs = (2000 \text{ lb})(120 \text{ ft})$$
$$= 2.4 \times 10^5 \text{ ft·lb}$$

Since 1 hp = 550 ft·lb/s, the power developed is

$$P = (60 \text{ hp})\left(\frac{550 \text{ ft·lb/s}}{1 \text{ hp}}\right) = 3.3 \times 10^4 \text{ ft·lb/s}$$

From Eq. (8-13),

$$P = \frac{Fs}{t}$$

so that

$$t = \frac{Fs}{P} = \frac{2.4 \times 10^5 \text{ ft·lb}}{3.3 \times 10^4 \text{ ft·lb/s}}$$

$$= 7.27 \text{ s}$$

Equation (8-12) can be solved for work: Work = Pt. Therefore, the *kilowatthour* (kW·h) unit used by electric companies in billing is a unit of *power* (kilowatt) times *time* (hour), or a unit of work. Quite reasonably, the bill is for the amount of work that has been done. However, the price per kilowatthour may also be determined by the peak power demand of the consumer.

8-8 ■ IMPULSE AND MOMENTUM

Energy and work are scalar quantities which say nothing about direction. The law of conservation of energy describes only the relationship between initial and final states; it says nothing about how the energies are distributed. For example, when two objects collide, we can say that the total energy before impact must equal the total energy after impact. But we need a new concept if we are to determine how the total energy is divided between the objects or even their relative directions after impact. The concepts of *impulse* and *momentum* will add a vector description to our discussion.

When a golf ball is driven from the ground, as in Fig. 8-9, a large average force **F** acts on the ball during a very short time Δt, causing the ball to accelerate from rest to a final velocity $\mathbf{v}_f$. It is extremely difficult to measure either the force or its duration, but their product **F** Δt can be determined from the resulting change in velocity of the golf ball. From Newton's second law we have

$$\mathbf{F} = m\mathbf{a} = m\,\frac{\mathbf{v}_f - \mathbf{v}_0}{\Delta t}$$

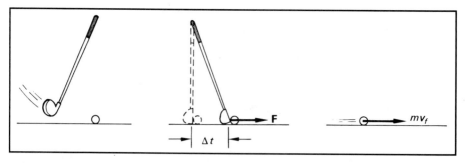

FIG. 8-9 When a golf club strikes the ball, a force F acting through the time interval t results in a change in its momentum.

Multiplying by Δt gives

$$\mathbf{F} \, \Delta t = m(\mathbf{v}_f - \mathbf{v}_0)$$

or

$$\boxed{\mathbf{F} \, \Delta t = m\mathbf{v}_f - m\mathbf{v}_0} \tag{8-15}$$

This equation is so useful in solving problems involving impact that the terms are given special names.

> The **impulse F** Δt *is a vector quantity equal in magnitude to the product of the force and the time interval in which it acts. Its direction is the same as that of the force.*

> The **momentum p** *of a particle is a vector quantity equal in magnitude to the product of its mass* m *and its velocity* **v**:

$$\boxed{\mathbf{p} = m\mathbf{v}}$$

Thus Eq. (8-15) can be stated verbally:

$$\text{Impulse } (\mathbf{F} \, \Delta t) = \text{change in momentum } (m\mathbf{v}_f - m\mathbf{v}_0)$$

The SI unit of impulse is the *newton-second* (N·s). The unit of momentum is the *kilogram-meter per second* (kg·m/s). It is convenient to distinguish these units even though they are actually the same:

$$\text{N·s} = \frac{\text{kg·m}}{\text{s}^2} \times \text{s} = \text{kg·m/s}$$

The corresponding USCS units are the *pound-second* (lb·s) and *slug-foot per second* (slug·ft/s).

EXAMPLE 8-13 A 3-kg sledgehammer is moving at 14 m/s as it strikes a steel spike. It is brought to a stop in 0.02 s. Determine the average force on the spike.

Solution Since $v_f = 0$, we have from Eq. (8-15)

$$F \, \Delta t = -mv_0$$

If we consider that the hammer is moving downward, we substitute $v_0 = -14$ m/s, giving

$$F = \frac{-mv_0}{\Delta t} = \frac{-(3 \text{ kg})(-14 \text{ m/s})}{0.02 \text{ s}}$$

$$= 2100 \text{ N}$$

This force, exerted on the hammer, is equal in magnitude but opposite in direction to the force exerted on the spike. It must be emphasized that the force found in this manner is an *average force*. At some instants the force will be much greater than 2100 N.

EXAMPLE 8-14   A 0.6-lb baseball moving toward the batter with a velocity of 44 ft/s is struck with a bat which causes it to move in a reversed direction with a velocity of 88 ft/s. (Refer to Fig. 8-10.) Find the impulse and the average force exerted on the ball if the bat is in contact with the ball for 0.01 s.

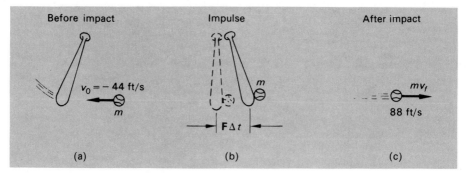

FIG. 8-10 Impact of a bat with a baseball.

Solution   Consider the final direction of motion as positive. Applying Eq. (8-15), we solve for the impulse as follows:

$$F \Delta t = mv_f - mv_0 = m(v_f - v_0)$$

since

$$m = \frac{W}{g} = \frac{0.6 \text{ lb}}{32 \text{ ft/s}} = 0.0188 \text{ slug}$$

$$F \Delta t = 0.0188 \text{ slug}[88 \text{ ft/s} - (-44 \text{ ft/s})]$$
$$= 0.0188 \text{ slug}(132 \text{ ft/s})$$
$$\text{Impulse} = F \Delta t = 2.48 \text{ lb·s}$$

The average force is then found by substituting $\Delta t = 0.01$ s:

$$F(0.01 \text{ s}) = 2.48 \text{ lb·s}$$

$$F = \frac{2.48 \text{ lb·s}}{0.01 \text{ s}} = 248 \text{ lb}$$

8-9 ■ LAW OF CONSERVATION OF MOMENTUM

Let us consider the *head-on collision* of masses m_1 and m_2, as shown in Fig. 8-11. We denote their velocities before impact as u_1 and u_2 and after impact as v_1 and v_2, respectively. The impulse of the force $\mathbf{F}_1$ acting on the right mass is

$$F_1 \, \Delta t = m_1 v_1 - m_1 u_1$$

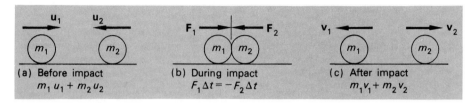

FIG. 8-11 Head-on collision of two masses.

Similarly, the impulse of the force $\mathbf{F}_2$ on the left mass is

$$F_2 \, \Delta t = m_2 v_2 - m_2 u_2$$

During the time Δt, $\mathbf{F}_1 = -\mathbf{F}_2$, so that

$$F_1 \, \Delta t = -F_2 \, \Delta t$$

or

$$m_1 v_1 - m_1 u_1 = -(m_2 v_2 - m_2 u_2)$$

and finally, after rearranging, we have

$$m_1 u_1 + m_2 u_2 = m_1 v_1 + m_2 v_2 \qquad (8\text{-}16)$$

Total momentum before impact = total momentum after impact

Thus we have derived a statement of the law of *conservation of momentum*:

The total linear momentum of colliding bodies before impact is equal to their total momentum after impact.

EXAMPLE 8-15 Assume that m_1 and m_2 of Fig. 8-11 have masses of 8 and 6 kg, respectively. The mass m_1 moves initially to the right with a velocity of 4 m/s and collides with m_2, moving to the left at 5 m/s. What is the total momentum before and after the collision?

Solution   We choose the direction to the right as positive, taking care to affix the proper sign to the velocities:

$$p_0(\text{before impact}) = m_1u_1 + m_2u_2$$
$$p_0 = (8 \text{ kg})(4 \text{ m/s}) + (6 \text{ kg})(-5 \text{ m/s})$$
$$= 32 \text{ kg·m/s} - 30 \text{ kg·m/s} = 2 \text{ kg·m/s}$$

The same total momentum must exist after impact, and so we can write

$$p_f = m_1v_1 + m_2v_2 = 2 \text{ kg·m/s}$$

If either v_1 or v_2 can be measured after the collision, the other can be determined from this relation.

EXAMPLE 8-16   A rifle weighs 8 lb and fires a bullet weighing 0.02 lb at a muzzle velocity of 2800 ft/s. Compute the recoil velocity if the rifle is freely suspended.

Solution   Since both the rifle m_1 and the bullet m_2 are initially at rest, the total momentum before firing must equal zero. The momentum is unaltered, and so it must also be zero after firing. Hence Eq. (8-16) gives

$$0 = m_1v_1 + m_2v_2$$
$$m_1v_1 = -m_2v_2$$

$$v_1 = -\frac{m_2v_2}{m_1}$$

$$= -\frac{(0.02 \text{ lb}/32 \text{ ft/s}^2)(2800 \text{ ft/s})}{8 \text{ lb}/32 \text{ ft/s}^2}$$

$$= -7 \text{ ft/s}$$

An interesting experiment which demonstrates the conservation of momentum can be performed with eight small marbles and a grooved track, as shown in Fig. 8-12. If one marble is released from the left, it will be stopped upon colliding with the others, and one on the right end will roll out with the same velocity. Similarly, when two, three, four, or five marbles are released from the left, the same number will roll out to the right with the same velocity, the others remaining at rest in the center.

You might reasonably ask why two marbles roll off in Fig. 8-12 instead of one with twice the velocity, since this condition would also conserve momentum. For example, if each marble has a mass of 50 g, and if two marbles approach from the left at a velocity of 20 cm/s, the total momentum before impact is 2000 g·cm/s. This same momentum could be achieved after impact if only one marble left, assuming it had a velocity of 40 cm/s. The answer lies in the fact that energy must be conserved. If one marble came off with twice the velocity, its kinetic energy would be much greater than that available from the other two. The kinetic energy put into the system would be

$$E_0 = \tfrac{1}{2}mv^2 = \tfrac{1}{2}(0.1 \text{ kg})(0.2 \text{ m/s})^2$$
$$= 2 \times 10^{-3} \text{ J}$$

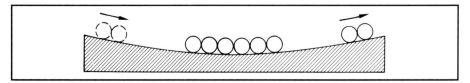

FIG. 8-12 Conservation of momentum.

The kinetic energy of one marble traveling at 40 cm/s is exactly twice this value:

$$E_f = \tfrac{1}{2}mv^2 = \tfrac{1}{2}(0.05 \text{ kg})(0.4 \text{ m/s})^2$$
$$= 4 \times 10^{-3} \text{ J}$$

Therefore, energy as well as momentum is important in describing impact phenomena.

Summary

The concepts of work, energy, power, and momentum have been discussed in this chapter. The major points to remember are summarized below:

- The *work* done by a force F acting through a distance s is found from the following equations (refer to Fig. 8-1):

$$\text{Work} = F_x s \qquad \text{work} = (F \cos \theta)s$$

SI unit: *joule* (J)   USCS unit: *foot-pound* (ft·lb)
- *Kinetic energy* E_k is the capacity for doing work as a result of motion. It has the same units as work and is found from

$$E_k = \tfrac{1}{2}mv^2 \qquad E_k = \frac{1}{2}\left(\frac{W}{g}\right)v^2$$

- Gravitational *potential energy* is the energy which results from the position of an object relative to the earth. Potential energy E_p has the same units as work and is found from

$$E_p = Wh \qquad E_p = mgh$$

where W or mg is the weight of the object and h is the height above some reference position.

■ Net work is equal to the change in kinetic energy:

$$\mathscr{F}s = \tfrac{1}{2}mv_f^2 - \tfrac{1}{2}mv_0^2$$

■ Conservation of mechanical energy under the action of a dissipative force F:

$$mgh = \tfrac{1}{2}mv_f^2 + \mathscr{F}_k s \qquad \text{initial } E_p = \text{final } E_k + (\text{work})_F$$

■ *Power* is the rate at which work is done:

$$P = \frac{\text{work}}{t} \qquad P = \frac{Fs}{t} \qquad P = Fv$$

SI unit: watt (W)   USCS unit: ft·lb/s
Other units 1 kW $= 10^3$ W   1 hp $= 550$ ft·lb/s

■ The *impulse* is the product of the average force F and the time interval Δt through which it acts:

$$\text{Impulse} = F\,\Delta t \qquad \text{SI units: N·s} \qquad \text{USCS units: lb·s}$$

■ The *momentum* of a particle is its mass times its velocity:

$$\text{Momentum } p = mv \qquad \text{SI units: kg·m/s} \qquad \text{USCS units: slug·ft/s}$$

■ The impulse is equal to the change in momentum:

$$F\,\Delta t = mv_f - mv_0 \qquad \text{N·s} = \text{kg·m/s (equivalent units)}$$

■ *Conservation of momentum:* The total momentum before impact is equal to the total momentum after impact:

$$m_1 u_1 + m_2 u_2 = m_1 v_1 + m_2 v_2$$

Questions

8-1. Define the following terms:

a. work
b. potential energy
c. kinetic energy
d. conservation of energy
e. conservation of momentum

f. power
g. horsepower
h. watt
i. impulse
j. momentum

8-2. Two teams are engaged in a tug of war. Is work done? When?

8-3. Whenever net work is done on a body, will the body undergo acceleration? Discuss.

8-4. A diver stands on a board 10 ft above the water. What kind of energy results from this position? What happens to this energy as the diver approaches the water? Is work done? If so, what is doing the work, and on what is the work done?

8-5. A hand compresses a spring. The _____ does negative work on the _____, and the _____ does positive work on the _____. If the compression takes place slowly and at a constant speed, is the net work zero? If so, how can there be potential energy after compression?

8-6. In stacking 8-ft boards, you lift an entire board at its center and lay it on the pile. Your helper lifts one end, rests it on the pile, and then lifts the other end. Compare the work done.

8-7. You first drive a nail into a wall by striking it horizontally with a hammer; then you drive a nail into the floor with the same hammer. The nail penetrates 2 mm in each case, and the material on the floor is identical to that on the wall. Is the average stopping force the same for each situation? What role does the weight of the hammer play in each case?

8-8. For an object of mass m which slides from rest at the top of a slope of height h, the final kinetic energy was given by

$$\tfrac{1}{2}mv^2 = mgh - \mathscr{F}_k s$$

Write a similar equation which expresses the minimum kinetic energy required at the bottom if the mass is to slide to the top of the slope.

8-9. A roller coaster at the fair boasts ''a maximum height of 100 ft with a maximum speed of 60 mi/h.'' Do you believe the advertisement? Explain.

8-10. A constant force is applied to an object producing constant acceleration. Is the power also constant? Discuss.

8-11. If you hold a weapon loosely when firing, it appears to give a greater kick than when you hold it tight against your shoulder. Explain. What effect does the weight of the weapon have?

8-12. A mortar shell explodes in midair. How is energy conserved? How is momentum conserved?

8-13. A father and daughter stand facing each other on a frozen pond. If the girl pushes the father backward, describe their relative motion and velocities. Would they differ if the father pushed the daughter?

Problems

8-1. A horizontal force of 20 N drags a small sled across the ground at constant speed. The speed is constant because the force of friction exactly balances the 20-N pull. If 42 m is covered, what is the work done by the pulling force? What is the work of the friction force? What is the total or net work?

Answer 840 J; −840 J; 0 J.

8-2. An external work of 400 ft·lb is applied in lifting a 30-lb motor. If all this work contributes to lifting the motor, how high will it be lifted?

8-3. How much work is done in lifting a 6-lb weight to a height of 2 ft? How much work is done in lifting a 9-kg mass 8 m?

Answer 12 ft·lb; 706 J.

8-4. A trunk is pulled 24 m across the floor by a rope making an angle θ with the horizontal, as shown in Fig. 8-13. The tension in the rope is 8 N. Compute the work done when θ is equal to (*a*) 0°, (*b*) 30°, (*c*) 60°.

8-5. The block in Fig. 8-13 has a mass of 2 kg, the angle θ is 60°, and the coefficient of kinetic friction is 0.1. What is the resultant work if the block is dragged for 4 m? What is the work of the friction force? What is the work of the 8-N force?

Answer 10.9 J; −7.84 J; 16 J.

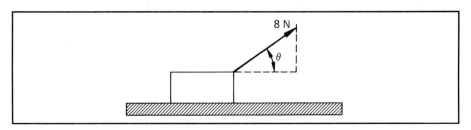

FIG. 8-13

8-6. A 10-kg block is pushed 8 m along a horizontal surface by a constant force of 26 N. If $\mu_k = 0.2$, what is the resultant work? What acceleration will the block receive?

8-7. A horizontal force of 60 lb drags a block along a horizontal surface for 8 ft. If the resultant work is 384 ft·lb, what was the force of kinetic friction?

Answer 12 lb.

8-8. An average force of 40 N shortens a coiled spring by 6 cm. What is the work done by the 40-N force? What is the potential energy?

8-9. A 2-kg block is pushed 6 m along a horizontal surface by a constant force of 16 N. If $\mu_k = 0.5$, what is the resultant work? How much work is done against friction?

Answer 37.2 J; −58.8 J.

8-10. What is the kinetic energy of a 6-g bullet at the instant its speed is 190 m/s? What is the kinetic energy of a 1200-kg car traveling at 80 km/h?

8-11. What is the kinetic energy of a 2400-lb automobile when its speed is 55 mi/h? What is the kinetic energy of a 9-lb ball when its speed is 40 ft/s?

Answer 244,000 ft·lb; 225 ft·lb.

8-12. What is the gravitational potential energy of a 16-lb hammer when it is lifted 20 ft? What is the potential energy of a 4-kg hammer at the same height?

8-13. The hammer in Fig. 8-14 has a mass of 3 kg and is traveling at 90 m/s at the instant it strikes the nail. If the nail is driven into the wall 4 mm, what was the average stopping force?

Answer −3.04 × 10⁶ N.

8-14. A 6-kg ball moving initially at 4 m/s is acted on by a constant force of 8 N for a distance of 16 m. What is the initial kinetic energy? How much work is done on the system? What is the final kinetic energy, and what is the final speed?

8-15. If a 2-N friction force opposes the motion in Prob. 8-14, what is the initial kinetic energy? What is the resultant work on the system? What is the final kinetic energy? What the final speed?

Answer 48 J; 96 J; 144 J; 6.93 m/s.

8-16. A 6-kg mass falls through a distance of 20 m. What is the loss in potential energy? What happens to this energy?

FIG. 8-14

8-17. A 10-kg mass is lifted to a height of 20 m. What are the potential energy, kinetic energy, and total energy at that height? Then the mass is released and falls freely. What are the potential energy, kinetic energy, and total energy when the mass is at the bottom of its fall? What is its final velocity based on its energy?

Answer 1960, 0, 1960 J; 0, 1960, 1960 J; 19.8 m/s.

8-18. In Prob. 8-17, what are the potential energy, the kinetic energy, the total energy, and the velocity when the ball is 5 m above the ground?

8-19. The hammer of a pile driver weighs 800 lb and falls 16 ft before striking the pile. The impact drives the pile 6 in. deeper into the ground. Based on energy considerations, what was the average force driving the pile?

Answer 25,600 lb.

8-20. A 10-kg sled traveling initially at 2 m/s is given a steady push P for a distance of 12 m. What must be the value of P in order to increase the speed to 4 m/s in this distance? First calculate from work-energy considerations; then check your answer from Newton's second law.

8-21. A ballistic pendulum is a laboratory device (Fig. 8-15) which might be used to calculate the velocity of a projectile. A 40-g ball is caught by a 500-g suspended mass. After impact, the two masses move together until they stop at a point 45 mm above the point of impact. From energy considerations, what was the velocity of the combined masses just after impact? (Neglect friction.)

Answer 0.939 m/s.

8-22. A 64-lb projectile leaves the ground with a kinetic energy of 4096 ft·lb. Use energy considerations to determine the maximum height the projectile will reach.

8-23. A 20-g bullet leaves a rifle barrel with a muzzle velocity of 1500 m/s. Calculate its kinetic energy. If the bullet penetrates a wooden block a distance of 1.2 cm, what is the average stopping force?

Answer 22,500 J; -1.88×10^6 N.

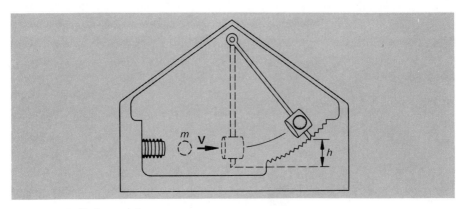

FIG. 8-15 Ballistic pendulum.

8-24. An automobile weighs 2400 lb. What is its mass? How much work is needed to lift this car to a height of 130 ft? What is the potential energy at that height? If the car is released and falls freely, what will be its velocity on striking the ground? Verify that the same velocity would be found from uniform acceleration equations.

8-25. A 20-kg sled is pushed up a 34° incline to a vertical height of 140 m above its initial position. What is the slope distance? Draw a free-body diagram of the forces acting on the sled. Neglecting friction, what is the minimum push **P** required to move the sled up the slope? How much work is done by the force **P?** What is the potential energy at the top?

8-26. Assume the same conditions as in Prob. 8-25 except that the coefficient of kinetic friction is 0.2. (*a*) What is the normal force exerted by the slope on the sled? (*b*) What is the force of kinetic friction? (*c*) What minimum push **P** was required? (*d*) What is the work done by the force **P?** (*e*) What is the resultant or net work if the speed was constant? (*f*) What is the potential energy at the top?

8-27. A sled weighing 20 lb slides from rest at the top of a hill 128 ft high. The slope of the hill is 37°, and a constant 4-lb friction force acts through the entire distance. Draw a free-body diagram of the forces acting on the sled. What is the initial potential energy? How much work is done against friction? What is the kinetic energy at the bottom? What is the velocity at the bottom?

Answer 2560 ft·lb; 851 ft·lb; 1709 ft·lb; 74.0 ft/s.

8-28. Do Prob. 8-27 for a 6-kg sled at the top of a hill 20 m high and 30 m long. Assume that the constant friction force is 12 N.

8-29. How much work is required by a force directed up the plane if a 4-kg box is to be moved up an incline 20 m long and 16 m high? The coefficient of kinetic friction is 0.3.

Answer 767 J.

8-30. What will be the kinetic energy of the box in Prob. 8-29 when it slides to the bottom of the plane after being released from the top?

8-31. A simple pendulum 1 m long has a wooden bob of mass 8 kg. (*a*) How much work is required to move the pendulum from its vertical position to a horizontal position? (*b*) Compute the kinetic energy and the velocity of the bob at the instant it passes through the lowest position in its path.

Answer (*a*) 78.4 J; (*b*) 78.4 J, 4.43 m/s.

8-32. A simple pendulum 2 ft long has a wooden bob which weighs 5 lb. As it passes through its lowest point, its velocity is measured to be 10 ft/s. What is its kinetic energy at that point? What is the maximum potential energy that can be attained? What is the maximum height to which the bob can rise?

8-33. A 5-lb hammer is moving horizontally at 25 ft/s when it strikes a nail. If the nail meets an average resistive force of 300 lb, compute the depth of penetration in inches.

Answer 1.95 in.

8-34. A water pump feeds water from a well into a watering trough 30 ft higher than the water level of the well. (Water weighs 62.4 lb/ft^3.) If the pump delivers 60 ft^3 of water each minute, how much work is done each minute?

8-35. A 40-kg mass is lifted at constant speed through a distance of 20 m in 3 s. What is the average power employed?

Answer 2.61 kW.

8-36. A power station conveyer belt lifts 500 tons/h of ore to a height of 90 ft. What average horsepower is required?

8-37. What is the maximum speed at which a 40-hp hoist can lift a 2-ton load?

Answer 5.5 ft/s.

8-38. A 20-kg mass is lifted at constant speed by an 800-W engine. Calculate the speed.

8-39. A 4-kg sled is dragged at constant speed by an engine with an output power of 1 kW. If $\mu_k = 0.6$, what time is required to drag the sled 16 m?

Answer 0.376 s.

8-40. A 200-lb man climbs an 800-ft slope in 7 h. How much work does he do in foot-pounds? In joules? What average power must he supply?

8-41. A 300-kg elevator is raised with a constant velocity through a vertical distance of 100 m in 2 min. What is the increase in potential energy? What is the useful output power of the hoist?

Answer 2.94 × 10^5 J; 2.45 kW.

8-42. What mass can a 3-kW engine drag along a level road at 15 m/s if the coefficient of friction is 0.1?

8-43. A 90-hp motor provides power for a small elevator in a hotel. If the weight of the elevator is 3000 lb, how much time is required to lift the elevator 200 ft at a constant speed?

Answer 12.1 s.

8-44. An elevator is powered by a 30-hp motor. The weight of the elevator and its load is 2000 lb. If the overall efficiency of the motor is 70 percent, what will be the upward speed of the elevator?

8-45. What is the maximum speed at which a 40-kW engine can hoist an 800-kg load?

Answer 5.10 m/s.

8-46. Compute the linear momentum of a 1000-kg car traveling at 80 km/h in a northerly direction.

8-47. What is the momentum of a 0.003-kg bullet moving at 600 m/s in a direction of 30° above the horizontal? What are the horizontal and vertical components of this momentum?

Answer 1.8 kg·m/s; 1.56 kg·m/s, 0.9 kg·m/s.

8-48. A 400-g rubber ball is dropped from a window which is 12 m from the pavement below. Find the speed of the ball just before impact. What is its momentum just before impact? If it leaves the pavement with a speed of 12 m/s, what is its momentum after impact? What is the total change in momentum? If the ball is in contact with the pavement for 0.01 s, what is the average force exerted on the ball?

8-49. A 2500-kg truck traveling at 40 km/h strikes a brick wall and comes to a stop in 0.2 s. (*a*) What is the impulse? (*b*) Find the average force on the truck during the crash.

Answer (*a*) -2.78×10^4 kg·m/s; (*b*) 1.39×10^5 lb.

8-50. A train travels with a speed of 60 mi/h and weighs 8×10^6 lb. (*a*) What impulse will the train exert while stopping? (*b*) If the train comes to a dead stop in 600 ft, how long will it take? (*c*) What is the required braking force?

8-51. A 0.2-kg baseball reaches the batter with a speed of 20 m/s. After the ball has been struck, it leaves the bat at 35 m/s in a reversed direction. If the ball exerts an average force of 8400 N, how long was it in contact with the bat?

Answer 0.00131 s.

8-52. A 24-g bullet is fired with a muzzle velocity of 900 m/s from a 5-kg rifle. Find the recoil velocity of the rifle. Find the ratio of the kinetic energy of the bullet to that of the rifle.

8-53. Two children, weighing 80 and 50 lb, are at rest on roller skates. If the larger child pushes the other so that the smaller one moves away at a speed of 6 mi/h, what will the velocity of the larger child be?

Answer 5.50 ft/s.

9

ROTATIONAL MOTION

As a result of completing this chapter, you should be able to

1. Demonstrate by definition and example your understanding of the concepts of *centripetal force* and *centripetal acceleration*.
2. Apply your knowledge of centripetal force and centripetal acceleration to the solution of problems similar to those in this text.
3. Define and apply the concepts of *frequency, period, angular speed,* and *angular acceleration*.
4. Write and apply the relationship between *linear* speed and *angular* speed.
5. Draw analogies relating angular motion parameters (θ, ω, and α) to linear motion parameters (s, v, and a) and solve problems involving uniform angular acceleration.
6. Solve simple problems involving the concepts of *rotational work* and *power*.

We have been considering only translational motion, in which an object's position is changing along a straight line. But it is possible for an object to move in a curved path, or to undergo rotational motion. For example, wheels, shafts, pulleys, gyroscopes, and many other mechanical devices rotate about their axes without translational motion. The generation and transmission of power are nearly always dependent on rotational motion of some kind. It is essential for you to be able to predict and control such motion. The concepts and formulas presented in this chapter are designed to provide you with these essential skills.

9-1 ■ MOTION IN A CIRCLE

Newton's first law tells us that all bodies moving in a straight line with constant speed will maintain their velocity unaltered unless acted on by an external force. The velocity of a body is a vector quantity consisting of both its speed and its direction. Just as a resultant force is required to change its speed, a resultant force must be applied to change its direction. Whenever this force acts in a direction other than the original direction of motion, the path of a moving particle is changed.

The simplest kind of two-dimensional motion occurs when a constant external force always acts at right angles to the path of a moving particle. In this case the resultant force will produce an acceleration which alters only the direction of motion, leaving the speed constant. This simple type of motion is referred to as *uniform circular motion*.

> **Uniform circular motion** *is motion in which there is no change in speed, only a change in direction.*

An example of uniform circular motion is afforded by swinging a rock in a circular path with a string, as shown in Fig. 9-1. As the rock revolves with constant speed, the inward force of the tension in the string constantly changes the direction of the rock, causing it to move in a circular path. If the string should break, the rock would fly off at a tangent perpendicular to the radius of its circular path.

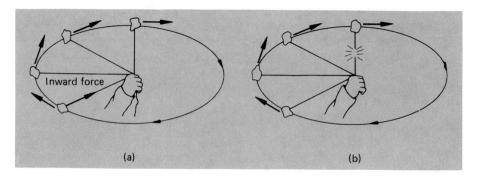

Inward force

(a)   (b)

FIG. 9-1 (*a*) The inward pull of the string on the rock causes it to move in a circular path. (*b*) If the string breaks, the rock will fly off at a tangent to the circle.

9-2 ■ CENTRIPETAL ACCELERATION

Newton's second law of motion states that a resultant force must produce an acceleration *in the direction of the force*. In uniform circular motion, a *central force* changes the velocity of a moving particle by continually changing its direction. Thus, it is possible for an object to undergo an acceleration (a change in velocity) even though it is moving with a constant circular speed. It can be shown through vector analysis that the direction of this acceleration is toward the center of the circular path. For an object which is caused to move in a circular path of radius R as in Fig. 9-2, the magnitude of the *centripetal acceleration* a_c is given by

$$a_c = \frac{v^2}{R}$$ *Centripetal Acceleration*   (9-1)

where v is the linear speed of the particle. The term *centripetal* means that the acceleration is always directed toward the center and perpendicular to the direction of the velocity at any instant.

The units of centripetal acceleration are the same as those for linear acceleration. For example, in SI, v^2/R would have the units

$$\frac{(m/s)^2}{m} = \frac{m^2/s^2}{m} = m/s^2$$

However, in the linear case, it is the *speed* that is changing. Only the *direction* changes in uniform circular motion.

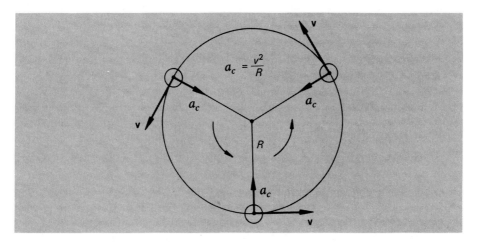

FIG. 9-2 Centripetal acceleration is always directed toward the center and perpendicular to the linear velocity.

EXAMPLE 9-1   A 3-kg body is tied to the end of a cord and whirled in a horizontal circle of radius 600 mm. If the body makes 3 complete revolutions (3 rev) every second, determine its linear speed and its centripetal acceleration.

Solution   If the body makes 3 rev/s, the time to travel one complete circle (a distance $2\pi R$) is $\frac{1}{3}$ s. Hence the linear speed is

$$v = \frac{2\pi R}{0.333 \text{ s}} = \frac{2\pi(0.6 \text{ m})}{0.333 \text{ s}}$$

$$= 11.3 \text{ m/s}$$

The centripetal acceleration, from Eq. (9-1), is

$$a_c = \frac{v^2}{R} = \frac{(11.3 \text{ m/s})^2}{0.6 \text{ m}}$$

$$= \frac{128 \text{ m}^2/\text{s}^2}{0.6 \text{ m}} = 213 \text{ m/s}^2$$

The procedure used to compute linear speed in the above example is so useful that it is worth remembering. If we define the *period* as the time for one complete revolution and designate it by the letter T, the linear speed can be computed by dividing the period into the circumference. Thus

$$v = \frac{2\pi R}{T} \tag{9-2}$$

Another useful parameter in engineering problems is the rotational speed, expressed in *revolutions per minute* (rpm) or *revolutions per second* (rev/s). This quantity is called the *frequency* of rotation and is given by the reciprocal of the period:

$$f = \frac{1}{T} \tag{9-3}$$

The validity of this relation is demonstrated by noting that the reciprocal of revolutions per second is seconds per revolution, or the period. Substitution of this definition into Eq. (9-2) yields an alternative equation for determining the linear speed:

$$v = 2\pi f R \tag{9-4}$$

For example, if the frequency is 1 rev/s and the radius is 1 ft, the linear speed would be 2π ft/s.

9-3 ■ CENTRIPETAL FORCE

The inward force necessary to maintain uniform circular motion is defined as the *centripetal force*. From Newton's second law of motion, the magnitude of this force must equal the product of mass and centripetal acceleration. Thus

$$F_c = ma_c = \frac{mv^2}{R} \qquad (9\text{-}5)$$

where m is the mass of an object moving with a speed v in a circular path of radius R. The units chosen for the quantities F_c, m, v, and R must be consistent for the system chosen. For example, the SI units for mv^2/R are

$$\frac{\text{kg·m}^2/\text{s}^2}{\text{m}} = \frac{\text{kg·m}}{\text{s}^2} = \text{N}$$

An inspection of Eq. (9-5) reveals that the inward force $\mathbf{F}_c$ is directly proportional to the square of the velocity of the moving object. This means that increasing the linear speed to twice its original value will require 4 times the original force. Similar reasoning will show that doubling the mass or halving the radius will require twice the original centripetal force.

For problems in which the rotational speed is expressed in terms of the frequency, the centripetal force can be determined from

$$F_c = \frac{mv^2}{R} = 4\pi^2 f^2 mR \qquad (9\text{-}6)$$

This relation results from substitution of Eq. (9-4) which expresses the linear speed in terms of the frequency of revolution.

EXAMPLE 9-2   A 4-kg ball is swung in a horizontal circle by a cord 2 m long. What is the tension in the cord if the period is 0.5 s?

Solution   The tension in the cord will be equal to the centripetal force necessary to hold the 4-kg body in a circular path. The linear speed is obtained by dividing the period into the circumference:

$$v = \frac{2\pi R}{T} = \frac{2\pi(2 \text{ m})}{0.5 \text{ s}} = 25.1 \text{ m/s}$$

from which the centripetal force is

$$F_c = \frac{mv^2}{R} = \frac{(4 \text{ kg})(25.1 \text{ m/s})^2}{2 \text{ m}}$$

$$= 1260 \text{ N}$$

EXAMPLE 9-3   Two 4-lb weights rotate about the center axis at 12 rev/s, as shown in Fig. 9-3. (a) What is the resultant force acting on each weight? (b) What is the tension in the rod?

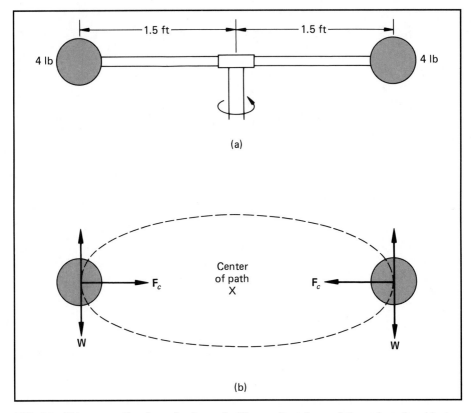

FIG. 9-3 Objects traveling in a circular path. The resultant force of the rod on the objects provides the necessary centripetal force. According to Newton's third law, the objects exert an equal and opposite reaction force, called the *centrifugal force*. These forces do not cancel because they act on different objects.

Solution (a)   The total downward force of the weights and rod is balanced by the upward force of the center support. Therefore, the resultant force acting on each revolving weight is a pull toward the center equal to the centripetal force. The mass of each weight is

$$m = \frac{W}{g} = \frac{4 \text{ lb}}{32 \text{ ft/s}^2} = 0.125 \text{ slug}$$

Substituting the given values of frequency, mass, and radius into Eq. (9-6), we obtain

$$F_c = 4\pi^2 f^2 mR = 4\pi^2 (12 \text{ rev/s})^2 (0.125 \text{ slug})(1.5 \text{ ft})$$
$$= 1066 \text{ lb}$$

The same calculations hold for the other weight.

Solution (b)   The resultant force just computed represents the centripetal force exerted *by* the rod *on* the weights. According to Newton's third law, there must be an equal and opposite reaction force exerted by the weight on the rod. Remember that although these forces are equal in magnitude and opposite in direction, they do not act on the same body. Because the outward force exerted on the rod is *fleeing the center,* it is sometimes referred to as the *centrifugal force.* It is this centrifugal force that causes the tension in the rod. Since it is equal in magnitude to the centripetal force, the tension in the rod must also be 1066 lb.

9-4 ■ FRICTION AND THE CENTRIPETAL FORCE

When an automobile is driven around a sharp turn on a perfectly level road, friction between the tires and the road provides centripetal force (see Fig. 9-4). If this centripetal force is not adequate, the car may slide off the road. The maximum value of the force of friction determines the maximum speed with which a car can negotiate a turn of a given radius.

EXAMPLE 9-4   What is the maximum speed at which an automobile can negotiate a curve of radius 300 ft without sliding if the coefficient of static friction is 0.7?

Solution   As the car increases its speed, the force of static friction required to hold it gets larger. Finally, the car attains a speed so great that the centripetal force equals the maximum force of static friction. At that instant

$$\mathscr{F}_s = F_c = \frac{mv^2}{R}$$

and since $\mathscr{F}_s = \mu_s \mathscr{N}$, we can write

$$\frac{mv^2}{R} = \mu_s \mathscr{N}$$

Applying the first condition for equilibrium to the vertical forces in Fig. 9-4 reveals that the normal force is equal to the weight of the car:

$$\mathscr{N} = W = mg$$

Hence

$$\frac{mv^2}{R} = \mu_s mg \qquad \text{or} \qquad v^2 = \mu_s gR$$

from which

$$\boxed{v = \sqrt{\mu_s gR}} \tag{9-7}$$

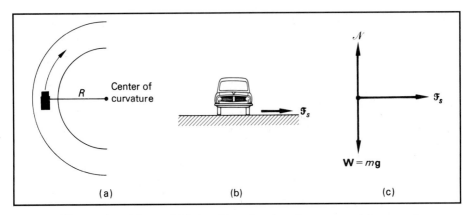

FIG. 9-4 The centripetal force of friction. Note that there is no outward force on the car.

Substituting known values for μ_s, g, and R, we can now compute the maximum speed:

$$v = \sqrt{(0.7)(32 \text{ ft/s}^2)(300 \text{ ft})} = 82 \text{ ft/s}$$

or approximately 56 mi/h.

*9-5 ■ BANKING OF CURVES

Now let us consider the effects of banking a turn to eliminate the need for a friction force. As seen from Fig. 9-5, a road can be banked in such a manner that the normal force **N** has vertical and horizontal components:

$$\mathcal{N}_x = \mathcal{N} \sin \theta \qquad \mathcal{N}_y = \mathcal{N} \cos \theta$$

The horizontal component $\mathbf{N}_x$ provides the necessary centripetal force. Therefore, if we represent the linear velocity by v and the radius of the turn by R, the banking angle θ required to eliminate the need for friction is obtained from

$$\mathcal{N} \sin \theta = \frac{mv^2}{R}$$

Also, since the vertical components are balanced, we write

$$\mathcal{N} \cos \theta = mg$$

Dividing the first equation by the second yields

$$\tan \theta = \frac{v^2}{R} \qquad\qquad (9\text{-}8)$$

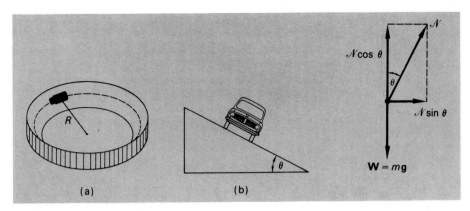

FIG. 9-5 Effects of banking a turn. The horizontal component of the normal force, $N \sin \theta$ provides the necessary centripetal acceleration.

EXAMPLE 9-5 Find the required banking angle for a curve of radius 160 m if the curve is to be negotiated at a speed of 80 km/h without the need of a friction force.

Solution The linear speed is first converted to meters per second:

$$v = 80 \text{ km/h} = 22.2 \text{ m/s}$$

Substituting into Eq. (9-8) yields

$$\tan \theta = \frac{v^2}{Rg} = \frac{(22.2 \text{ m/s})^2}{(160 \text{ m})(9.8 \text{ m/s}^2)} = 0.315$$

from which the required banking angle is

$$\theta = 17.5°$$

9-6 ■ ROTATION OF RIGID BODIES; ANGULAR DISPLACEMENT

Another type of circular motion exists when an entire body rotates about an axis. For example, wheels, drive shafts, and flywheels all use rotational effects to accomplish work. In such cases, it is often necessary to measure the amount of rotation, which we call *angular displacement*.

To understand what is meant by angular displacement, consider the rotating disk shown in Fig. 9-6. If point A rotates to point B as the disk turns on its axis, the angular displacement is denoted by the angle θ. There are several ways of measuring this angle. We are already familiar with the units of degrees and revolutions, which are related by the definition

$$1 \text{ rev} = 360°$$

Neither of these units is very useful in describing rotation of rigid bodies. A more applicable measure for angular displacement is the *radian* (rad). The definition of a radian may be seen from Fig. 9-7.

> An angle of **one radian** (*1 rad*) *is a central angle whose arc* s *is equal in length to the radius* R.

To calculate an angle in radians, we can use the following formula which results from the definition:

$$\theta = \frac{s}{R}$$

Angular Displacement   (9-9)

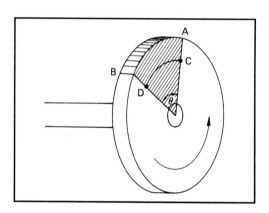

FIG. 9-6 Angular displacement θ is indicated by the shaded portion of the disk. The angular displacement is the same from C to D as it is from A to B for a rigid body.

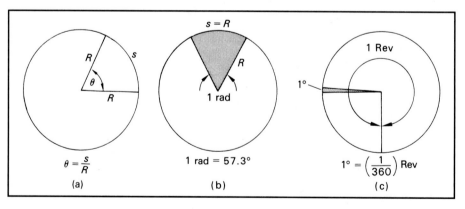

FIG. 9-7 The measure of angular displacement and a comparison of units.

Since the ratio of the arc distance s to the radius R is a length unit divided by a length unit, the *radian* is a dimensionless quantity.

Perhaps we can understand the size of a radian better by comparing it with the more familiar units of degrees. We can do this by considering an arc length s equal to the circumference of a circle $2\pi R$. Such an angle (360°) expressed in radians is found from Eq. (9-9):

$$\theta = \frac{2\pi R}{R} = 2\pi \text{ rad}$$

Hence, a complete revolution is an angle of 2π rad:

$$1 \text{ rev} = 360° = 2\pi \text{ rad}$$

from which we note that

$$1 \text{ rad} = \frac{360°}{2\pi} = 57.3° \qquad (9\text{-}10)$$

EXAMPLE 9-6 If the arc length is 6 ft and the radius is 10 ft, find the angular displacement in radians, degrees, and revolutions.

Solution Substituting directly into Eq. (9-9), we obtain

$$\theta = \frac{s}{R} = \frac{6 \text{ ft}}{10 \text{ ft}} = 0.6 \text{ rad}$$

Converting to degrees yields

$$\theta = (0.6 \text{ rad})\left(\frac{57.3°}{1 \text{ rad}}\right) = 34.4°$$

and since 1 rev = 360°,

$$\theta = (34.4°)\left(\frac{1 \text{ rev}}{360°}\right) = 0.0956 \text{ rev}$$

EXAMPLE 9-7 A point on the edge of a rotating disk or radius 8 m moves through an angle of 37°. Compute the length of the arc described by the point.

Solution Since Eq. (9-9) was defined for an angle measured in radians, we must first convert 37° to radian units:

$$\theta = (37°)\left(\frac{1 \text{ rad}}{57.3°}\right) = 0.646 \text{ rad}$$

The arc length is given by

$$s = R\theta = (8 \text{ m})(0.646 \text{ rad}) = 5.17 \text{ m}$$

The unit radian is dropped because it represents a ratio of length to length (m/m = 1).

9-7 ■ ANGULAR VELOCITY

The time rate of change in angular displacement is called the *angular velocity*. Thus, if an object rotates through an angle θ in a time t, its average angular velocity is given by

$$\bar{\omega} = \frac{\theta}{t} \qquad (9\text{-}11)$$

The symbol ω, the Greek letter *omega*, is used to denote rotational velocity. Although angular velocity may be expressed in *revolutions per minute* or *revolutions per second*, in most physical problems it is necessary to use *radians per second* to conform with convenient formulas. Since the rate of rotation in many technical problems is given in terms of the frequency of revolution, the following relation will be useful:

$$\omega = 2\pi f \qquad (9\text{-}12)$$

where ω is measured in *radians per second* and f is measured in *revolutions per second*.

EXAMPLE 9-8 Compute the average angular velocity of a long-playing phonograph record (33⅓ rpm).

Solution Note that no mention was made of the distance to the center of the record. The angular velocity depends only on the rate of rotation. The frequency of revolution is

$$f = \left(\frac{33\frac{1}{3} \text{ rev}}{\text{min}} \right)\left(\frac{1 \text{ min}}{60 \text{ s}} \right) = 0.555 \text{ rev/s}$$

or, by substituting into Eq. (9-12), the angular velocity is

$$\omega = \left(\frac{2\pi \text{ rad}}{\text{rev}} \right)\left(\frac{0.555 \text{ rev}}{\text{s}} \right) = 3.49 \text{ rad/s}$$

It is important to realize that the angular velocity discussed in this section represents an *average* velocity. The same distinction must be made between the average and the instantaneous angular velocities as that discussed in Chap. 6 for average and instantaneous linear velocities.

9-8 ■ ANGULAR ACCELERATION

Like linear motion, angular motion may be uniform or accelerated. The rate of rotation may increase or decrease under the influence of a resultant torque. For example, if the angular velocity changes from an initial value ω_0 to a final value ω_f in a time t, the average angular acceleration is given by

$$\alpha = \frac{\omega_f - \omega_0}{t}$$

The Greek letter α (*alpha*) denotes angular acceleration. A more useful form for this equation is

$$\omega_f = \omega_0 + \alpha t \tag{9-13}$$

A comparison of Eq. (9-13) with Eq. (6-4) for linear acceleration will show that their forms are identical if we draw analogies between angular and linear parameters.

Now that the concept of initial and final angular velocities has been introduced, we can express the average angular velocity in terms of its initial and final values:

$$\bar{\omega} = \frac{\omega_f + \omega_0}{2}$$

Substituting this equality for $\bar{\omega}$ in Eq. (9-11) yields a more useful expression for the angular displacement:

$$\theta = \bar{\omega}t = \frac{\omega_f + \omega_0}{2} t \tag{9-14}$$

This equation is also similar to an equation derived for linear motion. In fact, the equations for angular acceleration have the same basic form as those derived in Chap. 6 for linear acceleration if we draw the following analogies:

$$s \text{ (ft, m)} \leftrightarrow \theta \text{ (rad)}$$
$$v \text{ (ft/s, m/s)} \leftrightarrow \omega \text{ (rad/s)}$$
$$a \text{ (ft/s}^2, \text{ m/s}^2) \leftrightarrow \alpha \text{ (rad/s}^2)$$

Time, of course, is the same for both types of motion and is measured in seconds. Table 9-1 illustrates the similarities between angular and linear motion.

In applying these formulas, we must be careful to choose the appropriate units for each quantity. It is also important to choose a direction (clockwise or counterclockwise) as positive and to follow through consistently in affixing the appropriate sign to each quantity.

TABLE 9-1 COMPARISON OF LINEAR ACCELERATION AND ANGULAR ACCELERATION

Constant linear acceleration	Constant angular acceleration
$s = \bar{v}t = \dfrac{v_f + v_0}{2}\, t$	$\theta = \bar{\omega}t = \dfrac{\omega_f + \omega_0}{2}\, t$
$v_f = v_0 + at$	$\omega_f = \omega_0 + \alpha t$
$s = v_0 t + \frac{1}{2}at^2$	$\theta = \omega_0 t + \frac{1}{2}\alpha t^2$
$2as = v_f^2 - v_0^2$	$2\alpha\theta = \omega_f^2 - \omega_0^2$

EXAMPLE 9-9   A flywheel increases its rate of rotation from 6 to 12 rev/s in 8 s. What is its angular acceleration?

Solution   We will first compute the initial and final angular velocities:

$$\omega_0 = 2\pi f_0 = \left(\frac{2\pi \text{ rad}}{\text{rev}}\right)\left(\frac{6 \text{ rev}}{\text{s}}\right) = 12\pi \text{ rad/s}$$

$$\omega_f = 2\pi f_0 = \left(\frac{2\pi \text{ rad}}{\text{rev}}\right)\left(\frac{12 \text{ rev}}{\text{s}}\right) = 24\pi \text{ rad/s}$$

The angular acceleration is

$$\alpha = \frac{\omega_f - \omega_0}{t} = \frac{(24\pi - 12\pi) \text{ rad/s}}{8 \text{ s}}$$

$$= 1.5\pi \text{ rad/s}^2 = 4.71 \text{ rad/s}^2$$

EXAMPLE 9-10   A grinding disk rotating initially with an angular velocity of 6 rad/s receives a constant acceleration of 2 rad/s². (a) What angular displacement will it describe in 3 s? (b) How many revolutions will it make? (c) What is its final angular velocity?

Solution (a)   The angular displacement is given by

$$\theta = \omega_0 t + \tfrac{1}{2}\alpha t^2$$
$$= (6 \text{ rad/s})(3 \text{ s}) + \tfrac{1}{2}(2 \text{ rad/s}^2)(3 \text{ s})^2$$
$$= 18 \text{ rad} + (1 \text{ rad/s}^2)(9 \text{ s}^2)$$
$$= 27 \text{ rad}$$

Solution (b)   Since 1 rev $= 2\pi$ rad, we obtain

$$\theta = (27 \text{ rad}) \frac{1 \text{ rev}}{2\pi \text{ rad}} = 4.30 \text{ rev}$$

Solution (c)   The final velocity is equal to the initial velocity plus the change in speed. Thus

$$\omega_f = \omega_0 + \alpha t = 6 \text{ rad/s} + (2 \text{ rad/s}^2)(3 \text{ s})$$
$$= 12 \text{ rad/s}$$

9-9 ■ RELATION BETWEEN ANGULAR AND LINEAR MOTION

The *axis of rotation* of a rigid rotating body can be defined as that line of particles which remains stationary during rotation. This may be a line through the body, as with a spinning top, or it may be a line through space, as with a rolling hoop. In any case, our experience tells us that the farther a particle is from the axis of rotation, the greater its linear speed. This fact was expressed in Eq. (9-4) by the formula

$$v = 2\pi fR$$

where f is the frequency of rotation. We now derive a similar relation in terms of angular speed. The rotating particle in Fig. 9-8 turns through an arc s, which is given by

$$s = \theta R$$

from Eq. (9-9). If this distance is traversed in time t, the linear speed of the particle is given by

$$v = \frac{s}{t} = \frac{\theta R}{t}$$

Since $\theta/t = \omega$, the linear speed can be expressed as a function of the angular speed:

$$v = \omega R \qquad\qquad (9\text{-}15)$$

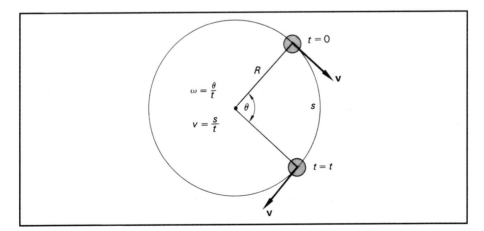

FIG. 9-8 The relation between angular speed and linear speed.

This result also follows from Eq. (9-12), in which the angular velocity is expressed as a function of the frequency of revolution. Since $s = \theta R$ is true only if θ is expressed in radians, $v = \omega R$ is true only if ω is in radians per unit time.

EXAMPLE 9-11   A drive shaft has an angular speed of 60 rad/s. At what distance from the axis should flyweights be positioned if they are to have a linear speed of 40 m/s?

Solution   Solving for R in Eq. (9-15), we obtain

$$ R = \frac{v}{\omega} = \frac{40 \text{ m/s}}{60 \text{ rad/s}} = 0.667 \text{ m} $$

9-10 ■ ROTATIONAL WORK AND POWER

Work was defined in Chap. 8 as the product of a displacement and a force. We now consider the work done in rotational displacement under the influence of a resultant torque. Consider a force **F** acting at the edge of a pulley of radius r, as shown in Fig. 9-9. The effect of such a force is to rotate the pulley through an angle θ while the point at which the force is applied moves through an arc distance s. The arc distance s is given by

$$ s = r\theta $$

Hence the work done by the force **F** is by definition

$$ \text{Work} = Fs = Fr\theta $$

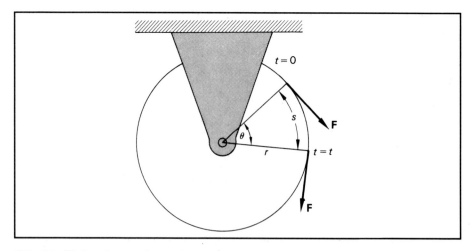

FIG. 9-9 Work and power in rotation.

but Fr is the torque τ due to the force **F,** so that we obtain

$$\boxed{\text{Work} = \tau\theta}$$

Rotational Work (9-16)

The angle θ must be expressed in radians for either system of units in order that the work can be expressed in foot-pounds or joules.

Mechanical energy is transmitted in the form of rotational work. When we speak of the power output of engines, we are concerned with the rate at which rotational work is done. Thus rotational power can be determined by dividing both sides of Eq. (9-16) by the time t required for the torque τ to cause a displacement θ:

$$\text{Power} = \frac{\text{work}}{t} = \frac{\tau\theta}{t}$$

(9-17)

Since θ/t is the average angular velocity $\bar{\omega}$, we write

$$\boxed{\text{Power} = \tau\bar{\omega}}$$

Rotational Power (9-18)

Notice the similarity of this equation to its analog, $P = Fv$, derived earlier for linear motion.

EXAMPLE 9-12 A wheel of radius 2 ft has a constant force of 12 lb applied at its rim. The wheel starts from rest and acquires a rotational speed of 600 rpm in 4 s. What average horsepower is developed?

Solution The average horsepower depends on the torque τ and the average angular speed $\bar{\omega}$. We will first find the torque by multiplying the force at the rim by the radius of the wheel:

$$\tau = Fr = (12 \text{ lb})(2 \text{ ft}) = 24 \text{ lb·ft}$$

Now the final angular speed is

$$\omega_f = 2\pi f = 2\pi\left(600 \,\frac{\text{rev}}{\text{min}}\right)\left(\frac{1 \text{ min}}{60 \text{ s}}\right) = 62.8 \text{ rad/s}$$

The average angular speed is one-half this value, since the initial angular speed was zero:

$$\bar{\omega} = \frac{\omega_0 + \omega_f}{2} = \frac{0 + 62.8 \text{ rad/s}}{2} = 31.4 \text{ rad/s}$$

Now the average power is found from Eq. (9-18):

$$\text{Power} = \tau\bar{\omega} = (24 \text{ lb·ft})(31.4 \text{ rad/s})$$
$$= 753.6 \text{ ft·lb/s}$$

Recalling that 1 hp = 550 ft·lb/s, we obtain finally

$$\text{Power} = (753.6 \text{ ft·lb/s})\left(\frac{1 \text{ hp}}{550 \text{ ft·lb/s}}\right) = 1.37 \text{ hp}$$

In this example, it was necessary to calculate an average angular speed. When a rigid body is rotating at a *constant* angular speed, Eq. (9-18) may be used directly.

Summary

In this chapter we have covered a large variety of physical relationships which contribute to our understanding of rotational motion. The major concepts and formulas are given below:

■ The linear speed v of an object in uniform circular motion can be calculated from the period T or the frequency f:

$$v = \frac{2\pi R}{T} \qquad v = 2\pi f R$$

■ The centripetal acceleration a_c is found from the linear speed, the period, or the frequency as follows:

$$a_c = \frac{v^2}{R} \qquad a_c = \frac{4\pi^2 R}{T^2} \qquad a_c = 4\pi^2 f^2 R$$

■ The centripetal force F_c is equal to the product of the mass m and the centripetal acceleration a_c. It is given by

$$F_c = \frac{mv^2}{R} \qquad F_c = 4\pi^2 f^2 mR$$

- The angle in radians is the ratio of the arc distance s to the radius R of the arc. Symbolically we write

$$\theta = \frac{s}{R} \qquad s = \theta R$$

The radian is a unitless ratio of two lengths.
- Angular velocity, which is the rate of angular displacement, can be calculated from θ or from the frequency of rotation:

$$\bar{\omega} = \frac{\theta}{t} \qquad \bar{\omega} = 2\pi f \qquad \textit{Average Angular Velocity}$$

- The maximum speed at which it is possible to negotiate a curve without slipping and the optimum banking angle for curves are given by

$$v = \sqrt{\mu_s g R} \qquad \tan \theta = \frac{v^2}{gR}$$

- Angular acceleration is the time rate of change in angular speed:

$$\alpha = \frac{\omega_f - \omega_0}{t} \qquad \textit{Angular Acceleration}$$

- By comparing θ to s, ω to v, and α to a, the following equations can be utilized for angular acceleration problems:

$$\theta = \frac{\omega_f - \omega_0}{2} t$$
$$\omega_f = \omega_0 + \alpha t$$
$$\theta = \omega_0 t + \tfrac{1}{2}\alpha t^2$$
$$2\alpha\theta = \omega_f^2 - \omega_0^2$$

When any three of the five parameters θ, α, t, ω_f, and ω_0 are given, the other two can be found from one of these equations. Choose a direction of rotation as being positive throughout your calculations.

■ When you are comparing linear speed v with angular speed ω, the following relationships should be remembered:

$$v = \omega R \qquad v = 2\pi fR$$

■ Rotational work and rotational power are found from

$$\text{Work} = \tau\theta \qquad \text{power} = \tau\overline{\omega}$$

Questions

9-1. Define the following terms:
    a. centripetal acceleration   **g.** radian
    b. centripetal force   **h.** angular displacement
    c. linear speed   **i.** angular acceleration
    d. angular speed   **j.** banking angle
    e. frequency   **k.** rotational work
    f. period   **l.** rotational power

9-2. A bicyclist leans to the side when negotiating a turn. Why? Describe with a free-body diagram the forces acting on the rider.

9-3. In negotiating a circular turn, a car hits a patch of ice and skids off the road. According to Newton's first law, the car will move forward in a direction tangent to the curve, not outward at right angles to it. Explain with diagrams.

9-4. If the force causing circular motion is directed toward the center of rotation, why is water thrown off clothes during the spin cycle of a washing machine?

9-5. When a ball tied at the end of a string is revolved in a circle at constant speed, the inward centripetal force is equal in magnitude to the outward centrifugal force. Does this represent a condition of equilibrium? Explain.

9-6. What factors contribute to the most desirable banking angles on roadways?

9-7. Does the centripetal force do work in uniform circular motion?

9-8. A motorcyclist negotiates a circular track at constant speed. What exerts the centripetal force, and on what does the force act? What exerts the centrifugal reaction force, and on what does it act?

9-9. State the angular analogies for the following translational equations:
    a. $v_f = v_0 + at$ **c.** power $= \text{work}/t = Fv$
    b. $s = v_0t + \frac{1}{2}at^2$ **d.** $s = \overline{v}t$

9-10. In a belt drive the input pulley is one-half the diameter of the output pulley. The linear speed of the single belt, of course, is the same for each pulley. Compare the rotational speeds of the input and output pulleys.

9-11. In the definitions of rotational work and power, why must the angular displacement be expressed in radians rather than in degrees?

9-12. A variable-speed industrial drill develops a constant power output of ½ hp. How is the output torque dependent on the rotational speed of the drill bit?

Problems

9-1. A 4-lb object is tied to a cord and swung in a horizontal circle of radius 3 ft. Neglect the effects of gravity, and assume a frequency of revolution of 80 rpm. Determine (a) the linear speed, (b) the centripetal acceleration, (c) the centripetal force, and (d) what happens if the cord breaks.

Answer (a) 25.1 ft/s; (b) 211 ft/s^2; (c) 26.3 lb; (d) the object moves at tangent.

9-2. Answer the same questions as in Prob. 9-1 for a 2-kg mass whirled in a circle of radius 1.5 m at 3 rev/s.

9-3. A drive shaft 60 mm in diameter rotates at 9 rev/s. What is the centripetal acceleration at the surface of the shaft? What is the linear speed of a point on the surface?

Answer 95.9 m/s^2; 1.70 m/s.

9-4. A merry-go-round revolves with a period of 6 s. How far from its center should you sit to experience a centripetal acceleration of 12 ft/s^2? If you weigh 128 lb, what is the centripetal force on you? What exerts this force? Is there an outward force on you?

9-5. Two 8-kg masses are attached to the end of a thin rod 400 mm long. The rod is supported in the middle and whirled in a circle. Assume that the rod can support a maximum tension of only 80 N. What is the maximum frequency of rotation in revolutions per minute?

Answer 67.5 rpm.

9-6. What must the speed of a satellite just above the surface of the earth be if it is to travel in a circular orbit about the earth? Assume that the radius of the earth is 4000 mi. What is the nature of the centripetal force in this case?

9-7. An electron revolves in an orbit about the nucleus of an atom. It follows a circular path of radius 6×10^{-11} m. If the mass of the electron is 9.11×10^{-31} kg and its linear speed is 3.2×10^6 m/s, compute the centripetal acceleration and the centripetal force.

Answer 1.71×10^{23} m/s^2; 1.56×10^{-7} N.

9-8. The breaking strength of a cord is 12 N. What is the maximum frequency of revolution if a 5-kg mass is to be moved in a horizontal circle of radius 300 mm? Now suppose that the cord is lengthened so that the mass moves in a circle of radius 500 mm. What must be the new frequency of rotation in order to produce the same tension in the cord?

9-9. A bus negotiates a curve of radius 487 m. What force will be exerted on an 80-kg passenger by the outside wall of the bus when it is traveling at 60 km/h? If $\mu_s = 0.5$, what is the maximum speed at which the bus can make the turn without slipping?

Answer 45.6 N; 48.8 m/s.

9-10. On a rainy day the coefficient of friction between the tires and the road is 0.4. What is the minimum radius of a curve in order that cars may travel at a speed of 80 km/h without slipping?

9-11. The speed limit for a curve of radius 800 ft is 55 mi/h. When the coefficient of friction is as low as 0.3, can this curve be negotiated at the speed limit? What is the maximum speed?

Answer Yes; 59.8 mi/h.

*9-12. Find the banking angle necessary for a curve of radius 600 ft when the speed limit is 55 mi/h.

*9-13. Find the required banking angle if a car is to make a 180° turn in a circular distance of 800 ft at 50 mi/h.

Answer 33.4°.

*9-14. A curve in a road 9 m wide has a radius of 96 m. How much higher should the outside edge be than the inside edge for an automobile to travel safely at a speed of 60 km/h?

9-15. A point near the edge of a rotating shaft of radius 3 ft moves through a distance of 2 ft. Compute the angular displacement (*a*) in radians. (*b*) in degrees, and (*c*) in revolutions.

Answer (*a*) 0.667 rad; (*b*) 38.2°; (*c*) 0.106 rev.

9-16. A point on the edge of a large wheel 8 m in diameter moves through an angle of 37°. Compute the length of the arc described by the point.

9-17. An electric motor turns at a frequency of 600 rpm. What is its angular speed? What is the angular displacement after 6 s?

Answer 62.8 rad/s; 377 rad.

9-18. A rotating pulley completes 12 rev in 4 s. Determine the average angular speed (*a*) in revolutions per second, (*b*) in revolutions per minute, and (*c*) in radians per second.

9-19. A rotating flywheel starts from rest and reaches a final rotational speed of 900 rpm in 4 s. Determine the angular acceleration and the angular displacement after 4 s.

Answer 23.6 rad/s²; 189 rad.

9-20. An electric motor rotating at 1900 rpm slows to 300 rev/min in 5 s when the power is turned off. Find (*a*) the angular acceleration and (*b*) the angular displacement during the 5 s.

9-21. A grinding stone rotating at 4 rev/s receives a constant angular acceleration of 3 rad/s². (*a*) What angular displacement will it describe in 3 s? (*b*) How many revolutions will it make? (*c*) What will its final angular speed be?

Answer (*a*) 88.9 rad; (*b*) 14.1 rev; (*c*) 34.1 rad/s.

9-22. What is the linear speed of a point on the surface of a rotating cylinder if the cylinder makes 10 complete revolutions in 20 s and the diameter of the cylinder is 3 m?

9-23. A cylindrical piece of metal stock 6 in. in diameter rotates in a lathe at 800 rpm. What is the linear velocity of the surface of the cylinder?

Answer 20.9 ft/s.

9-24. The proper tangential velocity for machining steel stock is about 2.3 ft/s. At how many revolutions per minute should a steel cylinder 3 in. in diameter be turned in a lathe?

9-25. A pulley 32 cm in diameter and rotating initially in 4 rev/s receives a constant angular acceleration of 2 rad/s². (*a*) Compute the angular speed after 8 s. (*b*) What is the angular displacement during this time? (*c*) What is the linear velocity of a belt wrapped around the pulley after 8 s?

Answer (*a*) 41.4 rad/s; (*b*) 266 rad; (*c*) 13.2 m/s.

9-26. An automobile engine develops 250 hp at 5000 rev/min. What is the mean torque exerted on the crankshaft?

9-27. A Prony brake, shown in Fig. 9-10, is a device for measuring power output. The engine, motor, or turbine turning the pulley must do work against the friction force. The magnitude of the friction force is the difference in scale readings ($F_2 - F_1$). Assume that the centers of the two vertical ropes are 18 in. apart and that the tensions are $F_2 = 325$ lb and $F_1 = 225$ lb when the motor is turning at 480 rev/min. What horsepower is being developed by the motor?

Answer 6.86 hp.

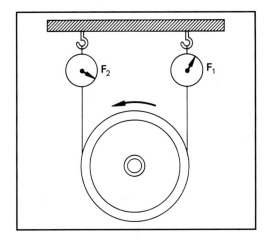

FIG. 9-10 The Prony brake.

9-28. What is the power developed by a motor as determined by the Prony brake if $F_2 =$ 500 N, $F_1 = 350$ N, $R = 250$ mm, and $f = 600$ rpm?

9-29. A high-speed industrial drill develops ½ hp at 1600 rpm. What torque is applied to the drill bit?

Answer 1.64 lb·ft.

9-30. What torque must be applied to develop 40 kW of power at 800 rpm?

9-31. A large commercial engine develops 2800 kW at 2400 rpm. If this engine is used to drive a winch whose drum diameter is 600 mm, how much weight can be lifted by a cable wrapped around the drum? With what speed will it be lifted?

Answer 37,136 N; 75.4 m/s.

10

SIMPLE
MACHINES

OBJECTIVES

As a result of completing this chapter, you should be able to

1. Describe a simple machine and its operation in general terms to the extent that *efficiency* and *conservation of energy* are explained.
2. Write and apply formulas for computing the efficiency of a simple machine in terms of work or power.
3. Distinguish by definition and example between *ideal* mechanical advantage and *actual* mechanical advantage.
4. Draw a diagram of each of the following simple machines and beside each diagram write a formula for computing the ideal mechanical advantage: lever, inclined plane, wedge, gears, pulley systems, wheel and axle, screw jack, belt drive.
5. Compute the mechanical advantage and the efficiency of each of the simple machines listed in the previous objective.

A simple machine is any device which transmits the application of a force into useful work. With a chain hoist we can transmit a small downward force into a very large upward force for lifting. In industry delicate samples of radioactive material are handled with machines that allow an applied force to be reduced significantly. Single pulleys may be used to change the direction of an applied force without affecting its magnitude. A study of machines and their efficiency is essential for the productive use of energy. In this chapter, you will become familiar with levers, gears, pulley systems, inclined planes, and other machines routinely used for many industrial applications.

10-1 ■ SIMPLE MACHINES AND EFFICIENCY

In a simple machine, input work is done by the application of a single force, and the machine performs output work by means of a single force. During any such operation, three processes occur, as shown in Fig. 10-1:

1. Work is supplied to the machine.
2. Work is done against friction.
3. Output work is done by the machine.

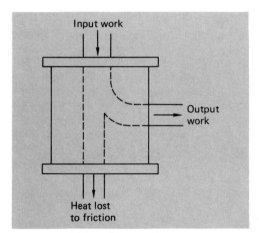

FIG. 10-1 Three processes occur in the operation of a machine: (1) the input of a certain amount of work, (2) the loss of energy in doing work against friction, and (3) the output of useful work.

According to the principle of conservation of energy, these processes are related as follows:

Input work = work against friction + output work

The amount of useful work performed by a machine can never be greater than the work supplied to it. There will always be some loss due to friction. For example, in pumping up a bicycle tire with a small hand pump, we exert a downward force on the plunger, causing air to be forced into the tire. That some of our input work is lost to friction can easily be seen by feeling how warm the wall of the hand pump becomes. The smaller we can make the friction loss in a machine, the greater the return for our effort. In other words, the effectiveness of a given machine can be measured by comparing its output work with the work supplied to it.

*The **efficiency** E of a machine is defined as the ratio of the work output to the work input:*

$$E = \frac{\text{work output}}{\text{work input}} \qquad \textit{Efficiency} \quad (10\text{-}1)$$

The efficiency as defined in Eq. (10-1) will always be a number between 0 and 1. Common practice is to express this fraction as a percentage by multiplying the decimal number by 100. For example, a machine which does 40 ft·lb of work when 80 ft·lb of work is supplied to it has an efficiency of 50 percent.

Another useful expression for efficiency can be found from the definition of power as work per unit time. We can write

$$P = \frac{\text{work}}{t} \qquad \text{or} \qquad \text{work} = Pt$$

The efficiency in terms of power input P_i and power output P_o is given by

$$E = \frac{\text{work output}}{\text{work input}} = \frac{P_o t}{P_i t}$$

or

$$E = \frac{\text{power output}}{\text{power input}} = \frac{P_o}{P_i} \qquad (10\text{-}2)$$

EXAMPLE 10-1   A 60-hp motor winds a cable around a drum. (*a*) If the cable lifts a 3-ton load of bricks to a height of 12 ft in 3 s, calculate the efficiency of the motor. (*b*) At what rate is work done against friction?

Solution (a)   First compute the output power:

$$P_o = \frac{Fs}{t} = \frac{(6000 \text{ lb})(12 \text{ ft})}{3 \text{ s}}$$

$$= (24{,}000 \text{ ft·lb/s})\left(\frac{1 \text{ hp}}{550 \text{ ft·lb/s}}\right)$$

$$= 43.6 \text{ hp}$$

The efficiency is then found from Eq. (10-2):

$$E = \frac{P_o}{P_i} = \frac{43.6 \text{ hp}}{60 \text{ hp}} = 0.727$$

$$= 72.7\%$$

Solution (b) The rate at which work is done against friction is the difference between the input power and the output power, or 16.4 hp.

10-2 ■ MECHANICAL ADVANTAGE

Simple machines like the lever, block and tackle, chain hoist, gears, inclined plane, and screw jack all play important roles in modern industry. We can illustrate the operation of any of these machines by the general diagram in Fig. 10-2. An input force F_i acts through a distance s_i, accomplishing the work $F_i s_i$. At the same time, an output force F_o acts through a distance s_o, performing the useful work $F_o s_o$.

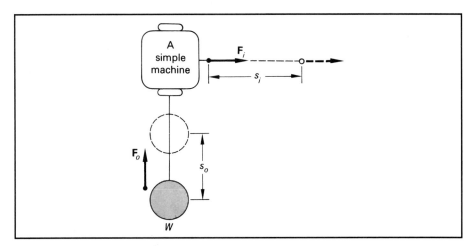

FIG. 10-2 During the operation of any simple machine, an input force F_i acts through a distance s_i while an output force F_o acts through a distance s_o.

The **actual mechanical advantage** M_A *of a machine is defined as the ratio of the output force* F_o *to the input force* F_i:

$$M_A = \frac{\text{output force}}{\text{input force}} = \frac{F_o}{F_i} \qquad (10\text{-}3)$$

An actual mechanical advantage greater than 1 indicates that the output force is greater than the input force. Although most machines have values of M_A greater than 1, this is not always the case. In handling small, fragile objects, it is sometimes desirable to make the output force smaller than the input force.

In the previous section, we noted that the efficiency of a machine increases as frictional effects become small. Applying the conservation-of-energy principle to the simple machine in Fig. 10-2 yields

$$\text{Work input} = \text{work against friction} + \text{work output}$$
$$F_i s_i = (\text{work})_{\mathscr{F}} + F_o s_o$$

The most efficient engine possible would realize no losses due to friction. We can represent this *ideal* case by setting $(\text{work})_{\mathscr{F}} = 0$ in the above equation. Thus

$$F_o s_o = F_i s_i$$

Since this equation represents an ideal case, we define the *ideal mechanical advantage* M_I as the *ideal* ratio F_o/F_i. Thus, M_I is given by

$$M_I = \left(\frac{F_o}{F_i}\right)_{\text{ideal}} = \frac{s_i}{s_o} \tag{10-4}$$

The **ideal mechanical advantage** *of a simple machine is equal to the ratio of the distance the input force moves to the distance the output force moves.*

The efficiency of a simple machine is the ratio of output work to input work. Therefore, for the general machine of Fig. 10-2 we have

$$E = \frac{F_o s_o}{F_i s_i} = \frac{F_o/F_i}{s_i/s_o}$$

Finally, utilizing Eqs. (10-3) and (10-4), we obtain

$$E = \frac{M_A}{M_I} \tag{10-5}$$

All the above concepts have been treated as they apply to a general machine. In the following sections we shall apply them to specific machines.

10-3 ■ LEVER

Possibly the oldest and most generally useful machine is the simple lever. A *lever* consists of any rigid bar pivoted at a certain point called the *fulcrum*. Figure 10-3

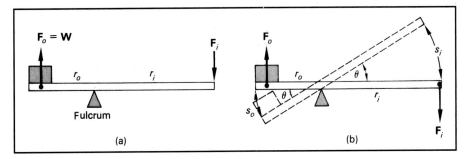

FIG. 10-3 Lever.

illustrates the use of a long rod to lift a weight **W**. We can calculate the ideal mechanical advantage of such a device in two ways. The first method involves the principle of equilibrium, and the second uses the principle of work, as discussed in the previous section. Since the equilibrium method is easier for the lever, we shall apply it first.

Because no translational motion is involved during the application of a lever, the condition for equilibrium is that the input torque equal the output torque:

$$F_i r_i = F_o r_o$$

The ideal mechanical advantage can be found from

$$M_I = \left(\frac{F_o}{F_i}\right)_{\text{ideal}} = \frac{r_i}{r_o} \qquad (10\text{-}6)$$

The ratio F_o/F_i is considered the *ideal* case because no friction forces are considered.

The same result is obtained from work considerations. Note from Fig. 10-3b that the force $\mathbf{F}_i$ moves through the arc distance s_i while the force $\mathbf{F}_o$ moves through the arc distance s_o. However, the two arcs are subtended by the same angle θ, and so we can write the proportion

$$\frac{s_i}{s_o} = \frac{r_i}{r_o}$$

Substitution into Eq. (10-4) will verify the result obtained from equilibrium considerations, that is, $M_I = r_i/r_o$.

EXAMPLE 10-2   An iron bar 3 m long is used to lift a 60-kg block. The bar is used as a lever, as shown in Fig. 10-3. The fulcrum is placed 80 cm from the block. What is the ideal mechanical advantage of the system, and what input force is required?

Solution   The distance $r_o = 0.8$ m, and the distance $r_i = 3$ m $- 0.8$ m $= 2.2$ m. Therefore, the ideal mechanical advantage is

$$M_I = \frac{r_i}{r_o} = \frac{2.2 \text{ m}}{0.8 \text{ m}} = 2.75$$

The output force in this case is equal to the weight of the 60-kg block ($W = mg$). Therefore, the required input force is given by

$$F_i = \frac{F_o}{M_I} = \frac{(60 \text{ kg})(9.8 \text{ m/s}^2)}{2.75}$$

$$= 214 \text{ N}$$

Before we leave the subject of the lever, we should note that very little of the input work is lost to friction forces. For all practical purposes the actual mechanical advantage for a simple lever is equal to its ideal mechanical advantage. Other examples of the lever are illustrated in Fig. 10-4.

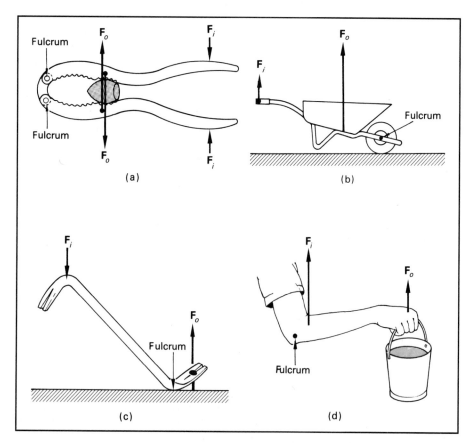

FIG. 10-4 The lever forms the operating principle of many simple machines.

10-4 ■ APPLICATIONS OF THE LEVER PRINCIPLE

A serious limitation of the elementary lever is the fact that it operates through a small angle. There are many ways of overcoming this restriction by allowing for continuous rotation of the lever arm. For example, the *wheel and axle* (Fig. 10-5) allows for the continued action of the input force F_i. By applying the reasoning described in Sec. 10-2 for a general machine, it can be shown that

$$M_I = \left(\frac{F_o}{F_i}\right)_{ideal} = \frac{R}{r} \qquad (10\text{-}7)$$

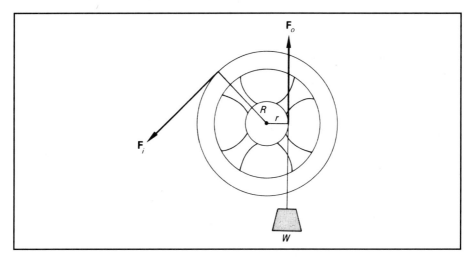

FIG. 10-5 Wheel and axle.

Thus the ideal mechanical advantage of a wheel and axle is the ratio of the radius of the wheel to the radius of the axle.

Another application of the lever concept is the use of *pulleys*. A single pulley, as shown in Fig. 10-6, is simply a lever whose input moment arm is equal to its output moment arm. From the principle of equilibrium, the input force will equal the output force, and the ideal mechanical advantage will be

$$M_I = \left(\frac{F_o}{F_i}\right)_{ideal} = 1 \qquad (10\text{-}8)$$

The only advantage of such a device lies in its ability to change the direction of an input force.

A single movable pulley (Fig. 10-7), however, has an ideal mechanical advantage of 2. Note that the two supporting ropes must each be shortened by 1 ft in order to lift the load through a distance of 1 ft. Therefore, the input force moves through a distance of 2 ft while the output force only moves a distance of 1 ft. Applying the principle of work, we have

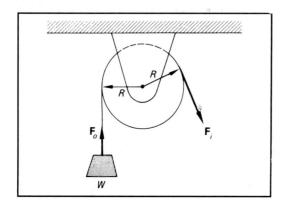

FIG. 10-6 A single fixed pulley serves only to change the direction of the input force.

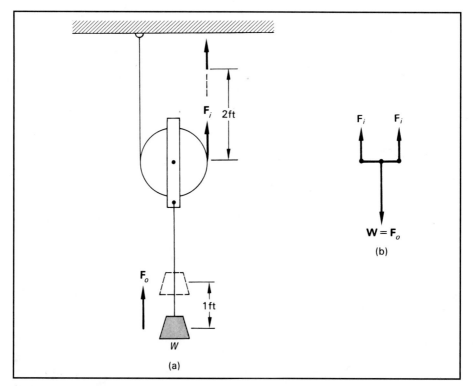

FIG. 10-7 A single movable pulley. (*a*) The input force moves through twice the distance that the output force travels. (*b*) The free-body diagram shows that $2F_i = F_o$.

$$F_i(2 \text{ ft}) = F_o(1 \text{ ft})$$

from which the ideal mechanical advantage is

$$M_I = \left(\frac{F_o}{F_i}\right)_{\text{ideal}} = 2 \tag{10-9}$$

The same result can be shown by constructing a free-body diagram, as in Fig. 10-7b. From the figure it is evident that

$$2F_i = F_o$$

or

$$M_I = \frac{F_o}{F_i} = 2$$

The latter method is usually applied to problems involving movable pulleys, since it allows one to associate M_I with the number of strands of a single rope supporting the movable pulley.

EXAMPLE 10-3 Calculate the ideal mechanical advantage of the block-and-tackle arrangement shown in Fig. 10-8.

Solution We first construct a free-body diagram, as shown in Fig. 10-8b. From the figure we note that

$$4F_i = F_o$$

from which

$$M_I = \left(\frac{F_o}{F_i} \right)_{ideal} = \frac{4F_i}{F_i} = 4$$

Note that the uppermost pulley serves only to change the direction of the input force. The same M_I would result if $\mathbf{F}_i$ were applied upward at point a.

10-5 ■ TRANSMISSION OF TORQUE

The simple machines discussed so far are used to transmit and apply forces in order to move loads. In most mechanical applications, however, work is done by transmitting torque from one drive to another. For example, the belt drive (Fig. 10-9) transmits the torque from a driving pulley to an output pulley. The mechanical advantage of such a system is the ratio of the torques between the output pulley and the driving pulley:

$$M_I = \frac{\text{output torque}}{\text{input torque}} = \frac{\tau_o}{\tau_i}$$

From the definition of torque, we can write this expression in terms of the radii of the pulleys:

$$M_I = \left(\frac{\tau_o}{\tau_i} \right)_{ideal} = \frac{F_o r_o}{F_i r_i}$$

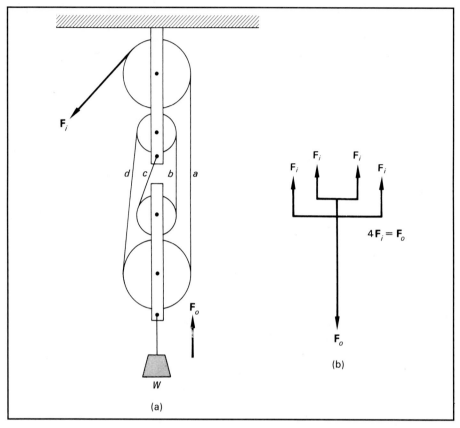

FIG. 10-8 The block and tackle. This arrangement has an ideal mechanical advantage of 4, since four strands support the movable block.

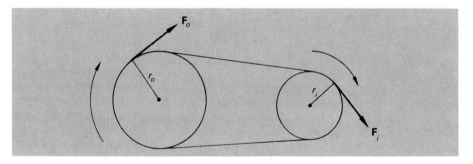

FIG. 10-9 Belt drive.

If there is no slippage between the belt and the pulleys, it is safe to say that the tangential input force F_i is equal to the tangential output force F_o. Thus

$$M_I = \frac{F_o r_o}{F_i r_i} = \frac{r_o}{r_i}$$

Since the diameters of pulleys are usually specified instead of their radii, a more convenient expression is

$$M_I = \frac{D_o}{D_i} \tag{10-10}$$

where D_i is the diameter of the driving pulley and D_o is the diameter of the output pulley.

Suppose we now apply the principle of work to the belt drive. Remember that work is defined in rotary motion as the product of torque τ and angular displacement θ. For the belt drive, assuming ideal conditions, the input work equals the output work. Thus

$$\tau_i \theta_i = \tau_o \theta_o$$

The power input must also be equal to the power output. Dividing the above equation by the time t required to rotate through the angles θ_i and θ_o, we obtain

$$\tau_i \frac{\theta_i}{t} = \tau_o \frac{\theta_o}{t} \quad \text{or} \quad \tau_i \omega_i = \tau_o \omega_o$$

where ω_i and ω_o are the angular speeds of the input and output pulleys. Note that the ratio τ_o / τ_i represents the ideal mechanical advantage. Therefore, we can add another expression to Eq. (10-10), to obtain

$$\boxed{M_I = \frac{D_o}{D_i} = \frac{\omega_i}{\omega_o}} \tag{10-11}$$

This important result shows that the mechanical advantage is achieved at the *expense* of rotary motion. In other words, if the mechanical advantage is 2, the input shaft must rotate with twice the angular speed of the output shaft. The ratio ω_i / ω_o is sometimes referred to as the *speed ratio*.

If the speed ratio is greater than 1, the machine produces an output torque which is greater than the input torque. As we have seen, this feat is accomplished at the expense of rotation. However, many machines are designed to increase the rotational output speed. In these cases, the speed ratio is less than 1, and the increased rotational speed is accomplished with reduced torque output.

EXAMPLE 10-4   Consider the belt drive illustrated in Fig. 10-9, where the diameter of the small driving pulley is 6 in. and the diameter of the driven pulley is 18 in. A 6-hp motor drives the input pulley at 600 rpm. Calculate the revolutions per minute and torque delivered to the driven wheel if the system is 75 percent efficient.

Solution   We first calculate the ideal mechanical advantage (100 percent efficiency) of the system. From Eq. (10-11),

$$M_I = \frac{D_o}{D_i} = \frac{18 \text{ in.}}{6 \text{ in.}} = 3$$

Since the efficiency is 75 percent, the actual mechanical advantage is given from Eq. (10-5):

$$M_A = EM_I = (0.75)(3) = 2.25$$

Now the actual mechanical advantage is the simple ratio of output torque τ_o to input torque τ_i. Recalling that the power in rotational motion is equal to the product of torque and angular velocity, we can solve for τ_i as follows:

$$\tau_i = \frac{P_i}{\omega_i} = \frac{(6 \text{ hp})[(550 \text{ ft·lb/s})/\text{hp}]}{(600 \text{ rev/min})(2\pi \text{ rad/rev})(1 \text{ min/60 s})}$$

$$= \frac{(6)(550 \text{ ft·lb/s})}{20\pi \text{ rad/s}} = 52.5 \text{ ft·lb}$$

Since $M_A = \tau_o/\tau_i$, the output torque is given by

$$\tau_o = M_A\tau_i = (2.25)(52.5 \text{ ft·lb})$$
$$= 118 \text{ ft·lb}$$

Assuming that the belt does not slip, it will move with the same linear velocity v around each pulley. Since $v = \omega r$, we can write the equality

$$\omega_i r_i = \omega_o r_o \qquad \text{or} \qquad \omega_i D_i = \omega_o D_o$$

from which

$$\omega_o = \frac{\omega_i D_i}{D_o} = \frac{(600 \text{ rpm})(6 \text{ in.})}{18 \text{ in.}} = 200 \text{ rpm}$$

Note that the ratio of ω_i to ω_o yields the ideal mechanical advantage and not the actual mechanical advantage. The difference between M_I and M_A is due to friction, both in the belt and in the shaft bearings. Since greater tension on the belt will result in greater friction forces, maximum efficiency is obtained by reducing the belt tension until it just prevents the belt from slipping on the pulleys.

Before leaving our discussion of the transmission of torque, we must consider the application of *gears*. A gear is simply a notched wheel which can transmit torque by meshing with another notched wheel, as shown in Fig. 10-10. A pair of meshing gears differs from a belt drive only in the sense that the gears rotate in opposite directions. The same relationships derived for the belt drive hold for gears:

$$M_I = \frac{D_o}{D_i} = \frac{\omega_i}{\omega_o} \qquad\qquad (10\text{-}12)$$

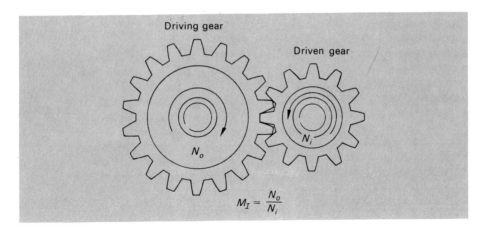

FIG. 10-10 Spur gears. The ideal mechanical advantage is the ratio of the number of teeth on the output gear to the number of teeth on the input gear.

A more useful expression makes use of the fact that the number of teeth N on the rim of a gear is proportional to its diameter D. Because of this dependence, the ratio of the number of teeth on the driven gear N_o to the number of teeth on the driving gear N_i is the same as the ratio of their diameters. Hence we can write

$$M_I = \frac{N_o}{N_i} = \frac{D_o}{D_i}$$

(10-13)

The use of gears avoids the problem of slippage, which is common with belt drives. It also conserves space and allows for a greater torque to be transmitted.

In addition to the *spur* gears illustrated in Fig. 10-10, there are several other types of gears. Four common types are worm gears, helical gears, bevel gears, and planetary gears. Examples of each are shown in Fig. 10-11. The same general relationships apply for all these gears.

10-6 ■ INCLINED PLANE

The only machines we have discussed so far involve application of the lever principle. A second fundamental machine is the *inclined plane*. Suppose you have to move a heavy load from the ground to a truck bed without hoisting equipment. You would probably select a few long boards and form a ramp from the ground to the bed of the truck. Experience has taught you that it takes less effort to push a load up a small elevation than it does to lift the load directly. Since a smaller input force results in the same output force, a mechanical advantage is realized. However, the smaller input force is accomplished at the cost of greater distance.

Consider the movement of a weight **W** up the inclined plane in Fig. 10-12. The slope angle θ is such that the weight must be moved through a distance s to reach a

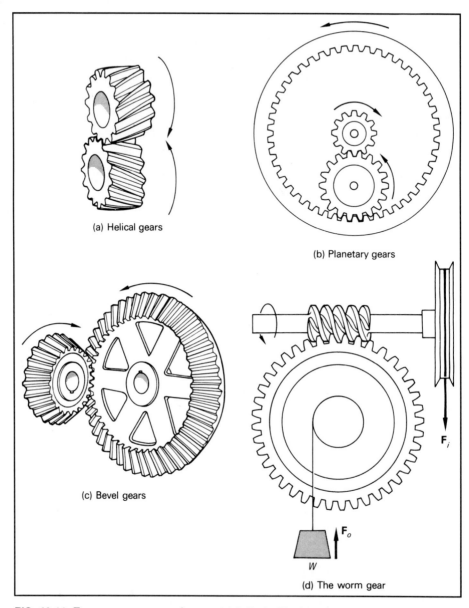

(a) Helical gears

(b) Planetary gears

(c) Bevel gears

(d) The worm gear

FIG. 10-11 Four common types of gears: (*a*) helical, (*b*) planetary, (*c*) bevel, (*d*) worm. (The spur gear, which is the most common type, is shown in Fig. 10-10.)

height h at the top of the incline. If we neglect friction, the work required to push the weight up the plane is the same as the work required to lift it up vertically. We can express this equality as

$$\text{Work input} = \text{work output}$$
$$F_i s = Wh$$

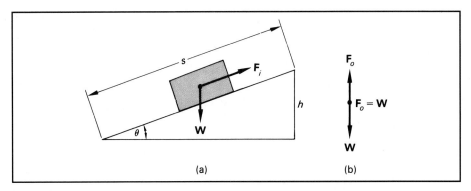

FIG. 10-12 The inclined plane. The input force represents the effort required to push the block up the plane; the output force is equal to the weight of the block.

where F_i is the input force and W is the output force. The ideal mechanical advantage will be the ratio of the weight to the input force. Stating this symbolically, we have

$$M_I = \frac{W}{F_i} = \frac{s}{h} \qquad (10\text{-}14)$$

EXAMPLE 10-5 The 200-lb wooden crate in Fig. 10-13 is to be raised to a loading platform 6 ft high. A ramp 12 ft long is used to slide the crate from the ground to the platform. Assume that the coefficient of friction is 0.3. (*a*) What is the ideal mechanical advantage of the ramp? *(*b*) What is the actual mechanical advantage (optional)?

Solution (a) The ideal mechanical advantage, from Eq. (10-14), is

$$M_I = \frac{s}{h} = \frac{12 \text{ ft}}{6 \text{ ft}} = 2$$

This value represents the mechanical advantage of the ramp if it were frictionless.

Solution (b) The actual mechanical advantage is the ratio of the weight lifted to the required input force, considering friction. Applying the first condition for equilibrium to the free-body diagram (Fig. 10-13*b*), we find that the normal force $\mathcal{N}$ is given by

$$\mathcal{N} = W_y = (200 \text{ lb})(\cos 30°) = 173 \text{ lb}$$

from which the friction force must be

$$\mathcal{F} = \mu \mathcal{N} = (0.3)(173 \text{ lb}) = 51.9 \text{ lb}$$

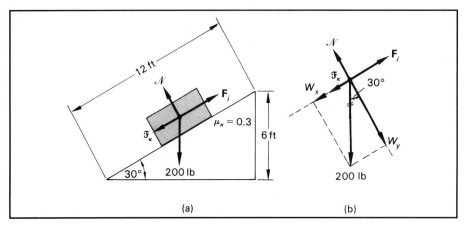

FIG. 10-13 Comparing the actual mechanical advantage of an inclined plane.

Summing the forces along the plane, we obtain

$$F_i - \mathcal{F} - W_x = 0$$

But $W_x = (200 \text{ lb})(\sin 30°) = 100 \text{ lb}$, so that we have

$$F_i - 51.9 \text{ lb} - 100 \text{ lb} = 0$$
$$F_i = 51.9 \text{ lb} + 100 \text{ lb} = 152 \text{ lb}$$

We can now compute the actual mechanical advantage:

$$M_A = \frac{F_o}{F_i} = \frac{200 \text{ lb}}{152 \text{ lb}} = 1.32$$

It is left as an exercise for the student to show that the efficiency of the ramp is 66 percent.

10-7 ■ APPLICATIONS OF THE INCLINED PLANE

Many machines apply the principle of the inclined plane. The simplest is the *wedge* (Fig. 10-14), which is actually a double inclined plane. In the ideal case, the mechanical advantage of a wedge of length L and thickness t is given by

$$M_I = \frac{L}{t} \qquad (10\text{-}15)$$

This equation is a direct consequence of the general relation expressed by Eq. (10-14). The ideal mechanical advantage is always much greater than the actual mechanical

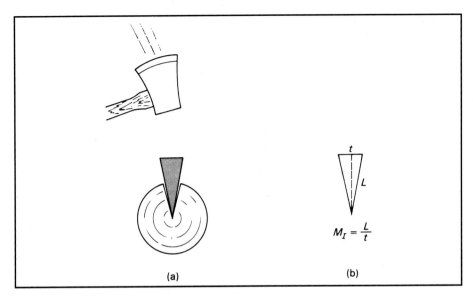

$$M_I = \frac{L}{t}$$

(a)   (b)

FIG. 10-14 The wedge is actually a double inclined plane.

advantage because of the large friction forces between the surfaces in contact. The wedge finds its application in axes, knives, chisels, planers, and all other cutting tools. A cam is a kind of rotary wedge which is used to lift valves in internal-combustion engines.

One of the most useful applications of the inclined plane is the *screw*. This principle can be explained by examining a common tool known as the *screw jack* (Fig. 10-15). The threads are essentially an inclined plane wrapped continuously around a cylindrical shaft. When the input force F_i turns through a complete revolution $(2\pi R)$, the output force F_o will advance through the distance p. This distance p is actually the distance between two adjacent threads, and it is called the *pitch* of the screw. The ideal mechanical advantage is the ratio of the input distance to the output distance:

$$M_I = \frac{s_i}{s_o} = \frac{2\pi R}{p} \tag{10-16}$$

The screw is an example of a very inefficient machine, but in this case it is usually an advantage since friction forces are needed to hold the load in place while the input force is not being applied.

Summary

A simple machine has been defined as a device which converts a single input force F_i to a single output force F_o. In general, the input force moves through a distance s_i, and the output force moves through a distance s_o. The purpose is to accomplish useful work in a manner suited to a particular application. The major concepts are given below:

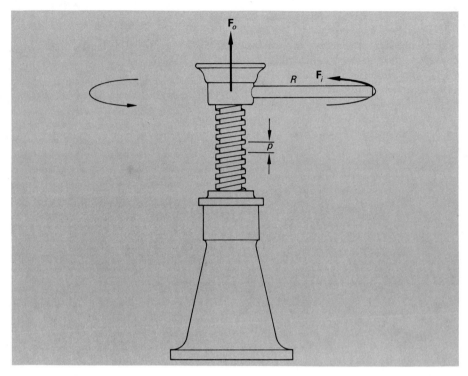

FIG. 10-15 Screw jack.

■ A simple machine is a device which converts a single input force F_i to a single output force F_o. The input force moves through a distance s_i, and the output force moves a distance s_o. There are two mechanical advantages:

$$M_A = \frac{F_o}{F_i} \qquad \text{actual mechanical advantage (friction considered)}$$

$$M_I = \frac{s_o}{s_i} \qquad \text{ideal mechanical advantage (assumes no friction)}$$

■ The efficiency of a machine is a ratio of output work to input work. It is normally expressed as a percentage and can be calculated from any of the following relations:

$$E = \frac{\text{work output}}{\text{work input}} \qquad E = \frac{\text{power output}}{\text{power input}} \qquad E = \frac{M_A}{M_I}$$

■ The ideal mechanical advantages for a number of simple machines are given below.

$$M_I = \left(\frac{F_o}{F_i}\right)_{\text{ideal}} = \frac{r_i}{r_o} \qquad \qquad \textit{Lever}$$

$$M_I = \left(\frac{F_o}{F_i}\right)_{ideal} = \frac{R}{r}$$

Wheel and Axle

$$M_I = \frac{D_o}{D_i} = \frac{\omega_i}{\omega_o}$$

Belt Drive

$$M_I = \frac{W}{F_i} = \frac{s}{h}$$

Inclined Plane

$$M_I = \frac{L}{t}$$

Wedge

$$M_I = \frac{N_o}{N_i} = \frac{D_o}{D_i}$$

Gears

$$M_I = \frac{s_i}{s_o} = \frac{2\pi R}{p}$$

Screw Jack

Questions

10-1. Define the following terms:
   a. machine
   b. efficiency
   c. lever
   d. pulley
   e. gears
   f. wedge
   g. screw
   h. pitch
   i. inclined plane
   j. wheel and axle
   k. actual mechanical advantage
   l. ideal mechanical advantage
   m. belt drive

10-2. What is meant by *useful work* or *output work?* What is meant by *input work?* Write the general relationship between input work and output work.

10-3. Two jacks are operated simultaneously to lift the front end of a car. Immediately afterward it is noted that the left jack feels warmer than the right one. Which jack is more efficient? Explain.

10-4. A machine may alter the magnitude and/or the direction of an input force. (*a*) Name several examples in which both changes occur. (*b*) Give examples in which only the magnitude of the input force is altered. (*c*) Give some examples in which only the direction is altered.

10-5. A machine lifts a load through a vertical distance of 4 ft while the input force moves through a distance of 2 ft. Would this machine be helpful in lifting large weights? Explain.

10-6. A bicycle can be operated in three gear ranges. In *low range* the pedals describe 2 complete revolutions while the rear wheel turns through 1 revolution. In *medium range* the pedals and the wheels turn at the same rate. In *high range* the rear wheel of the bicycle completes 2 revolutions for every complete pedal revolution. Discuss the advantages and disadvantages of each range.

10-7. What happens to the ideal mechanical advantage if a simple machine is operated in reverse? What happens to its efficiency?

10-8. Give several examples of machines which have an actual mechanical advantage less than 1.

10-9. Why do buses and trucks often use larger steering wheels than those found on automobiles? What principle is used?

10-10. Draw diagrams of pulley systems which have ideal mechanical advantages of 2, 3, and 5.

10-11. Usually the road to the top of a mountain winds around the mountain instead of going straight up the side. Why? If we neglect friction, is more work required to reach the top along the spiral road? Is more power required? If we consider friction, would it require less work to drive straight up the side of the mountain? Explain.

Problems

10-1. A 25 percent efficient machine performs external work of 200 J. What input work is required?

Answer 800 J.

10-2. During the operation of a 300-hp engine, energy is lost at the rate of 200 hp because of friction. What are the useful output power and the efficiency of the engine?

10-3. A 60-W motor lifts a 2-kg mass to a height of 4 m in 3 s. (*a*) Compute the output power. (*b*) What is the efficiency of the motor? (*c*) What is the rate at which work is done against friction?

Answer (*a*) 26.1 W; (*b*) 43.6%; (*c*) 33.9 W.

10-4. One edge of a 200-lb safe is lifted with a 4-ft steel rod. What input force is required at the end of the rod if a fulcrum is placed 6 in. from the safe? (To lift one edge, a force equal to one-half the weight of the safe is required.)

10-5. A frictionless machine lifts a 200-lb load through a distance of 10 ft. If the input force moves through a distance of 30 ft, what is the ideal mechanical advantage of the machine? What is the magnitude of the input force?

Answer 3; 66.7 lb.

10-6. What would be the required input force if the machine in Prob. 10-5 were 60 percent efficient?

10-7. A 60-N weight is lifted in the three different ways shown in Fig. 10-16. Compute the ideal mechanical advantage and the required input force for each application.

Answer (*a*) 2, 30 N; (*b*) 3, 20 N; (*c*) 0.33, 180 N.

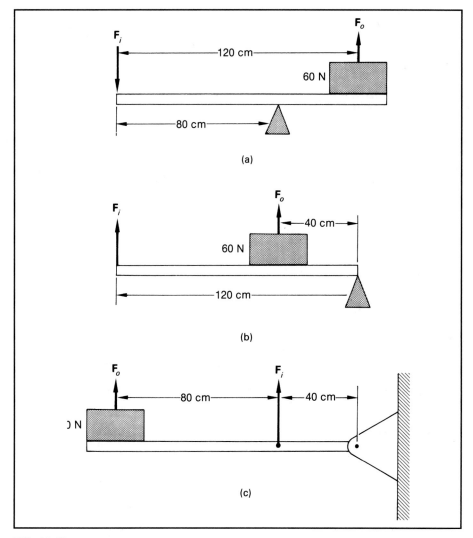

FIG. 10-16

10-8. A 20-kg mass is to be lifted with a rod 2 m long. If you can exert a downward force of 40 N on one end of the rod, where should you place a block of wood to act as a fulcrum?

10-9. A wheel 0.2 m in diameter is attached to an axle with a diameter of 6 cm. If a weight of 400 N is attached to the axle, what force must be applied to the rim of the wheel to lift the weight at constant speed? Neglect friction.

Answer 120 N.

10-10. What is the mechanical advantage of a screwdriver used as a wheel and axle if its blade is 0.3 in. wide and the handle diameter is 0.8 in.?

10-11. Determine the force F required to lift a 200-N load W with each of the four pulley systems shown in Fig. 10-17.

Answer (*a*) 100 N; (*b*) 40 N; (*c*) 50 N; (*d*) 50 N.

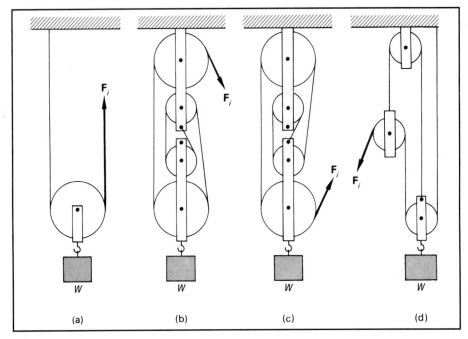

FIG. 10-17

10-12. The *chain hoist* (Fig. 10-18) is a combination of the wheel and axle and the block and tackle. Show that the ideal mechanical advantage of such a device is given by

$$M_I = \frac{2R}{R - r}$$

10-13. A pair of step pulleys (Fig. 10-19) makes it possible to change output speeds merely by shifting the belt. If an electric motor turns the input pulley at 2000 rpm, find the possible angular speeds of the output shaft. The pulley diameters are 4, 6, and 8 in.

Answer (*a*) Small-input pulley; 2000, 1333, 1000 rpm;
(*b*) middle-input pulley: 3000, 2000, 1500 rpm;
(*c*) large-input pulley; 4000, 2667, 2000 rpm.

10-14. An 8-hp motor drives the input pulley of a belt drive at 600 rpm. Compute the revolutions per minute and torque delivered to the driven pulley if the system is 60 percent efficient. The diameters of the input and output pulleys are 4 and 8 in., respectively.

10-15. A worm drive similar to that shown in Fig. 10-11 has n teeth in the gear wheel. (If $n = 80$, one complete turn of the worm will advance the wheel one-eightieth of a revolution.) Derive an expression for the ideal mechanical advantage of the worm gear in terms of the radius of the input pulley R, the radius of the output shaft r, and the number of teeth n in the gear wheel.

Answer $M_I = nR/r$.

10-16. The worm drive of Prob. 10-15 has 80 teeth in the gear wheel. If the radius of the input wheel is 18 in. and the radius of the output shaft is 3 in., what input force is required to lift a 2-ton load? Assume an efficiency of 80 percent.

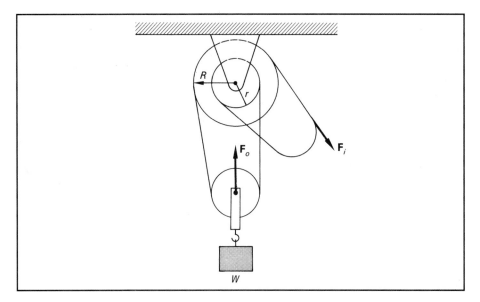

FIG. 10-18 Chain hoist.

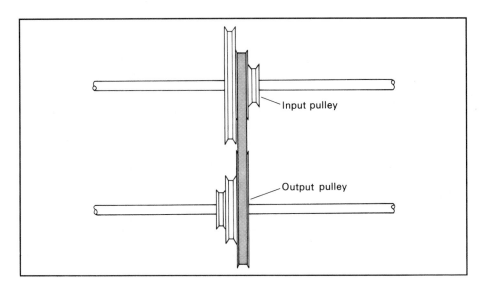

FIG. 10-19 Step pulley.

*10-17. A 37° inclined plane forms a loading ramp for a 40-kg crate. The ramp is 8.2 m long, and the coefficient of friction is 0.2. (*a*) What is the ideal mechanical advantage of the ramp? (*b*) What is the actual mechanical advantage? (*c*) What is the efficiency?

Answer (*a*) 1.66; (*b*) 1.31; (*c*) 79%.

10-18. The lever of a screw jack is 24 in. long. If the screw has six threads per inch, what is its ideal mechanical advantage? If the jack is 15 percent efficient, what force is needed to lift 2000 lb?

10-19. A wrench with a 6-in. handle acts to tighten a ¼-in.-diameter bolt having 10 threads per inch. What is the pitch of the bolt? Compute the ideal mechanical advantage. If an input force of 20 lb results in a 600-lb force on the nut, what is the efficiency?

Answer 0.1 in.; 377; 7.96%.

10-20. A shaft rotating at 800 rpm delivers a torque of 240 N·m to an output shaft which is rotating at 200 rpm. If the efficiency of the machine is 70 percent, compute the output torque. What is the output power in horsepower?

11

PROPERTIES
OF SOLIDS

As a result of completing this chapter, you should be able to

1. **Demonstrate by example and discussion your understanding of the following terms:** *elasticity, compression, elastic limit, stress, strain, shearing, tension, hardness, malleability, ductility, ultimate strength.*
2. **Write and apply formulas for calculating Young's modulus and the shear modulus.**

Until now we have been discussing objects in motion or at rest. The objects have been assumed to be rigid and absolutely solid. But we know that wire can be stretched, that rubber tires will compress, and that bolts will sometimes break. A more complete understanding of nature requires a study of the mechanical properties of matter. The concepts of elasticity, tension, and compression are analyzed in this chapter. As the number of kinds of alloys increases and the demands on them become greater, our knowledge of such concepts becomes more important. For example, the stress placed on spaceships or on cables in modern bridges is of a magnitude unheard of a few years ago.

11-1 ■ *ELASTIC PROPERTIES OF MATTER*

We define an *elastic* body as one which returns to its original size and shape when a deforming force is removed. The quicker a body returns to its original shape, the more elastic the body is. Rubber bands, golf balls, trampolines, diving boards, footballs, and springs are common examples of elastic bodies. Putty, dough, and clay are examples of inelastic bodies. For all elastic bodies, we shall find it convenient to establish a cause-and-effect relationship between a deformation and the deforming forces.

Consider the coiled spring of length *l* shown in Fig. 11-1. We can study its elasticity by adding successive weights and observing the increase in length. A 2-lb weight lengthens the spring by 1 in.; a 4-lb weight lengthens the spring by 2 in.; and a 6-lb weight lengthens the spring by 3 in. Evidently, there is a direct relationship between the elongation of a spring and the applied force.

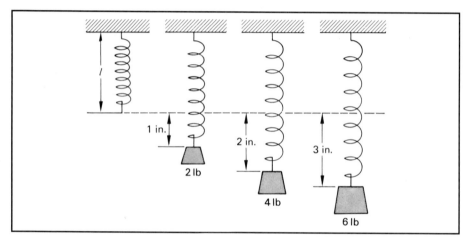

FIG. 11-1 Uniform elongation of a spring.

Robert Hooke first stated this relationship in connection with the invention of a balance spring for a clock. In general, he found that a force **F** acting on a spring (Fig. 11-2) produces an elongation *s* that is directly proportional to the magnitude of the force. *Hooke's law* can be written

$$F = ks \qquad (11\text{-}1)$$

The proportionality constant *k* varies extremely with the type of material and is called the *spring constant*. For the example illustrated in Fig. 11-1 the spring constant is

$$k = \frac{F}{s} = 2 \text{ lb/in.}$$

Hooke's law is by no means restricted to coiled springs; it applies to the deformation of all elastic bodies. In order to make the law more generally applicable, it will be convenient to define the terms *stress* and *strain*. Stress refers to the *cause* of an elastic deformation, whereas strain refers to the *effect,* i.e., the deformation itself.

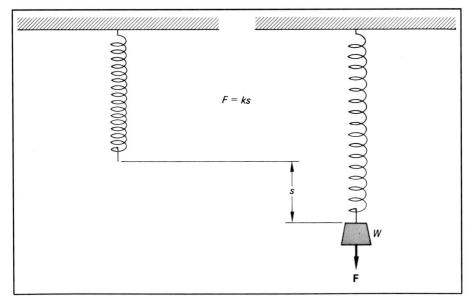

$$F = ks$$

FIG. 11-2 Relation between a stretching force F and the elongation it produces.

Three common types of stresses and their corresponding deformations are shown in Fig. 11-3. A *tensile stress* occurs when equal and opposite forces are directed away from each other. A *compressive stress* occurs when equal and opposite forces are directed toward each other. A *shearing stress* occurs when equal and opposite forces do not have the same line of action.

The effectiveness of any force producing a stress is highly dependent upon the area over which the force is distributed. For this reason, a more complete definition of stress can be stated as follows:

> **Stress** *is the ratio of an applied force to the area over which it acts, e.g., newtons per square meter or pounds per square inch.*

As mentioned earlier, the term *strain* must represent the effect of a given stress. The general definition of strain might be as follows:

> **Strain** *is the relative change in the dimensions or shape of a body as the result of an applied stress.*

In the case of a tensile or compressive stress, the strain may be considered a change in length per unit length. A shearing stress, however, may alter only the shape of a body without changing its dimensions. Shearing strain is usually measured in terms of an angular displacement.

The *elastic limit* is the maximum stress a body can experience without becoming permanently deformed. For example, an aluminum rod whose cross-sectional area is 1 in.^{2} will become permanently deformed by the application of a tensile force greater than 19,000 lb. This does not mean that the aluminum rod will break at this point; it means only that the rod will not return to its original size. In fact, the tension can be

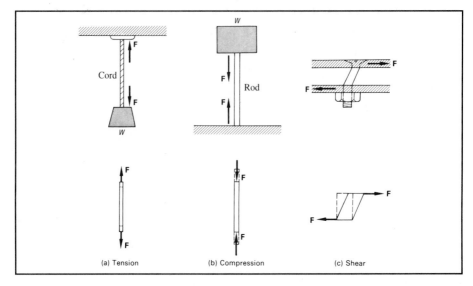

FIG. 11-3 Three common stresses shown with their corresponding deformations: (*a*) tension, (*b*) compression, and (*c*) shear.

increased to about 21,000 lb before the rod breaks. It is this property of metals which allows them to be drawn out into wires of smaller cross sections. The greatest stress a wire can withstand without breaking is known as its *ultimate strength*.

If the elastic limit of a material is not exceeded, we can apply Hooke's law to any elastic deformation. Within the limits of a given material, it has been experimentally verified that the ratio of a given stress to the strain it produces is a constant. In other words, the stress is directly proportional to the strain. *Hooke's law* states the following:

> *Provided that the elastic limit is not exceeded, an elastic deformation (strain) is directly proportional to the magnitude of the applied force per unit area (stress).*

If we call the proportionality constant the *modulus of elasticity,* we can write Hooke's law in its most general form:

$$\text{Modulus of elasticity} = \frac{\text{stress}}{\text{strain}} \qquad (11\text{-}2)$$

In the following sections we shall discuss the specific applications of this fundamental relation.

11-2 ■ YOUNG'S MODULUS

In this section we consider longitudinal stresses and strains as they apply to wires, rods, or bars. For example, in Fig. 11-4 a force **F** is applied to the end of a wire of cross-sectional area A. The longitudinal stress is given by

$$\text{Longitudinal stress} = \frac{F}{A}$$

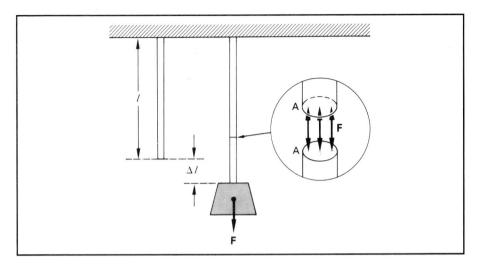

FIG. 11-4 Computing Young's modulus for a wire of cross section A. The elongation Δl is exaggerated for clarity.

The metric unit for stress is the *newton per square meter,* which is defined as the *pascal* (Pa):

$$1 \text{ Pa} = 1 \text{ N/m}^2$$

The USCS unit for stress is the *pound per square inch* (lb/in.^{2}). Since the pound per square inch remains in common use, it will be helpful to compare it with the SI unit:

$$1 \text{ lb/in.}^2 = 6895 \text{ Pa} = 6.895 \text{ kPa}$$

The effect of such a stress is to stretch the wire, i.e., to increase its length. Hence the longitudinal strain can be represented by the change in length per unit length. We can write

$$\text{Longitudinal strain} = \frac{\Delta l}{l}$$

where l is the original length and Δl is the elongation. Experimentation has shown that a comparable decrease in length occurs for a compressive stress. The same equations will apply whether we are discussing an object under tension or an object under compression.

If we define the longitudinal modulus of elasticity as *Young's modulus Y,* we can write Eq. (11-2) as

$$\text{Young's modulus} = \frac{\text{longitudinal stress}}{\text{longitudinal strain}}$$

$$Y = \frac{F/A}{\Delta l/l} = \frac{Fl}{A \, \Delta l} \qquad (11\text{-}3)$$

The units of Young's modulus are the same as the units of stress, i.e., pounds per square inch or pascals. This follows since the longitudinal strain is a unitless quantity. Representative values for some of the most common materials are listed in Tables 11-1 and 11-2.

TABLE 11-1 ELASTIC CONSTANTS FOR VARIOUS MATERIALS IN SI UNITS

Material	Young's modulus Y, MPa*	Shear modulus S, MPa	Elastic limit, MPa	Ultimate strength, MPa
Aluminum	68,900	23,700	131	145
Brass	89,600	35,300	379	455
Copper	117,000	42,300	159	338
Iron	89,600	68,900	165	324
Steel	207,000	82,700	248	489

* 1 MPa = 10^6 Pa.

TABLE 11-2 ELASTIC CONSTANTS FOR VARIOUS MATERIALS IN USCS UNITS

Material	Young's modulus Y, lb/in.^{2}	Shear modulus S, lb/in.^{2}	Elastic limit, lb/in.^{2}	Ultimate strength, lb/in.^{2}
Aluminum	10×10^6	3.44×10^6	19,000	21,000
Brass	13×10^6	5.12×10^6	55,000	66,000
Copper	17×10^6	6.14×10^6	23,000	49,000
Iron	13×10^6	10×10^6	24,000	47,000
Steel	30×10^6	12×10^6	36,000	71,000

EXAMPLE 11-1   A telephone wire 120 m long and 2.2 mm in diameter is stretched by a force of 380 N. What is the longitudinal stress? If the length after stretching is 120.10 m, what is the longitudinal strain? Determine Young's modulus for the wire.

Solution   The cross-sectional area of the wire is

$$A = \frac{\pi D^2}{4} = \frac{\pi (2.2 \times 10^{-3} \text{ m})^2}{4} = 3.8 \times 10^{-6} \text{ m}^2$$

Thus

$$\text{Stress} = \frac{F}{A} = \frac{380 \text{ N}}{3.8 \times 10^{-6} \text{ m}^2}$$

$$= 100 \times 10^6 \text{ N/m}^2 = 100 \text{ MPa}$$

$$\text{Strain} = \frac{\Delta l}{l} = \frac{0.10 \text{ m}}{120 \text{ m}} = 8.3 \times 10^{-4}$$

$$Y = \frac{\text{stress}}{\text{strain}} = \frac{100 \text{ MPa}}{8.3 \times 10^{-4}} = 120,000 \text{ MPa}$$

EXAMPLE 11-2   What is the maximum load which can be hung from a steel wire ¼ in. in diameter if its elastic limit is not to be exceeded? Determine the increase in length under this load if the original length is 3 ft.

Solution   From Table 11-2, the elastic limit for steel is 36,000 lb/in.^{2}. Since this value represents the limiting stress, we write

$$\frac{F}{A} = 36,000 \text{ lb/in.}^2$$

where A is given by

$$A = \frac{\pi D^2}{4} = \frac{\pi (0.25 \text{ in.})^2}{4} = 0.0491 \text{ in.}^2$$

Thus the limiting load is

$$
\begin{aligned}
F &= (36,000 \text{ lb/in.}^2)(A) \\
&= (36,000 \text{ lb/in.}^2)(0.0491 \text{ in.}^2) = 1770 \text{ lb}
\end{aligned}
$$

The increase in length under such a load is found from Eq. (11-3) as follows:

$$
\begin{aligned}
\Delta l &= \frac{l}{Y} \frac{F}{A} = \left(\frac{36 \text{ in.}}{30 \times 10^6 \text{ lb/in.}^2}\right)(36,000 \text{ lb/in.}^2) \\
&= 0.0432 \text{ in.}
\end{aligned}
$$

11-3 ■ SHEAR MODULUS

Compressive and tensile stresses produce a slight change in volume as a result of altered dimensions. As mentioned earlier, a shearing stress alters only the shape of a body, leaving its volume unchanged. For example, consider the parallel noncurrent forces acting on the cube in Fig. 11-5. The applied force causes each successive layer of atoms to slip sideways, much like the pages of a book under similar stress. The interatomic forces restore the block to its original shape when the stress is relieved.

The shearing stress is defined as the ratio of the tangential force **F** to the area A over which it is applied. The shearing strain is defined as the angle ϕ (in radians), which is called the *shearing angle* (refer to Fig. 11-5b). Applying Hooke's law, we can now define the *shear modulus S* as follows:

$$S = \frac{\text{shearing stress}}{\text{shearing strain}} = \frac{F/A}{\phi} \qquad (11\text{-}4)$$

The angle ϕ is usually so small that it is approximately equal to tan ϕ. Making use of this fact, we can rewrite Eq. (11-4) in the form

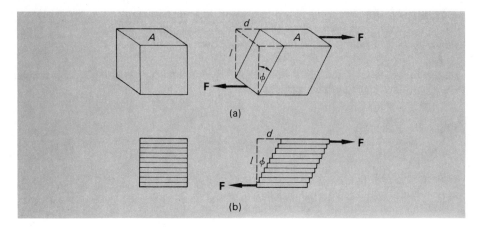

FIG. 11-5 Shearing stress and shearing strain.

$$S = \frac{F/A}{\tan \phi} = \frac{F/A}{d/l} \qquad (11\text{-}5)$$

Since the value of S is an indication of the rigidity of a body, it is sometimes referred to as the *modulus of rigidity*.

EXAMPLE 11-3   A steel stud (Fig. 11-6) 1 in. in diameter projects 1.5 in. out from the wall. If the end of the bolt is subjected to a shearing force of 8000 lb, compute its downward deflection.

Solution   The cross-sectional area is

$$A = \frac{\pi D^2}{4} = \frac{\pi (1 \text{ in.})^2}{4} = 0.785 \text{ in.}^2$$

If we represent the downward deflection by d, we can solve as follows:

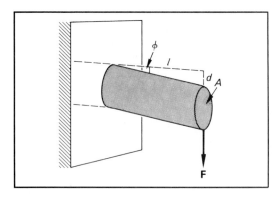

FIG. 11-6 The downward deflection of a stud is an example of shearing strain.

$$S = \frac{F/A}{d/l} = \frac{Fl}{Ad}$$

$$d = \frac{Fl}{AS} = \frac{(8000 \text{ lb})(1.5 \text{ in.})}{(0.785 \text{ in.}^2)(12 \times 10^6 \text{ lb/in.}^2)}$$

$$= 1.27 \times 10^{-3} \text{ in.}$$

You should show that this deflection results in a shearing angle of 8.47×10^{-4} rad.

11-4 ■ OTHER PHYSICAL PROPERTIES OF METALS

In addition to the elasticity, the tensile strength, and the shearing strength, metals have other important properties. A solid consists of molecules arranged so closely together that they attract each other very strongly. This attraction, called *cohesion,* gives a solid a definite shape and size. It also affects its usefulness to industry as a working material. Such properties as hardness, ductility, malleability, and conductivity must be understood before metals are chosen for specific industrial applications. (See Fig. 11-7.)

Hardness is an industrial term used to describe the ability of metals to resist forces that tend to penetrate them. Hard materials resist being scratched, worn away, penetrated, or otherwise damaged physically. Some metals, such as sodium and potassium, are very soft, while iron and steel are two of the hardest metals. The hardness of metals is tested with machines that push a cone-shaped diamond point into test materials. The penetration is measured, and a hardness reading is taken directly from a dial.

Two other special properties of materials are *ductility* and *malleability*. The meaning of each of these terms can be seen from Fig. 11-7. Ductility is defined as the ability of a metal to be drawn out into a wire. Tungsten and copper are very ductile. Malleability is the property which enables us to hammer or bend metals into any desired shape or to roll them into sheets. Most metals are malleable, gold being the most malleable.

Conductivity refers to the ability of metals to permit the flow of heat or electricity. The best conductors are silver, copper, gold, and aluminum, in that order. More will be said about this property in later chapters.

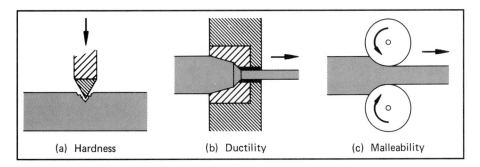

(a) Hardness (b) Ductility (c) Malleability

FIG. 11-7 Illustrating the working properties of metals: (*a*) A hard metal resists penetration, (*b*) a ductile metal can be drawn into a wire, and (*c*) a malleable metal can be rolled into sheets.

Summary

In industry, we must utilize materials effectively and for appropriate situations. Otherwise, the result will be metal failures that produce costly damage or serious injury to employees. In this chapter, we have discussed the elastic properties of matter and some of the formulas used to predict the effects of stress on certain solids. The following points will summarize this chapter.

- According to *Hooke's law,* an elastic body will deform or elongate an amount s under the application of a force F. The constant of proportionality k is the *spring constant:*

$$F = ks \qquad k = \frac{F}{s} \qquad\qquad \textit{Hooke's Law}$$

- *Stress* is the ratio of an applied force to the area over which it acts. *Strain* is the relative change in dimensions which results from the stress. For example,

$$\text{Longitudinal stress} = \frac{F}{A} \qquad \text{longitudinal strain} = \frac{\Delta l}{l}$$

- *The modulus of elasticity* is the constant ratio of stress to strain:

$$\text{Modulus of elasticity} = \frac{\text{stress}}{\text{strain}}$$

- *Young's modulus Y* is for longitudinal deformations:

$$Y = \frac{F/A}{\Delta l/l} \qquad \text{or} \qquad Y = \frac{Fl}{A\,\Delta l} \qquad\qquad \textit{Young's Modulus}$$

- A shearing strain occurs when an angular deformation ϕ is produced:

$$S = \frac{F/A}{\tan \phi} \qquad \text{or} \qquad S = \frac{F/A}{d/l} \qquad\qquad \textit{Shear Modulus}$$

- *Hardness* is the ability of materials to resist penetration; *ductility* is the ability to be drawn into a wire; *malleability* is the ability to be rolled into sheets; and *conductivity* is the ability to carry heat or electricity.

Questions

11-1. Define the following terms:
a. elasticity
b. Hooke's law
c. spring constant
d. tensile stress
e. compressive stress
f. shear stress
g. strain
h. elastic limit
i. ultimate strength
j. Young's modulus
k. shear modulus

11-2. Explain clearly the relationship between stress and strain.

11-3. Two wires have the same length and cross-sectional area but are not of the same material. Each wire is hung from the ceiling with a 2000-lb weight attached. The wire on the left stretches twice as far as the one on the right. Which has the greater Young's modulus?

11-4. Does Young's modulus depend on the length and cross-sectional area? Explain.

11-5. Two wires, A and B, are made of the same material and subjected to the same loads. Discuss their relative elongations when (a) wire A is twice as long as wire B and has twice the diameter of wire B and (b) wire A is twice as long as wire B and has one-half the diameter of wire B.

11-6. After studying the various elastic constants given in Tables 11-1 and 11-2, would you say it was usually easier to stretch a material or to shear a material? Explain.

11-7. A 400-lb weight is evenly supported by three wires of the same dimensions, one of copper, one of aluminum, and one of steel. Which wire experiences the greatest stress? Which experiences the least stress? Which wire experiences the greatest strain? Which experiences the least strain?

11-8. Discuss the various stresses resulting when a machine screw is tightened.

11-9. Give several practical examples of longitudinal, shearing, and volume strains.

Problems

11-1. A vertical coil spring supports an 8-lb weight. If the spring stretches 2 in., what is the spring constant? What weight would be required to stretch the spring 5 in.?

Answer 4 lb/in.; 20 lb.

11-2. A 10-kg mass is supported by a spring whose constant is 12 N/m. Compute the elongation of the spring under this load.

11-3. A coil spring is used to support a 5-kg mass. If the spring stretches 7 mm, what is the spring constant in newtons per meter? What mass must be hung from this spring to stretch it 12 mm?

Answer 7000 N/m; 8.57 kg.

11-4. A spring scale is designed so that the scale platform depresses a distance of 5 mm when a load of 80 g is placed on the platform. What is the spring constant? What mass will cause the scales to depress a distance of 8 mm?

11-5. A rod 4 m long and 0.5 cm^2 in cross-sectional area is stretched 0.2 cm under a tension of 12,000 N. (a) What is the applied stress? (b) What is the strain? (c) What is Young's modulus for this material?

Answer (a) 2.4×10^8 N/m^2; (b) 2000; (c) 480,000 MPa.

11-6. A 6-kg mass is hung from the end of a 0.8-mm-diameter steel wire which is 1.6 m long. How far will the wire stretch under this load?

11-7. A no. 18 copper wire has a diameter of 0.04 in. and is originally 10 ft long. (*a*) What is the greatest load that can be supported by this wire without exceeding its elastic limit? (*b*) Compute the change in length of the wire under this load. (*c*) What is the maximum load that can be supported by this wire without breaking? (*d*) What is the maximum elongation?

Answer (a) 28.9 lb; (*b*) 0.162 in.;
(*c*) 61.6 lb; (*d*) 0.346 in.

11-8. An aluminum rivet 0.5 in. in diameter is subjected to a stress of 4500 lb/in.^{2}. What force is applied to the rivet? If the rivet is initially 12 in. long, how far will it stretch?

11-9. What is the minimum diameter of a brass rod if it is to support a 400-N load without exceeding the elastic limit?

Answer 1.16 mm.

11-10. A solid cylindrical column is 12 ft high and 6 in. in diameter. What will be its decrease in length when it is supporting a load of 90 tons?

11-11. A cube of aluminum 6 cm on a side is subjected to a shearing force of 80,000 N. Compute the deflection *d* of the cube and the shearing angle ϕ in radians.

Answer 5.62 × 10^{-5} rad.

11-12. Two sheets of aluminum on an aircraft wing are to be held together by aluminum rivets of cross-sectional area 0.25 in.^{2}. The shearing stress on each rivet must not exceed one-tenth of the elastic limit for aluminum. How many rivets are needed if each rivet supports the same fraction of the total shearing force of 25,000 lb?

11-13. The twisting of a cylindrical shaft (Fig. 11-8) through an angle θ is an example of a shearing strain. An analysis of the situation shows that the angle of twist in radians is given by

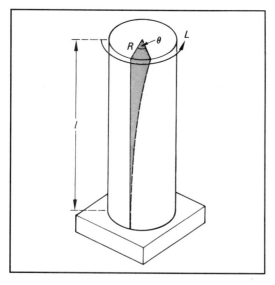

FIG. 11-8 A torque τ applied at one end of a solid cylinder causes it to twist through an angle θ.

$$\theta = \frac{2Ll}{2\pi SR^4}$$

where L is the applied torque, l is the length of the cylinder, R is the radius, and S is the shear modulus for the material.

Answer 7.64×10^{-3} rad.

11-14. An engine delivers 140 hp at 800 rpm to an 8-ft solid shaft 2 in. in diameter. Find the angle of twist in the drive shaft (refer to Prob. 11-13).

12

FLUIDS

As a result of completing this chapter, you should be able to

1. Compute the *weight* or *mass density* of a solid or contained fluid of regular shape when given its weight or mass.
2. Define and apply the concepts of *fluid pressure* and *buoyant force* to the solution of physical problems similar to the examples in this text.
3. Write and illustrate with drawings the four basic principles of fluid pressure summarized in Sec. 12-4.
4. Define *absolute pressure, gauge pressure,* and *atmospheric pressure,* and demonstrate by examples your understanding of the relationship among these terms.
5. Write and apply formulas for calculating the mechanical advantage of a *hydraulic press* in terms of input and output forces or areas.
6. Write and apply a formula for predicting the *rate of flow* of a fluid in terms of its velocity and cross-sectional area.
7. Discuss the change in velocity of a fluid as the cross-sectional area of a pipe changes.

Liquids and gases are called *fluids* because they flow freely and fill their containers. In this chapter you will learn that fluids may exert forces on the walls of their containers. Such forces acting on definite surface areas create a condition of *pressure*. A hydraulic press utilizes fluid pressure to lift heavy loads. The structure of water basins, dams, and large oil tanks is determined largely by pressure considerations. The design of boats, submarines, and weather balloons must take into account the pressure and density of the surrounding fluid. The motion of fluids is also important. For example, control of the injection of gasoline by carburetors determines the efficiency and smooth operation of many engines. In aircraft, fuel must be accurately and automatically metered at all engine speeds and loads regardless of changes in altitude or throttle position. An understanding of the principles of pressure, density, buoyant force, and fluid flow is essential in many industrial applications.

12-1 ■ FLUIDS

Matter exists in three phases: solid, liquid, and gaseous. As discussed in the preceding chapter, the molecules of solids have such a strong attraction for one another that the solids have a definite shape or volume. In liquids, the greater molecular separation allows successive layers to flow past each other. Thus a liquid has a definite volume but assumes the shape of its container. The molecules of a gas are usually so far apart that they exert little attraction on one another. Gases have neither a definite shape nor a definite volume. The individual molecules move freely until they fill their container. Because liquids and gases can *flow*, they are referred to as *fluids*. We shall learn that since fluids have weight, they can exert pressure and force. The forces can be exerted by fluids at rest or by fluids in motion.

12-2 ■ DENSITY

Before we discuss the statics and dynamics of fluids, it is important to understand the relation of a body's weight to its volume. For example, we refer to lead or iron as *heavy*, whereas wood or cork is considered *light*. What we really mean is that a block of wood is lighter than a block of lead *of similar size*. The terms *light* and *heavy* are comparative terms. As illustrated in Fig. 12-1, it is possible for a block of lead to weigh the same as a block of wood if their relative size differs greatly. However, 1 ft^3 of lead weighs more than 16 times as much as 1 ft^3 of wood.

The quantity which relates a body's weight to its volume is known as its *weight density*.

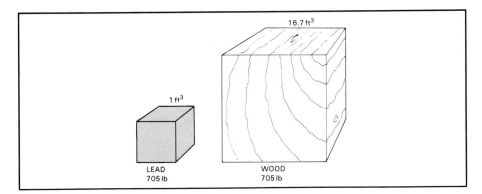

FIG. 12-1 A comparison of weight and volume in lead and wood.

The **weight density** D *of a body is defined as the ratio of its weight* W *to its volume* V. *Units are the newton per cubic meter* (N/m^3) *and the pound per cubic foot* (lb/ft^3):

$$D = \frac{W}{V} \qquad W = DV \qquad \text{(12-1)}$$

Thus, if a 20-lb object occupies a volume of 4 ft³, it has a weight density of 5 lb/ft³.

As mentioned in Chap. 7, the weight of a body is not constant but varies according to location. A more useful relation for density takes advantage of the fact that *mass* is a universal constant, independent of gravity.

The **mass density** ρ *of a body is defined as the ratio of its mass* m *to its volume* V:

$$p = \frac{m}{V} \qquad m = \rho V \qquad \text{(12-2)}$$

The units of mass density are the ratio of a mass unit to a volume unit, i.e., grams per cubic centimeter, kilograms per cubic meter, or slugs per cubic foot.

The relation between weight density and mass density is found by recalling that $W = mg$. Thus

$$D = \frac{mg}{V} = \rho g \qquad \text{(12-3)}$$

In USCS units, matter is usually described in terms of its weight. For this reason, weight density is more often used when one is working with this system of units. In SI units, mass is the more convenient quantity, and the mass density is preferred. Table 12-1 lists the weight densities and mass densities of some common substances.

EXAMPLE 12-1 A cylindrical tank for gasoline is 3 m long and 1.2 m in diameter. How many kilograms of gasoline will the tank hold?

Solution First we find the volume:

$$V = \pi r^2 h = \pi(0.6 \text{ m})^2(3 \text{ m}) = 3.39 \text{ m}^3$$

Substituting the volume and mass density into Eq. (12-1), we obtain

$$m = \rho V = (680 \text{ kg/m}^3)(3.39 \text{ m}^3) = 2310 \text{ kg}$$

TABLE 12-1 DENSITY AND WEIGHT DENSITY

Substance	D, lb/ft^3	ρ g/cm^3	ρ kg/m^3
Solids:			
Aluminum	169	2.7	2,700
Brass	540	8.7	8,700
Copper	555	8.89	8,890
Glass	162	2.6	2,600
Gold	1204	19.3	19,300
Ice	57	0.92	920
Iron	490	7.85	7,850
Lead	705	11.3	11,300
Oak	51	0.81	810
Silver	654	10.5	10,500
Steel	487	7.8	7,800
Liquids:			
Alcohol	49	0.79	790
Benzene	54.7	0.88	880
Gasoline	42	0.68	680
Mercury	850	13.6	13,600
Water	62.4	1.0	1,000
Gases (0°C):			
Air	0.0807	0.00129	1.29
Hydrogen	0.0058	0.000090	0.090
Helium	0.0110	0.000178	0.178
Nitrogen	0.0782	0.00125	1.25
Oxygen	0.0892	0.00143	1.43

12-3 ■ PRESSURE

The effectiveness of a given force often depends upon the area over which it acts. For example, a woman wearing narrow heels will do much more damage to floors than she would with flat heels. Even though she exerts the same downward force in each case, with the narrow heels her weight is spread over a much smaller surface area. The *normal force per unit area* is called *pressure*. Symbolically, the pressure P is given by

$$P = \frac{F}{A}$$

(12-4)

where A is the area over which the perpendicular force F is applied. The unit of pressure is the ratio of any force unit to a unit of area. Examples are newtons per square meter and pounds per square inch. In SI units, the newton per square meter is renamed the *pascal* (Pa). The *kilopascal* (kPa) is the most appropriate measure for fluid pressure.

$$1 \text{ kPa} = 1000 \text{ N/m}^2 = 0.145 \text{ lb/in.}^2$$

EXAMPLE 12-2   A golf shoe has 10 cleats, each having an area of 0.01 in.^{2} in contact with the floor. Assume that in walking there is one instant when all 10 cleats support the entire weight of a 180-lb person. What is the pressure exerted by the cleats on the floor? Express the answer in SI units.

Solution   The total area in contact with the floor is 0.1 in.^{2} (10 × 0.01 in.^{2}). Substitution into Eq. (12-4) yields

$$P = \frac{F}{A} = \frac{180 \text{ lb}}{0.1 \text{ in.}^2} = 1800 \text{ lb/in.}^2$$

Converting to SI units, we obtain

$$P = (1800 \text{ lb/in.}^2)\left(\frac{1 \text{ kPa}}{0.145 \text{ lb/in.}^2}\right) = 12{,}400 \text{ kPa}$$

As the area of the shoe in contact with the floor decreases, the pressure becomes larger. It is easy to see why this factor must be considered in floor construction.

12-4 ■ FLUID PRESSURE

There is a very significant difference between the way a force acts on a fluid and on a solid. Since a solid is a rigid body, it can withstand the application of a force without a significant change in shape. A liquid, however, can sustain a force only at an enclosed surface or boundary. If a fluid is not restrained, it will flow under a shearing stress instead of being deformed elastically.

> *The force exerted by a fluid on the walls of its container must always act perpendicular to the walls.*

It is this characteristic property of fluids that makes the concept of pressure so useful. Holes bored in the bottom and sides of a barrel of water (Fig. 12-2) demonstrate that the force exerted by the water is everywhere perpendicular to the surface of the barrel.

A moment's reflection will show the student that a liquid also exerts an upward pressure. Anyone who tries to keep a rubber float under the surface of water is immediately convinced of the existence of an upward pressure. In fact, we find that

> *Fluids exert pressure in all directions.*

Figure 12-3 shows a liquid under pressure. The forces acting on the face of the piston, the walls of the enclosure, and the surfaces of a suspended object are shown in the figure.

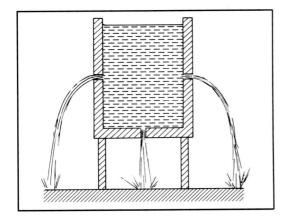

FIG. 12-2 The forces exerted by a fluid on the walls of its container are perpendicular at every point.

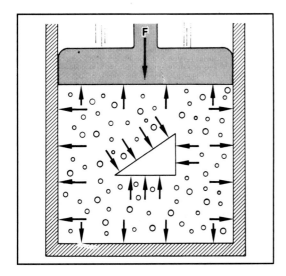

FIG. 12-3 Fluids exert pressure in all directions.

Just as larger volumes of solid objects exert greater forces against their supports, fluids exert greater pressure at increasing depths. The fluid at the bottom of a container is always under greater pressure than that near the surface. This is due to the weight of the overlying liquid. However, we must point out a distinct difference between the pressure exerted by solids and that exerted by liquids. A solid object can exert only a *downward* force due to its weight. At any particular depth in a fluid, the pressure is the same in all directions. If this were not true, the fluid would flow under the influence of a resultant pressure until a new condition of equilibrium was reached.

Since the weight of the overlying fluid is proportional to its density, the pressure at any depth is also proportional to the density of the fluid. This can be seen by considering a rectangular column of water extending from the surface to a depth h, as shown in Fig. 12-4. The weight of the entire column acts on the surface area A at the bottom of the column. From Eq. (12-1), we can write the weight of the column as

$$W = DV = DAh$$

where D is the weight density of the fluid. The pressure (weight per unit area) at the depth h is given by

$$P = \frac{W}{A} = Dh$$

or, in terms of mass density,

$$\boxed{P = Dh = \rho gh} \qquad (12\text{-}5)$$

The fluid pressure at any point is directly proportional to the density of the fluid and to the depth below the surface of the fluid.

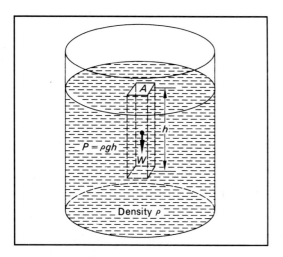

FIG. 12-4 Relationship of pressure, density, and depth.

EXAMPLE 12-3   The water pressure in a certain house is 20 lb/in.^{2}. How high must the water level in a reservoir be above the point of release in the house?

Solution   The weight density of water is 62.4 lb/ft^3. The pressure is 20 lb/in.^{2}. To avoid inconsistency in units, we convert the pressure to units of pounds per square foot:

$$P = (20 \text{ lb/in.}^2)\left(\frac{144 \text{ in.}^2}{1 \text{ ft}^2}\right) = 2800 \text{ lb/ft}^2$$

Solving for h in Eq. (12-5), we have

$$h = \frac{P}{D} = \frac{2880 \text{ lb/ft}^2}{62.4 \text{ lb/ft}^3} = 46 \text{ ft}$$

The above example made no mention of the size or shape of the reservoir containing the supply of water. Additionally, it gave no information about the path of the water or the size of the pipes connecting the reservoir to the home. Can we assume that our answer is correct when it is based only upon the difference in water levels? Doesn't the shape or area of a container have any effect on liquid pressure? In order to answer these questions, we must recall some of the characteristics of fluids already discussed.

Consider a series of vessels of different areas and shapes interconnected as shown in Fig. 12-5. It would seem at first glance that the greater volume of water in vessel B should develop a greater pressure at the bottom than vessel D. The effect of such a difference in pressure would then be to force the liquid to rise higher in vessel D. However, when the vessels are filled with liquid, the levels are the same for each vessel.

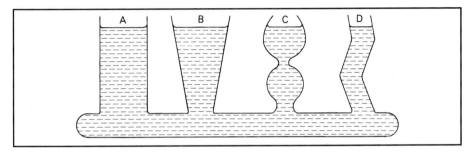

FIG. 12-5 Water seeks its own level, indicating that the pressure is independent of the area or shape of the container.

Part of the problem in understanding this paradox results from confusing the terms *pressure* and *total force*. Since pressure is measured in terms of a unit area, we do not consider the *total area* when we solve problems involving pressure. For example, in vessel A the area of the liquid at the bottom of the vessel is much greater than the area at the bottom of vessel D. This means that the liquid in vessel A will exert a greater *total force* on the bottom than the liquid in vessel D. But the greater force is applied over a larger area, so that the pressure is the same in both vessels.

If the bottoms of vessels B, C, and D have the same area, we can say that the total forces are also equal at the bottoms of these containers. (Of course, the pressures are equal at any particular depth.) You may wonder how the total forces can be equal when vessels B and C contain a greater volume of water. The additional water in each case is supported by upward components of forces exerted by the walls of the container on the fluid. (See Fig. 12-6.) When the walls of a container are vertical, the forces acting on the sides have no upward components. The total force at the bottom of a container is therefore equal to the weight of a straight column of water above the base area.

EXAMPLE 12-4 Assume that the vessels in Fig. 12-5 are filled with gasoline until the fluid level is 1 ft above the base of each vessel. The areas at the bases of vessels A and B are 20 and 10 in.^{2}, respectively. Compute the pressure and the total force at the base of each container.

Solution The pressure is the same at the base of either container and is given by

$$P = Dh = (42 \text{ lb/ft}^3)(1 \text{ ft}) = 42 \text{ lb/ft}^2$$

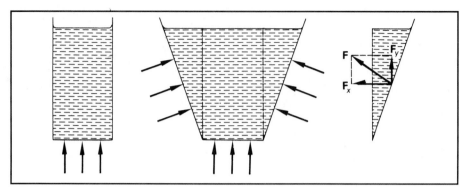

FIG. 12-6 The pressure at the bottom of each vessel is a function of only the depth of the liquid and is the same in all directions. Since the area at the bottom is the same for both vessels, the total force exerted on the bottom of each container is also the same.

The total force in each case is the product of the pressure and base area ($F = PA$). Thus

$$F_A = (42 \text{ lb/ft}^2)(20 \text{ in.}^2)\left(\frac{1 \text{ ft}^2}{144 \text{ in.}^2}\right) = 5.8 \text{ lb}$$

$$F_B = (42 \text{ lb/ft}^2)(10 \text{ in.}^2)\left(\frac{1 \text{ ft}^2}{144 \text{ in.}^2}\right) = 2.9 \text{ lb}$$

Before we consider other applications of fluid pressure, let us summarize the principles discussed in this section for fluids at rest:
1. The forces exerted by a fluid on the walls of its container are always perpendicular.
2. The fluid pressure is directly proportional to the depth of the fluid and to its density.
3. At any particular depth, the fluid pressure is the same in all directions.
4. Fluid pressure is independent of the shape or area of its container.

12-5 ■ MEASURING PRESSURE

The pressure discussed in the previous section is due only to the fluid itself and can be calculated from Eq. (12-5). Unfortunately, this is usually not the case. Any liquid in an open container, for example, is subjected to atmospheric pressure in addition to the pressure of its own weight. Since the liquid is relatively incompressible, the external pressure of the atmosphere is transmitted equally throughout the volume of the liquid. This fact, first stated by the French mathematician Blaise Pascal (1623–1662), is called *Pascal's law*. Generally, it can be stated as follows:

> *An external pressure applied to an enclosed fluid is transmitted uniformly throughout the volume of the liquid.*

Most devices which measure pressure directly actually measure the difference between the *absolute pressure* and *atmospheric pressure*. The result is called the *gauge pressure:*

Absolute pressure = gauge pressure + atmospheric pressure

Atmospheric pressure at sea level is 101.3 kPa, or 14.7 lb/in.^{2}. Because atmospheric pressure enters into so many calculations, we often use a pressure unit of *one at-mopshere* (atm), defined as the average pressure exerted by the atmosphere at sea level, that is, 14.7 lb/in.^{2}.

A common device for measuring gauge pressure is the open-tube *manometer* (muh-nom'-uh-ter), shown in Fig. 12-7. The manometer consists of a U-shaped tube containing a liquid, usually mercury. When both ends of the tube are open, the mercury seeks its own level because 1 atm of pressure is exerted at each of the open ends. When one end is connected to a pressurized chamber, the mercury will rise in the open tube until the pressures are equalized. The difference between the two levels of mercury is a measure of the gauge pressure, i.e., the difference between the absolute pressure in the chamber and atmospheric pressure at the open end. The manometer is used so often in laboratory situations that atmospheric pressures and other pressures are often expressed in *centimeters of mercury* or *inches of mercury*.

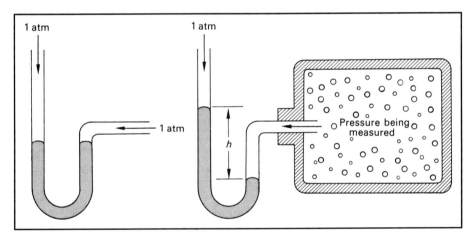

FIG. 12-7 Open-tube manometer. Pressure is measured by the height h of the mercury column.

Atmospheric pressure is usually measured in the laboratory with a mercury barometer. The principle of its operation is shown in Fig. 12-8. A glass tube, closed at one end, is filled with mercury. The open end is covered, and the tube is inverted in a bowl of mercury. When the open end is uncovered, the mercury flows out of the tube until the pressure exerted by the column of mercury exactly balances the atmospheric pressure acting on the mercury in the bowl. Since the pressure in the tube above the column of mercury is zero, the height of the column above the level of mercury in the bowl indicates the atmospheric pressure. At sea level an atmospheric pressure of 14.7 lb/in.^{2} will cause the level of the mercury in the tube to stabilize at a height of 76 cm, or 30 in.

In summary, we can write the following equivalent measures of atmospheric pressure:

$$1 \text{ atm} = 101.3 \text{ kPa} = 14.7 \text{ lb/in.}^2 = 76 \text{ cm of mercury}$$
$$= 30 \text{ in. of mercury} = 2116 \text{ lb/ft}^2$$

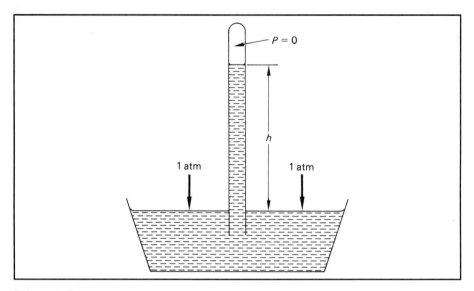

FIG. 12-8 Barometer.

EXAMPLE 12-5 The mercury manometer is used to measure the pressure of a gas inside a tank. (Refer to Fig. 12-7.) If the difference between the two mercury levels is 36 cm, what is the absolute pressure inside the tank?

Solution The gauge pressure is 36 cm of mercury, and atmospheric pressure is 76 cm of mercury. Thus the absolute pressure is found from Eq. (12-5):

Absolute pressure = 36 cm + 76 cm = 112 cm of mercury

The pressure in the tank is equivalent to the pressure that would be exerted by a column of mercury 112 cm high:

$$
\begin{aligned}
P &= Dh = \rho g h \\
&= (13{,}600 \text{ kg/m}^3)(9.8 \text{ m/s}^2)(1.12 \text{ m}) \\
&= 1.49 \times 10^5 \text{ N/m}^2 = 149 \text{ kPa}
\end{aligned}
$$

You should verify that this absolute pressure is also 21.6 lb/in.^{2}, or 1.47 atm.

12-6 ■ HYDRAULIC PRESS

The most universal application of Pascal's law is found with the hydraulic press, shown in Fig. 12-9. According to Pascal's principle, a pressure applied to the liquid in the left column will be transmitted undiminished to the liquid in the column at the right. Thus, if an input force F_i acts upon a piston of area A_i, it will cause an output force F_o to act on a piston of area A_o so that

Input pressure = output pressure

$$
\frac{F_i}{A_i} = \frac{F_o}{A_o} \tag{12-6}
$$

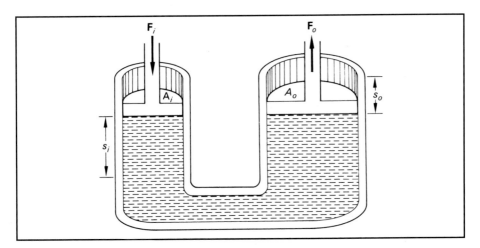

FIG. 12-9 Hydraulic press.

The ideal mechanical advantage of such a device is equal to the *ideal* ratio of the output force to the input force. Symbolically, we write

$$M_I = \frac{F_o}{F_i} = \frac{A_o}{A_i} \tag{12-7}$$

A small input force can be multiplied to yield a much larger output force simply by having the output piston much larger in area than the input piston. The output force is given by

$$F_o = F_i \frac{A_o}{A_i} \tag{12-8}$$

According to the methods developed in Chap. 10 for simple machines, the input work must equal the output work if we neglect friction. If the input force F_i travels through a distance s_i while the output force F_o travels through a distance s_o, we can write

Input work = output work
$$F_i s_i = F_o s_o$$

This relation leads to another useful expression for the ideal mechanical advantage of a hydraulic press:

$$M_I = \frac{F_o}{F_i} = \frac{s_i}{s_o} \tag{12-9}$$

Notice that the mechanical advantage is gained at the expense of input distance. For this reason, most applications utilize a system of valves to permit the output piston to be raised by a series of short input strokes.

EXAMPLE 12-6   The smaller and larger pistons of a hydraulic press have diameters of 2 and 24 in., respectively. (*a*) What input force is required in order to deliver a total output force of 2000 lb at the larger piston? (*b*) How far must the smaller piston travel in order to lift the larger piston 1 in.?

Solution (a)   The mechanical advantage is

$$M_I = \frac{A_o}{A_i} = \frac{\pi d_o^2/4}{\pi d_i^2/4} = \left(\frac{d_o}{d_i}\right)^2$$

$$= \left(\frac{24 \text{ in.}}{2 \text{ in.}}\right)^2 = 12^2 = 144$$

Neglecting friction, the required input force is given by

$$F_i = \frac{F_o}{M_I} = \frac{2000 \text{ lb}}{144} = 13.9 \text{ lb}$$

Solution (b)   Applying Eq. (12-9), we can compute the input distance:

$$s_i = M_I s_o = (144)(1 \text{ in.}) = 144 \text{ in.}$$

The principle of the hydraulic press is found in many engineering and mechanical devices. Power steering, the hydraulic jack, shock absorbers, and automobile braking systems are a few common examples.

12-7 ■ ARCHIMEDES' PRINCIPLE

Anyone familiar with swimming and other water sports has observed that objects seem to lose weight when they are submerged in water. In fact, an object may even float on the surface because of the upward pressure exerted by the water. An ancient Greek mathematician Archimedes (287–212 B.C.) first studied the buoyant force exerted by fluids. *Archimedes' principle* can be stated as follows:

> *An object which is completely or partly submerged in a fluid experiences an upward force equal to the weight of the fluid displaced.*

Archimedes' principle can be demonstrated by studying the forces a fluid exerts on a suspended object. Consider a disk of area A and height H which is completely submerged in a fluid, as shown in Fig. 12-10. Recall that the pressure at any depth h in a fluid is given by

$$P = \rho g h$$

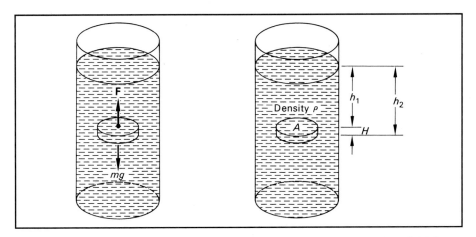

FIG. 12-10 The buoyant force exerted on the disk is equal to the weight of the fluid that the disk displaces.

where ρ is the mass density of the fluid and g is the acceleration of gravity. Of course, if we wish to represent the absolute pressure within the fluid, we must add the external pressure exerted by the atmosphere. The total downward pressure P_1 on top of the disk in Fig. 12-10 is therefore

$$P_1 = P_a + \rho g h_1 \qquad \text{downward}$$

where P_a is atmospheric pressure and h_1 is the depth at the top of the disk. Similarly, the upward pressure P_2 on the bottom of the disk is

$$P_2 = P_a + \rho g h_2 \qquad \text{upward}$$

where h_2 is the depth at the bottom of the disk. Since h_2 is greater than h_1, the pressure on the bottom of the disk will exceed the pressure at the top, resulting in a net upward force. If we represent the downward force by F_1 and the upward force by F_2, we can write

$$F_1 = P_1 A \qquad F_2 = P_2 A$$

The net upward force exerted *by* the fluid *on* the disk is called the *buoyant force* and is given by

$$\begin{aligned} F_B &= F_2 - F_1 = A(P_2 - P_1) \\ &= A(P_a + \rho g h_1 - P_a - \rho g h_2) \\ &= A\rho g(h_1 - h_2) = A\rho g H \end{aligned}$$

where $H = h_1 - h_2$ is the height of the disk. Finally, if we recall that the volume of the disk is $V = AH$, we obtain the important result

$$F_B = V\rho g = mg \tag{12-10}$$

Buoyant force = weight of displaced fluid

which is Archimedes' principle.

In applying this result, it must be recalled that Eq. (12-10) allows us to compute only the *buoyant force* due to the difference in pressures. It does not represent the resultant force. A submerged body will sink if the weight of the fluid it displaces (the buoyant force) is less than the weight of the body. If the weight of the displaced fluid is exactly equal to the weight of the submerged body, it will neither sink nor rise. In this instance, the body will be in equilibrium. If the weight of the displaced fluid exceeds the weight of a submerged body, the body will rise to the surface and float. When the floating body comes to equilibrium at the surface, it will displace its own weight of liquid. Figure 12-11 demonstrates this point, using an overflow can and a beaker to catch the fluid displaced by a wooden block.

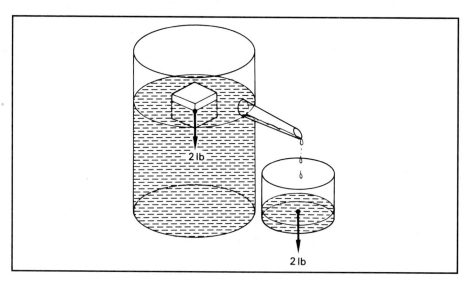

FIG. 12-11 A floating body displaces its own weight of fluid.

EXAMPLE 12-7 A cork float has a volume of 2 ft³ and a density of 15 lb/ft³. (*a*) What volume of the cork is beneath the surface when the cork floats in water? (*b*) What downward force is required to submerge the cork completely?

Solution (a) The floating cork will displace a volume of water equal to its own weight, which is

$$W = DV = (15 \text{ lb/ft}^3)(2 \text{ ft}^3) = 30 \text{ lb}$$

Since water has a weight density of 62.4 lb/ft³, the volume of water displaced is

$$V = \frac{W}{D} = \frac{30 \text{ lb}}{62.4 \text{ lb/ft}^3} = 0.481 \text{ ft}^3$$

Thus the volume of the cork under the water is also 0.481 ft³.

Solution (b) In order to submerge the cork, a downward force F must be exerted, in addition to the weight W of the cork, such that their sum equals the buoyant force F_B. Symbolically

$$F + W = F_B$$

The required force F is therefore equal to the difference between the buoyant force and the weight of the cork:

$$F = F_B - W$$

The buoyant force in this case can be found by computing the weight of 2 ft³ of water (the amount of water displaced when the cork is entirely submerged). We obtain

$$F_B = DV = (62.4 \text{ lb/ft}^3)(2 \text{ ft}^3) = 124.8 \text{ lb}$$

The force F required to submerge the block is

$$F = 124.8 \text{ lb} - 30 \text{ lb} = 94.8 \text{ lb}$$

EXAMPLE 12-8 A weather balloon is to operate at an altitude where the density of air is 0.9 kg/m³. At this altitude, the balloon has a volume of 20 m³ and is filled with hydrogen ($\rho_H = 0.09$ kg/m³). If the balloon bag weighs 118 N, what load can it support at this level?

Solution The buoyant force is equal to the weight of the displaced air. Thus

$$F_B = \rho g V = (0.9 \text{ kg/m}^3)(9.8 \text{ m/s}^2)(20 \text{ m}^3) = 176 \text{ N}$$

The weight of 20 m³ of hydrogen is

$$W_H = \rho_H g V = (0.09 \text{ kg/m}^3)(9.8 \text{ m/s}^2)(20 \text{ m}^3) = 17.6 \text{ N}$$

The load supported is

$$\begin{aligned} W_L &= F_B - W_H - W_B \\ &= 176 \text{ N} - 17.6 \text{ N} - 118 \text{ N} = 40.4 \text{ N} \end{aligned}$$

Large balloons can maintain a condition of equilibrium at any altitude by adjusting their weight and changing the buoyant force. The weight can be lightened by releasing the ballast provided for that purpose. The buoyant force can be decreased by releasing gas from the balloon or increased by pumping more gas into the flexible balloon. Hot-air balloons use the lower density of heated air to provide their buoyancy.

12-8 ■ FLUIDS IN MOTION

In addition to the study of fluids at rest, we need to understand some basic principles of fluids in motion. Unfortunately, the mathematical difficulties encountered with fluid flow are formidable. However, we will analyze simple fluid motion by making a few assumptions and generalizations. For a more detailed treatment you should refer to a fluid dynamics textbook.

In studying the dynamics of fluids, we shall assume that all fluids in motion exhibit *streamline flow:*

> **Streamline flow** *is the motion of a fluid in which every particle in the fluid follows the same path (past a particular point) as that followed by previous particles.*

Figure 12-12 illustrates the *streamlines* of air flowing past two stationary obstacles. Note that the streamlines break down as air passes over the second obstacle, setting up whirls and eddies. These little whirlpools represent *turbulent flow* and absorb much of the fluid energy, increasing the frictional drag through the fluid.

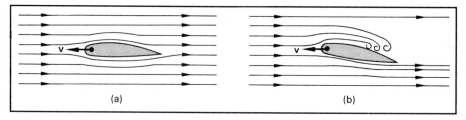

FIG. 12-12 Streamline and turbulent flow of a fluid.

We shall further consider that fluids are incompressible and have essentially no internal friction. Under these conditions, we can make certain predictions about the rate of fluid flow through a pipe or other container:

> The **rate of flow** *is defined as the volume of fluid that passes a certain cross section per unit of time.*

In order to express this rate quantitatively, we shall consider a liquid flowing through the pipe of Fig. 12-13 with an average speed v. During a time interval t, each particle in the stream moves through a distance vt. The volume V flowing through a cross section A is given by

$$V = Avt$$

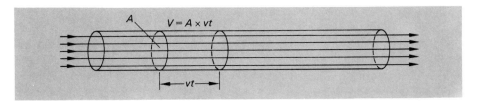

FIG. 12-13 Computing the rate of flow of a fluid through a pipe.

Thus the rate of flow (volume per unit time) can be calculated from

$$R = \frac{Avt}{t} = vA$$

(12-11)

Rate of flow = velocity × cross section

The units of R express the ratio of a volume unit to a time unit. Common examples are cubic feet per second, cubic meters per second, liters per second, and gallons per minute.

If the fluid is incompressible and we ignore the effects of internal friction, the rate of flow R will remain constant. This means that a variation in the pipe cross section, as illustrated in Fig. 12-14, will result in a change in the speed of the liquid so that the product vA remains constant. Symbolically, we write

$$R = v_1A_1 = v_2A_2$$

(12-12)

A liquid will flow faster through a narrow section of pipe and more slowly through a broad section. It is this principle which causes water to flow more rapidly when the banks of a small stream suddenly come closer together.

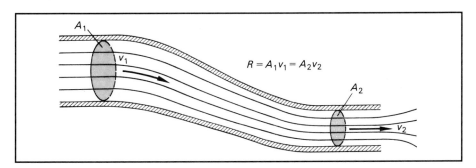

FIG. 12-14 In streamline flow, the product of the fluid velocity and the cross-sectional area of the pipe is constant at any point.

EXAMPLE 12-9 Water flows through a rubber hose 1 in. in diameter at a speed of 4 ft/s. What must the diameter of the nozzle be if water is to emerge with a velocity of 20 ft/s? What is the rate of flow in gallons per minute?

Solution   The rate of flow is constant, so that $A_1v_1 = A_2v_2$. Since the area A is proportional to the square of the diameter, we have

$$d_1^2v_1 = d_2^2v_2 \qquad \text{or} \qquad d_2^2 = \frac{v_1}{v_2}\, d_1^2$$

from which

$$d_2^2 = \left(\frac{4 \text{ ft/s}}{20 \text{ ft/s}}\right)(1 \text{ in.})^2 = 0.2 \text{ in.}^2$$

or

$$d_2 = 0.447 \text{ in.}$$

In order to determine the rate of flow, we must first determine the area of the 1-in. hose:

$$A_1 = \frac{\pi d_1^2}{4} = \left(\frac{\pi}{4}\right)(1 \text{ in.})^2 = 0.785 \text{ in.}^2$$

$$= (0.785 \text{ in.}^2)\left(\frac{1 \text{ ft}^2}{144 \text{ in.}^2}\right) = 5.45 \times 10^{-3} \text{ ft}^2$$

The rate of flow is $R = A_1v_1$, so that

$$R = (5.45 \times 10^{-3} \text{ ft}^2)(4 \text{ ft/s}) = 0.0218 \text{ ft}^3/\text{s}$$

Recalling that $1 \text{ ft}^3 = 7.48$ gal and that $1 \text{ min} = 60$ s, we can express this rate in the desired units:

$$R = (0.0218 \text{ ft}^3/\text{s})\left(\frac{7.48 \text{ gal}}{1 \text{ ft}^3}\right)\left(\frac{60 \text{ s}}{1 \text{ min}}\right) = 9.78 \text{ gal/min}$$

The same rate would be obtained by considering the product A_2v_2.

12-9 ■ PRESSURE AND VELOCITY

We have noted that a fluid's speed increases when the fluid flows through a constriction. An increase in speed can result only through the presence of an accelerating force. In order to accelerate the liquid as it enters the constriction, the pushing force from the large cross section must be greater than the resisting force from the constriction. In other words, the pressure at points A and C in Fig. 12-15 must be greater than the pressure at B. The tubes inserted into the pipe above these points clearly indicate the difference in pressure. The fluid level in the tube above the restriction is lower than the level in the adjacent areas. If h is the difference in height, the pressure differential is given by

$$P_A - P_B = \rho g h \qquad (12\text{-}13)$$

This assumes that the pipe is horizontal and that no pressure changes are introduced because of a change in potential energy.

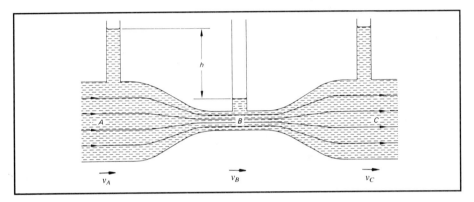

FIG. 12-15 The increased velocity of a fluid flowing through a constriction causes a drop in pressure.

The above example, as illustrated in Fig. 12-15, shows the principle of the *venturi meter*. From a determination of the difference in pressure, this device makes it possible to calculate the velocity of water in a horizontal pipe.

The venturi effect has many other applications for both liquids and gases. The carburetor in an automobile uses the venturi principle to mix gasoline vapor and air. Air passing through a constriction on its way to the cylinders creates a low-pressure area as its speed increases. The decrease in pressure is used to draw fuel into the air column, where it is readily vaporized.

Figure 12-16 shows two methods you can use to demonstrate the decrease in pressure that results from an increase in velocity. The simplest example consists of blowing air past the top surface of a sheet of paper, as shown in Fig. 12-16a. The pressure in the airstream above the paper will be reduced. This allows the excess pressure on the bottom to force the paper upward.

A second demonstration requires a hollow spool, a cardboard disk, and a pin (Fig. 12-16b). The pin is driven through the cardboard disk and placed in one end of the hollow spool. If you blow through the open end of the spool, you will find that the disk becomes more tightly pressed to the other end. One might expect the cardboard to fly off immediately. The explanation is that air blown into the spool must escape through the narrow space between the disk and the end of the spool. This action creates a low-pressure area, allowing the external atmospheric pressure to push the disk tightly against the spool.

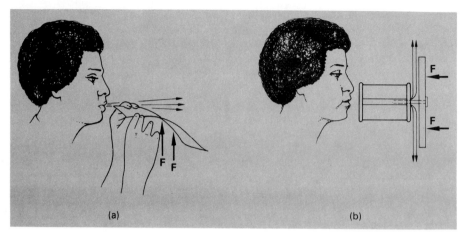

FIG. 12-16 Demonstrations of the decrease in pressure resulting from an increase in fluid speeds.

Summary

We have studied the effects of fluids at rest and in motion. Density, pressure, buoyant force, and other concepts have been defined in order to study the effects of fluids on their surroundings. The following ideas have been discussed in this chapter:

- A very important physical property of matter is its *density*. The weight density D and the mass density ρ are defined as follows:

$$\text{Weight density} = \frac{\text{weight}}{\text{volume}} \qquad \boxed{D = \frac{W}{V}} \qquad \text{N/m}^3 \text{ or lb/ft}^3$$

$$\text{Mass density} = \frac{\text{mass}}{\text{volume}} \qquad \boxed{\rho = \frac{m}{V}} \qquad \text{kg/m}^3 \text{ or slugs/ft}^3$$

- Since $W = mg$, the relationship between D and ρ is

$$\boxed{D = \rho g} \qquad \text{weight density} = \text{mass density} \times \text{gravity}$$

- Important points to remember about fluid pressure:
  a. The forces exerted by a fluid on the walls of its container are always perpendicular.
  b. The fluid pressure is directly proportional to the depth of the fluid and to its density:

$$P = \frac{F}{A} \qquad P = Dh \qquad P = \rho gh$$

c. At any particular depth, the fluid pressure is the same in all directions.
d. Fluid pressure is independent of the shape or area of its container.
- Pascal's law states that *an external pressure applied to an enclosed fluid is transmitted uniformly throughout the volume of the liquid.*
- When one is measuring fluid pressure, it is essential to distinguish between *absolute* pressure and *gauge* pressure:

$$\begin{aligned} \text{Absolute pressure} &= \text{gauge pressure} + \text{atmospheric pressure} \\ \text{Atmospheric pressure} &= 1 \text{ atm} = 1.013 \times 10^5 \text{ N/m}^2 \\ &= 1.013 \times 10^5 \text{ Pa} = 14.7 \text{ lb/in.}^2 \\ &= 76 \text{ cm of mercury} \end{aligned}$$

- Applying Pascal's law to the hydraulic press gives the following for the ideal advantage:

$$M_I = \frac{F_o}{F_i} = \frac{s_i}{s_o} \qquad \textit{Ideal Mechanical Advantage for Hydraulic Press}$$

- Archimedes' principle: *An object which is completely or partly submerged in a fluid experiences an upward force equal to the weight of the fluid displaced:*

$$F_B = mg \qquad \text{or} \qquad F_B = V\rho g \qquad \textit{Buoyant Force}$$

- The *rate of flow* is defined as the volume of fluid that passes a certain cross section A per unit of time t. In terms of fluid velocity v, we write

$$R = \frac{V}{t} = vA \qquad \textit{Rate of flow}$$

- For an incompressible fluid flowing through pipes in which the cross sections vary, the rate of flow is constant:

$$v_1 A_1 = v_2 A_2 \qquad d_1^2 v_1 = d_2^2 v_2$$

Questions

12-1. Define the following terms:

a. weight density
b. mass density
c. pressure
d. total force
e. Pascal's law
f. absolute pressure
g. gauge pressure

h. manometer
i. Archimedes' principle
j. buoyant force
k. streamline flow
l. turbulent flow
m. rate of flow
n. venturi effect

12-2. Make a list of the units for weight density and the similar units for mass density.

12-3. Which is numerically larger, the weight density of an object or its mass density?

12-4. The density of water is given in Table 12-1 as 62.4 lb/ft^3. In performing an experiment with water on the surface of the moon, would you trust this value? Explain.

12-5. Which is heavier, 870 kg of brass or 3.5 ft^3 of copper?

12-6. Why are dams so much thicker at the bottom than at the top? Does the pressure exerted on the dam depend on the length of the reservoir perpendicular to the dam?

12-7. A large block of ice floats in a bucket of water so that the level of the water is at the top of the bucket. Will the water overflow when the ice melts? Explain.

12-8. A tub of water rests on weighing scales which indicate 40 lb total weight. Will the total weight increase when a 5-lb fish is floating on the surface of the water? Discuss.

12-9. Suppose that an iron block supported by a string is completely submerged in the tub of Question 12-8. How will the reading on the scales be affected?

12-10. A boy just learning to swim finds that he can float on the surface more easily after inhaling air. He also observes that he can hasten his descent to the bottom of the pool by exhaling air on the way down. Explain his observations.

12-11. A toy sailboat filled with pennies floats in a small tub of water. If the pennies are thrown into the water, what happens to the water level in the tub?

12-12. Is it more difficult to hold a cork float barely under the surface than it is to hold it at a depth of 5 ft? Explain.

12-13. Is it possible to construct a barometer using water instead of mercury? How high will the column of water be if the external pressure is 1 atm?

12-14. Discuss the operation of a submarine and a weather balloon. Why will a balloon rise to a definite height and stop? Will a submarine sink to a particular depth and stop if no changes are made after submerging?

12-15. Why does the flow of water from a faucet decrease when someone turns on another faucet in the same building?

12-16. Two rowboats moving parallel to each other in the same direction are drawn together. Explain.

12-17. Explain what would happen in a modern jet airliner at high speed if a hijacker fired a bullet through the window or broke open an escape hatch.

12-18. During high-velocity windstorms or hurricanes, the roofs of houses are sometimes blown off without otherwise damaging the homes. Explain with the use of diagrams.

12-19. A small child knocks a balloon over the heating duct in her home and is surprised to find that the balloon remains suspended above the duct, bobbing from one side to the other. Explain.

12-20. What conditions would determine the maximum lift capacity of a streamlined aircraft wing? Draw figures to justify your answer.

Problems

12-1. What volume does 0.4 kg of alcohol occupy? What is the weight of this volume?

Answer 5.06×10^{-4} m^3; 3.92 N.

12-2. An unknown substance has a volume of 20 ft^3 and weights 3370 lb. Considering its density, what might the substance be?

12-3. What volume of water has the same weight as 1 ft^3 of lead? Compute the mass density of the water in slugs per cubic feet.

Answer 11.1 ft^3; 1.95 slugs/ft^3.

12-4. An open U-shaped tube containing mercury is 1 cm^2 in cross section. What volume of water must be poured into the left tube in order to cause the mercury in the right tube to rise 1 cm above its original position?

12-5. Find the pressure in kilopascals due to a column of mercury 600 mm high. What is this pressure in pounds per square inch? In atmospheres?

Answer 80 kPa; 11.6 lb/in.^{2}; 0.789 atm.

12-6. A submarine dives to a depth of 120 ft and levels off. The interior of the submarine is maintained at atmospheric pressure. What total force is exerted against a hatch 2 ft wide and 3 ft long?

12-7. A 20-kg piston rests on a sample of gas in a cylinder 80 mm in diameter. What is the gauge pressure of the gas? What is the absolute pressure?

Answer 39 kPa; 140.3 kPa.

12-8. A water pressure gauge indicates a pressure of 50 lb/in.^{2} at the foot of a building. What is the maximum height to which the water will rise in the building?

12-9. The gauge pressure in an automobile tire is 28 lb/in.^{2}. If the wheel supports 1000 lb, what area of the tire is in contact with the ground?

Answer 35.7 in.^{2}.

12-10. A cylindrical water tank 50 ft high and 20 ft in diameter is filled with water. (*a*) What is the water pressure on the bottom of the tank? (*b*) What is the total force on the bottom? (*c*) What is the pressure in a water pipe which is located 90 ft below the water surface?

12-11. The areas of the small and large pistons in a hydraulic press are 122 and 448 mm^2, respectively. What is the ideal mechanical advantage of the press? What force must be exerted in order to lift a 150-kg mass? If the mass is lifted a distance of 200 mm, what distance does the input force move?

Answer 4; 368 N; 800 mm.

12-12. The small and large pistons of a hydraulic press have diameters of 4 and 38 in., respectively. (*a*) What input force is needed to develop a lifting force of 40,000 lb at the larger piston? (*b*) How far must the small piston move in order to raise the large piston 5 in.?

12-13. A force of 50 N is applied to the small piston of a hydraulic press. The small piston has a diameter of 60 mm. What must be the diameter of the large piston if it is to lift a 20-kg mass?

Answer 119 mm.

12-14. The inlet pipe which supplies air pressure to operate a hydraulic lift is 20 mm in diameter. The output piston is 320 mm in diameter. What air pressure must be used to lift an 1800-kg car?

12-15. The area of the piston in a force pump is 10 in.^{2}. What force is required to raise water with the piston to a height of 100 ft?

Answer 433 lb.

12-16. A piece of cast iron weighs 24 lb in air. It has an apparent weight of only 21 lb when submerged in water. What is the buoyant force? What is the density of this object?

12-17. A 64-lb metal block has a volume of 0.2 ft³. The block is suspended from a cord and submerged in oil (D = 48 lb/ft³). Find the buoyant force and the tension in the cord.

Answer 9.6 lb; 54.4 lb.

12-18. A stone of unknown composition weighs 82 lb in air. Its apparent weight is 74 lb when submerged in water. What are the volume and density of the stone?

12-19. A 20-kg balloon is filled with 80 m³ of hydrogen. What force is required to hold it down?

Answer 745 N.

12-20. A balloon 40 m in diameter is filled with helium. What total mass can be lifted by this balloon in air of density 0.9 kg/m³?

12-21. A block of wood weighs 16 lb in air. A lead sinker which has an apparent weight of 28 lb in water is attached to the wood, and both are submerged in water. If their combined apparent weight in water is 18 lb, find the density of the wooden block.

Answer 38.4 lb/ft³.

*12-22. The floor of a river barge is 18 ft wide and 70 ft long. How much deeper will it sink into the water if a 400,000-lb load of coal is placed in the barge?

12-23. A block of wood which has a volume of 120 cm³ is found to have a mass of 100 g. Will it float in water? Will it float in gasoline?

Answer Floats in water; sinks in gasoline.

12-24. Water flows from a terminal 3 cm in diameter and has an average speed of 2 m/s. What is the rate of flow in cubic meters per minute?

12-25. Gasoline flows through a 1-in.-diameter hose at an average speed of 5 ft/s. What is the rate of flow in cubic feet per second? How many minutes are required to fill a 20-gal tank? (1 ft³ = 7.481 gal.)

Answer 0.0273 ft³/s; 1.63 min.

12-26. What must be the area of a hose if it is to deliver 8 L of oil in 1 min with an exit speed of 2 m/s? (1 L = 0.001 m³.)

12-27. Oil is forced through a pipe at the rate of 6 gal/min. What should the diameter of a connecting tube be if an exit velocity of 4 ft/s is desired?

Answer 0.784 in.

12-28. Water flowing at 6 m/s through a 60-mm pipe is suddenly channeled into a 30-mm pipe. What is the velocity in the small pipe? Is the rate of flow greater, smaller, or the same in the smaller pipe?

13

TEMPERATURE AND EXPANSION

As a result of completing this chapter, you should be able to

1. **Demonstrate your understanding of the Celsius, Fahrenheit, kelvin, and Rankine temperature scales by converting from specific temperatures on one scale to corresponding temperatures on another scale.**
2. **Distinguish between specific temperatures and temperature intervals and convert an interval on one scale to the equivalent interval on another scale.**
3. **Write formulas for linear expansion, area expansion, and volume expansion and be able to apply them to the solution of problems similar to those given in this chapter.**

The concept of temperature is familiar to us as an indication of the sensation of ''hot'' or ''cold.'' However, in industry, we need to have a more precise definition, one that we can measure or use in a practical way. The SI unit for temperature is the *kelvin* (K), but you will also work with three other temperature scales. For example, room temperature may be given as 293 K, 20°C, 68°F, or 528°R. In this chapter, you will become familiar with each of these units. You will also be able to predict the changes in the dimensions of an object as a result of changes in temperature. Anyone who has worked with materials which are exposed to changing temperatures can appreciate the need for such information. Construction engineers must constantly make allowances for expansion through the use of expansion joints or rollers. Mechanics will sometimes preheat nuts so that they will form a tight fit on the bolt after cooling. Some liquids will overflow their containers unless allowances are made for the greater expansion of the liquid. An understanding of temperature and expansion is important for many technical applications.

13-1 ■ TEMPERATURE AND THERMAL ENERGY

Until now we have been concerned primarily with the causes and effects of external motion. A brick resting on the floor is in translational and rotational equilibrium. However, a much closer study reveals that the brick is active internally. Figure 13-1 shows a simplified model of a solid. Individual molecules are held together by elastic forces analogous to the springs in the figure. These molecules move back and forth about their equilibrium positions. Thus, there is both potential and kinetic energy associated with the molecular motion. Since this internal energy is related to the hotness or coldness of a body, it is often referred to as *thermal energy:*

Thermal energy *is the total energy of an object, i.e., the sum of its molecular kinetic and potential energies.*

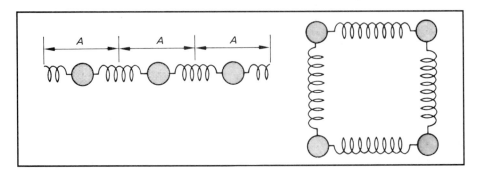

FIG. 13-1 A simplified model of a solid in which the individual molecules are held together by elastic forces.

When two objects with different temperatures are placed in contact, energy is transferred from one to the other. For example, suppose hot coals are dropped into a container of water, as shown in Fig. 13-2. Thermal energy will be transferred from the coals to the water until the system reaches a stable condition, called *thermal equilibrium.* When they are touched, the coals and the water produce similar sensations and there is no more transfer of thermal energy.

Such changes in thermal energy states cannot be satisfactorily explained in terms of classical mechanics alone. Therefore, all objects must have a new fundamental property which determines whether they will be in thermal equilibrium with other objects. This property is called *temperature.* In our example, the coals and the water are said to have the same temperature when the transfer of energy is zero.

Two objects are said to be in **thermal equilibrium** *if and only if they are at the same temperature.*

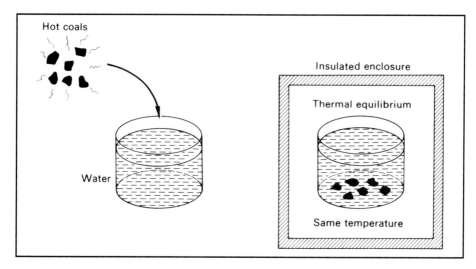

FIG. 13-2 Thermal equilibrium.

Once we establish a means of measuring temperature, we have a necessary and sufficient condition for thermal equilibrium. The transfer of thermal energy which is due only to a difference in temperature is defined as *heat:*

> **Heat** *is defined as the transfer of thermal energy which is due to a difference of temperature.*

13-2 ■ MEASUREMENT OF TEMPERATURE

Temperature is usually determined by measuring some mechanical, optical, or electrical quantity which varies with temperature. For example, most substances expand as their temperature increases. If a change in any dimension can be shown to have a one-to-one correspondence with changes in temperature, the variation can be calibrated to measure temperature. A device calibrated in this way is called a *thermometer*. The temperature of another object can then be measured by placing the thermometer in close contact with it and allowing the two to reach thermal equilibrium. The temperature indicated by a number on the graduated thermometer also corresponds to the temperature of the surrounding objects.

> A **thermometer** *is a device which, through marked scales, can give an indication of its own temperature.*

A thermometer is constructed by establishing standard temperatures that are easily duplicated. This is usually done by selecting lower and upper fixed points (temperatures). Two convenient temperatures are the *ice point* and the *steam point*.

> *The lower fixed point* **(ice point)** *is the temperature at which water and ice coexist in equilibrium at a pressure of 1 atm. The upper fixed point* **(steam point)** *is the temperature at which water and steam coexist in equilibrium at a pressure of 1 atm.*

The scale most widely used in industrial nations is the *Celsius scale*. The number 0 is assigned to the ice-point temperature, and the number 100 is assigned to the steam-point temperature. The 100 divisions between 0 and 100 represent temperatures between the ice point and the steam point. Each division or unit on the scale is called a *degree* (°). For example, we may write room temperature as 20°C (read "twenty degrees Celsius").

Another scale which is widely used in the United States is the *Fahrenheit scale*. Even though this unit is being gradually phased out, modern industrial measurements are often given in degrees Fahrenheit. This scale was based on different fixed temperatures and, therefore, results in a different set of numbers for corresponding temperatures on the Celsius scale. It will be noted that 0° and 100°C correspond to temperatures of 32 and 212°F, respectively.

We can compare the two scales by calibrating ordinary mercury-in-glass thermometers. This type of thermometer makes use of the fact that liquid mercury expands with increasing temperature. It consists of an evacuated glass capillary tube with a reservoir of mercury at the bottom and a closed top. Since the mercury expands more than the glass tube, the mercury column rises in the tube. The mercury will continue to rise until the mercury, the glass, and its surroundings are in thermal equilibrium.

Suppose we make two ungraduated thermometers and place them in a mixture of ice and water, as in Fig. 13-3. After allowing the mercury column to stabilize, we mark 0°C on one thermometer and 32°F on the other thermometer. Next, we place the same two thermometers directly above boiling water, allowing the mercury columns to stabilize at the steam point. Again we mark the two thermometers, inscribing 100 and 212°F next to the mercury levels. These marks will be directly above those made for the ice point. The level of mercury is the same in each thermometer. Thus the only difference between the two thermometers is how they are graduated. There are 100 divisions, or Celsius degrees (C°), between the ice point and steam point on the Celsius thermometer. There are 180 divisions, or Fahrenheit degrees (F°), on the Fahrenheit thermometer. Thus 100 Celsius degrees represent the same temperature interval as 180 Fahrenheit degrees. Symbolically,

$$100 \text{ C}° = 180 \text{ F}° \quad \text{ or } \quad 5 \text{ C}° = 9 \text{ F}° \tag{13-1}$$

In this text, the degree mark (°) is placed after the C or F to indicate that the numbers correspond to temperature *intervals* and not to *specific* temperatures. In other words, 20 C° is read "twenty Celsius degrees" and corresponds to a *difference* between two temperatures on the Celsius scale. The symbol 20°C, however, will refer to a specific mark on the Celsius thermometer. For example, suppose a heated piece of metal cools from 78 to 26°C. These numbers correspond to specific temperatures, as indicated by the height of a column of mercury. However, they represent a temperature interval of

$$\Delta t = 78°C - 26°C = 52 \text{ C}°$$

The symbol Δt is used to denote a change in temperature.

The physics which treats the transfer of thermal energy is nearly always concerned with changes in temperature. Thus it often becomes necessary to convert a temperature interval from one scale to the corresponding interval on the other scale. This can best be accomplished by recalling from Eq. (13-1) that an interval of 5 C° is equivalent to an interval of 9 F°. The appropriate conversion factors can be written as

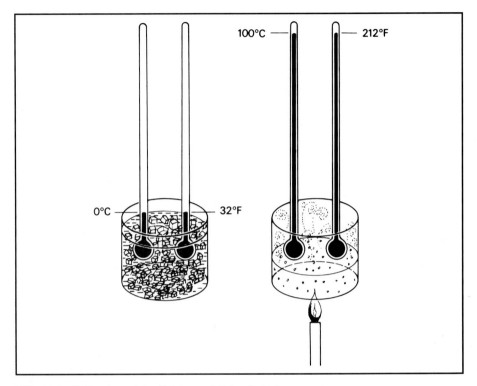

FIG. 13-3 Calibration of the Celsius and Fahrenheit thermometers.

$$\frac{5\ C^\circ}{9\ F^\circ} = 1 = \frac{9\ F^\circ}{5\ C^\circ} \tag{13-2}$$

When one is converting Fahrenheit to Celsius degrees, the factor on the left should be used; when one is converting Celsius to Fahrenheit degrees, the factor on the right should be used.

EXAMPLE 13-1 During a 24-h period, a steel rail varies in temperature from 20°F at night to 70°F in the middle of the day. Express this change in temperature in Celsius degrees.

Solution The temperature interval is

$$\Delta t = 70°F - 20°F = 50\ F°$$

In order to convert the interval from Fahrenheit to Celsius degrees, we choose the conversion factor which will cancel the Fahrenheit units. Thus

$$\Delta t = 50\ F° \times \frac{5\ C°}{9\ F°} = 27.8\ C°$$

Remember that Eq. (13-2) applies to temperature *intervals*. It can be used only when you are working with *differences* in temperatures. It is another matter entirely to find the temperature on the Fahrenheit scale which corresponds to the same temperature on the Celsius scale. Using the mathematics of ratio and proportion, we can derive the following relationships to convert a particular number on one scale to the corresponding number on the other scale:

$$t_C = \frac{5}{9}(t_F - 32) \tag{13-3}$$

$$t_F = \frac{9}{5}t_C + 32 \tag{13-4}$$

EXAMPLE 13-2   The melting point of lead is 330°C. What is the corresponding temperature on the Fahrenheit scale?

Solution   Substitution into Eq. (13-4) yields

$$t_F = \frac{9}{5}t_C + 32 = \frac{9}{5}(330) + 32$$

$$= 594 + 32 = 626°F$$

It is important to recognize that the t_F and t_C of Eqs. (13-3) and (13-4) represent identical temperatures. The numbers are different because the origin of each scale was at a different point and the degrees are of different size. What these equations tell us is the relationship between the *numbers* which are assigned to specific temperatures on two *different* scales.

13-3 ■ ABSOLUTE TEMPERATURE SCALE

It has probably occurred to you that the Celsius and Fahrenheit scales have a very serious limitation. Neither 0°C nor 0°F is a true zero of temperature. For temperatures much below the ice point, a negative temperature results. This certainly does not mean a lack or absence of temperature. Even more serious is the fact that a formula involving temperature as a variable will not work with the Celsius and Fahrenheit scales. For example, the volume of a gas is known to vary directly with temperature. If the temperature is doubled, the volume will also be doubled if the pressure is constant. This relationship can be expressed as

$$V = kt \tag{13-5}$$

where k is the proportionality constant and t is the temperature. Certainly the volume of a gas is not zero at 0°C or at 0°F. Neither will the volume be negative for negative temperatures. Yet, these conclusions might be drawn from Eq. (13-5).

This example provides a clue for establishing an *absolute scale*. If we can determine the temperature at which the volume of a gas under constant pressure becomes zero, we can establish a true zero of temperature. Suppose we measure the volume of a certain gas, first at the ice point and then at the steam point. These two points can be plotted on a graph, as in Fig.13-4, with the volume as the vertical axis and the temperature as the horizontal axis. The points A and B correspond to the gas volume at temperatures of 0 and 100°C, respectively. A straight line through these points, extended both to the left and to the right, provides a description of the change in volume as a function of temperature. Note that we can extend the line indefinitely to the right, indicating that there is no upper limit to temperature. However, we cannot extend the line indefinitely to the left because it will eventually cross the temperature axis. At this theoretical point, the gas would have zero volume. Further extension of the line would indicate a negative volume, which is meaningless. Therefore, the point at which the line intercepts the temperature axis is called the *absolute zero* of temperature. (Actually, any real gas would liquefy before reaching this point.)

If the preceding experiment is performed for several different gases, the slopes of the curves will vary slightly. But the temperature intercept will always be the same and near −273°C. Application of mathematics to controlled experiments has established that the absolute zero of temperature is −273.15°C. In this text, we shall assume that it is −273°C without fear of significant error. Conversion of this temperature to the Fahrenheit scale, by using Eq. (13-4), shows that absolute zero is −460°F:

$$\text{Absolute zero} = -273°\text{C} = -460°\text{F} \qquad (13\text{-}6)$$

An absolute temperature scale has as its zero point the absolute zero of temperature. One such scale was developed by Lord Kelvin (1824–1907). The standard interval on this scale, the *kelvin,* has been adopted by the international (SI) metric system as the base unit for the measurement of temperature. The interval on the *kelvin scale* represents the same change in temperature as the Celsius degree. Thus, an interval of 5 K (read "five kelvins") is exactly the same as a change in temperature of 5 C°. Since zero on the kelvin scale corresponds to −273°C, we may write

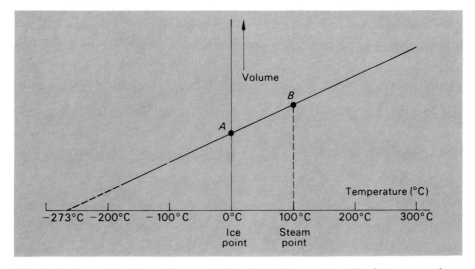

FIG. 13-4 Change in volume of a gas as a function of temperature. Absolute zero can be defined by extending the line until it reaches zero volume.

$$T_K = t_C + 273 \qquad (13\text{-}7)$$

For example, 0°C would correspond to 273 K, and 100°C would correspond to 373 K. (See Fig. 13-5). Hereafter we will use the symbol T for absolute temperature and the symbol t for other temperatures.

A second absolute scale which remains in use for some industrial applications is the *Rankine scale*. This scale has its zero point at -460°F. The degrees on this scale are identical in size with Fahrenheit degrees. The relationship between the temperature in degrees Rankine (°R) and the corresponding temperature in degrees Fahrenheit is

$$T_R = t_F + 460 \qquad (13\text{-}8)$$

For example, 0°F corresponds to 460°R, and 212°F corresponds to 672°R (see Fig. 13-5).

Remember that Eqs. (13-7) and (13-8) apply for specific temperatures. If we are concerned with a change in temperature or a difference in temperature, the absolute change or difference is the same in kelvins as it is in Celsius degrees. It is helpful to recall that

$$1 \text{ K} = 1 \text{ C}° \qquad 1 \text{ R}° = 1 \text{ F}° \qquad (13\text{-}9)$$

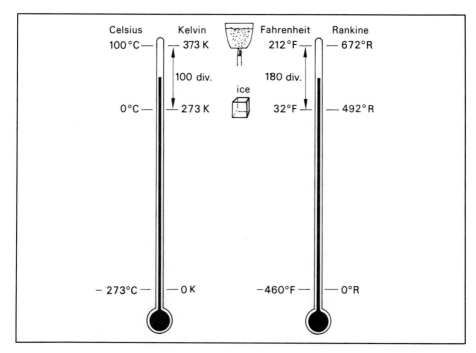

FIG. 13-5 A comparison of the four common temperature scales.

EXAMPLE 13-3 A mercury-in-glass thermometer may not be used at temperatures below −40°C. This is because mercury freezes at this temperature. (*a*) What is the freezing point of mercury on the kelvin scale? (*b*) What is the difference between this temperature and the freezing point of water? Express the answer in kelvins.

Solution (a) Direct substitution of −40°C into Eq. (13-7) yields

$$T_K = -40°C + 273 = 233 \text{ K}$$

Solution (b) The difference in the freezing points is

$$\Delta t = 0°C - (-40°C) = 40 \text{ C°}$$

Since the size of the kelvin is identical to that of the Celsius degree, the difference is also 40 K.

At this point you may ask why we still retain the Celsius and Fahrenheit scales. When working with heat, one is nearly always concerned with changes in temperature. In fact, there must be a change in temperature in order for heat to be transferred. Otherwise, the system would be in thermal equilibrium. Since the kelvin and Rankine scales are based on the same intervals as the Celsius and Fahrenheit scales, it makes no difference which scale is used for temperature intervals. However, if a formula calls for a specific temperature rather than a temperature difference, the absolute scale must be used.

13-4 ■ LINEAR EXPANSION

The most common effect produced by temperature changes is a change in size. With a few exceptions, all substances increase in size with rising temperature. The atoms in a solid are held together in a regular pattern by electric forces. At any temperature the atoms vibrate with a certain frequency and amplitude. As the temperature is increased, the amplitude (maximum displacement) of the atomic vibrations increases. This results in an overall change in the dimensions of the solid.

A change in any *one* dimension of a solid is called *linear expansion*. It is found experimentally that an increase in a single dimension, for example the length of a rod, is dependent on the original dimension and the change in temperature. Consider, for example, the rod in Fig. 13-6. The initial length is L_0, and the initial temperature is t_0. When the rod is heated to a temperature t, its new length will be L. Thus a change in temperature $\Delta t = t - t_0$ has resulted in a change in length $\Delta L = L - L_0$. The proportional change in length ΔL is given by

$$L - L_0 = \alpha L_0 (t - t_0)$$

or

$$\boxed{\Delta L = \alpha L_0 \Delta t} \tag{13-10}$$

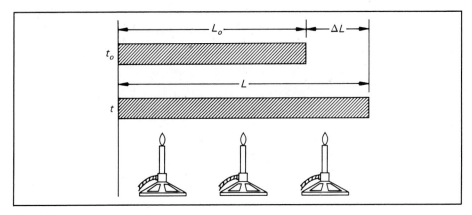

FIG. 13-6 Linear expansion.

where α is a proportionality constant called the *coefficient of linear expansion*. Since an increase in temperature does not produce the same change in length for all materials, the coefficient α is a property of the material. Solving Eq. (13-10) for α, we obtain

$$\alpha = \frac{\Delta L}{L_0 \, \Delta t} \qquad \text{\textit{Coefficient of} \\ \textit{Linear Expansion}} \quad (13\text{-}11)$$

The coefficient of linear expansion of a substance can be defined as the *change in length per unit length per degree change in temperature*. Since Eq. (13-11) involves a change in length per unit length ($\Delta L/L_0$), the units of α are in inverse degrees, that is $1/C°$ or $1/F°$. Expansion coefficients for many common materials are given in Table 13-1.

EXAMPLE 13-4 An iron pipe is 300 m long at room temperature (20°C). If the pipe is to be used as a steam pipe, how much allowance must be made for expansion, and what will be the new length of the pipe?

Solution The temperature of steam is 100°C, and α_{iron} ($1.2 \times 10^{-5}/C°$) is found from Table 13-1. The change in length is found by direct substitution into Eq. (13-10):

$$\Delta L = \alpha L_0 \, \Delta t = (1.2 \times 10^{-5}/C°)(300 \text{ m})(100°C - 20°C)$$
$$= (1.2 \times 10^{-5}/C°)(300 \text{ m}) \, (80 \text{ C°}) = 0.288 \text{ m}$$

The new length L is found by adding the increase ΔL to the original length L_0. Thus

$$L = L_0 + \Delta L = 300.29 \text{ m}$$

We can see from this example that the new length may be calculated by the following relation:

$$L = L_0 + \alpha L_0 \, \Delta t \qquad (13\text{-}12)$$

TABLE 13-1 LINEAR EXPANSION COEFFICIENTS

Substance	α	
	$10^{-5}/\text{C}°$	$10^{-5}/\text{F}°$
Aluminum	2.4	1.3
Brass	1.8	1.0
Concrete	0.7–1.2	0.4–0.7
Copper	1.7	0.94
Glass, Pyrex	0.3	0.17
Iron	1.2	0.66
Lead	3.0	1.7
Silver	2.0	1.1
Steel	1.2	0.66
Zinc	2.6	1.44

Remember, when you are calculating ΔL, that the units of α must be consistent with the units for Δt.

Linear expansion has both useful and destructive properties when applied to physical situations. The destructive effects require engineers to use expansion joints or rollers to make allowances for expansion. The predictable expansion of some materials, however, can be used to open or close switches at certain temperatures. Such devices are called *thermostats*.

Probably the most common application of the principle of linear expansion is the bimetallic strip. This device, shown in Fig. 13-7, consists of two flat strips of different metals welded or riveted together. The strips are fused so that they are the same length at a chosen temperature t_0. If we heat the strip, causing a rise in temperature, the material with the larger expansion coefficient will expand more. For example, a brass-iron strip will bend in an arc toward the iron side. When the source of heat is removed, the strip will gradually return to its original position. Cooling the strip below the initial temperature will cause the strip to bend in the other direction. This results because the material with the higher coefficient of expansion also *decreases* in length at a faster

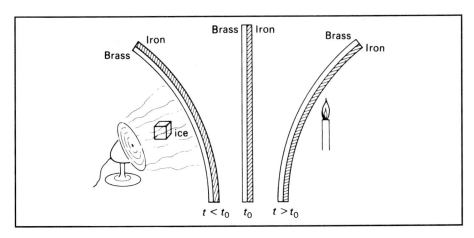

FIG. 13-7 Bimetallic strip.

rate. The bimetallic strip has many useful applications, from thermostatic control systems to blinking lights. Since the expansion is in direct proportion to an increase in temperature, the bimetallic strip can also be used as a thermometer.

13-5 ■ AREA EXPANSION

Linear expansion is by no means restricted to the length of a solid. Any line drawn through the solid will increase in length per unit length at the rate given by its expansion coefficient α. For example, in a solid cylinder the length, diameter, and even a diagonal drawn through the solid will all increase their dimensions in the same proportion. In fact, the expansion of a surface is exactly analogous to a photographic enlargement (see Fig. 13-8). Notice also that if the material contains a hole, the area of the hole expands at the same rate as it would if it were filled with material.

It can be shown that changes in area as a result of temperature changes can be predicted by an equation very similar to that derived for linear expansion. The change in area ΔA is proportional to the initial area A_0 and to the change in temperature. We may write

$$\Delta A = 2\alpha A_0 \, \Delta t \qquad (13\text{-}13)$$

Notice that the coefficient of area expansion is twice the coefficient of linear expansion α. If we label the coefficient of area expansion by γ, we have

$$\gamma = 2\alpha \qquad (13\text{-}14)$$

where γ (gamma) is the change in area per unit initial area per degree change in temperature. Using this definition, we may write the following formulas for area expansion:

$$\Delta A = \gamma A_0 \, \Delta t \qquad (13\text{-}15)$$
$$A = A_0 + \gamma A_0 \, \Delta t \qquad (13\text{-}16)$$

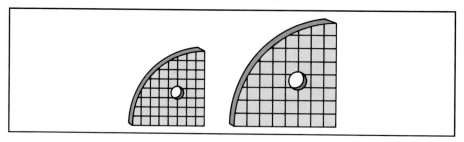

FIG. 13-8 Thermal expansion is similar to a photographic enlargement. Note that the hole gets larger in the same proportion as the material.

EXAMPLE 13-5 A brass disk has a hole 80 mm in diameter punched in its center at 20°C. If the disk is placed in boiling water, what will be the new area of the hole?

Solution We first compute the area of the hole at 20°C:

$$A_0 = \frac{\pi D^2}{4} = \frac{\pi(80 \text{ mm})^2}{4} = 5027 \text{ mm}^2$$

Now the area expansion coefficient is

$$\gamma = 2\alpha = (2)(1.8 \times 10^{-5}/\text{C}°) = 3.6 \times 10^{-5}/\text{C}°$$

The increase in the area of the hole is found from Eq. (13-15) as follows:

$$\Delta A = \gamma A_0 \, \Delta t$$
$$= (3.6 \times 10^{-5}/\text{C}°)(5027 \text{ mm}^2)(100°\text{C} - 20°\text{C})$$
$$= 14.5 \text{ mm}^2$$

The new area is found by adding this increase to the original area [see Eq. (13-16)]:

$$A = A_0 + \Delta A$$
$$= 5027 \text{ mm}^2 + 14.5 \text{ mm}^2 = 5041.5 \text{ mm}^2$$

13-6 ■ VOLUME EXPANSION

The expansion of heated material is the same in all directions. Therefore, the volume of a liquid, gas, or solid will have a predictable increase in volume with a rise in temperature. Reasoning similar to that of the previous sections will give us the following formulas for volume expansion:

$$\Delta V = \beta V_0 \, \Delta t \qquad (13\text{-}17)$$
$$V = V_0 + \beta V_0 \, \Delta t \qquad (13\text{-}18)$$

The symbol β (beta) is the *volume expansion coefficient*. It represents the *change in volume per unit volume per degree change in temperature*. For solid materials it is approximately 3 times the linear expansion coefficient:

$$\beta = 3\alpha \qquad (13\text{-}19)$$

When working with solids, we can compute β from the table of linear expansion coefficients (Table 13-1). For different liquids, the volume expansion coefficients are listed in Table 13-2. The molecular separation in gases is so great that they all expand at approximately the same rate. Volumetric expansion of gases will be discussed later.

TABLE 13-2 VOLUME EXPANSION COEFFICIENTS

Liquid	β	
	$10^{-4}/C°$	$10^{-4}/F°$
Alcohol, ethyl	11	6.1
Benzene	12.4	6.9
Glycerin	5.1	2.8
Mercury	1.8	1.0
Water	2.1	1.2

EXAMPLE 13-6   A Pyrex glass bulb is filled with 50 cm³ of mercury at 20°C. What volume will overflow if the system is heated uniformly to a temperature of 60°C? Refer to Fig. 13-9.

Solution   The inside volume of the glass bulb is 50 cm³ initially and will increase according to Eq. (13-17). Remember that $\beta_G = 3\alpha_G$. At the same time, the mercury will increase in volume according to the value of β_M. The overflow will, therefore, be the difference between the two expansions.

$$\begin{array}{ccc} \text{Volume} \\ \text{overflow} \end{array} = \begin{array}{c} \text{volume increase} \\ \text{in mercury} \end{array} - \begin{array}{c} \text{volume increase} \\ \text{in glass} \end{array}$$

$$V_{\text{overflow}} = \Delta V_M - \Delta V_G$$
$$= \beta_M V_M\, \Delta t - \beta_G V_G\, \Delta t$$

We will compute the volume increases separately:

$$\Delta V_M = \beta_M V_M\, \Delta t = (1.8 \times 10^{-4}/C°)(50\text{ cm}^3)(40\text{ C}°) = 0.36\text{ cm}^3$$
$$\Delta V_G = 3\alpha_G V_G\, \Delta t = 3(0.3 \times 10^{-5}/C°)(50\text{ cm}^3)(40\text{ C}°) = 0.018\text{ cm}^3$$

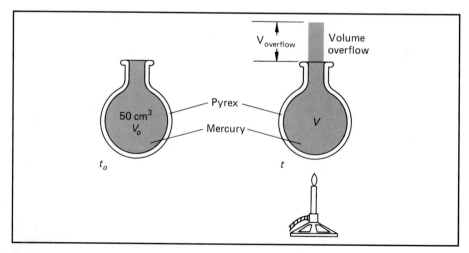

FIG. 13-9 The volume overflow is found by subtracting the change in volume of the glass from the change in volume of the liquid.

Thus the volume overflow is

$$V_{\text{overflow}} = \Delta V_M - \Delta V_G = 0.36 \text{ cm}^3 - 0.018 \text{ cm}^3$$
$$= 0.342 \text{ cm}^3$$

Summary

We have seen that because of the existence of four commonly used temperature scales, temperature conversions are important. You have also studied one very important effect of changes in temperature of materials, a change in the physical dimensions. The major concepts are summarized below:

- There are four temperature scales that you should be thoroughly familiar with. These scales are compared in Fig. 13-5, giving values for the steam point, the ice point, and absolute zero on each scale. It is very important for you to distinguish between a temperature interval Δt and a specific temperature t. For temperature intervals:

$$\frac{5 \text{ C}°}{9 \text{ F}°} = 1 = \frac{9 \text{ F}°}{5 \text{ C}°} \qquad 1 \text{ K} = 1 \text{ C}° \qquad 1 \text{ R}° = 1 \text{ F}° \qquad \textit{Temperature Intervals}$$

- For specific temperatures you must correct for the interval difference, but you must also correct for the fact that different numbers are assigned for the same temperatures:

$$t_C = \frac{5}{9}(t_F - 32) \qquad t_F = \frac{9}{5}t_C + 32 \qquad \textit{Specific Temperatures}$$

$$T_K = t_C + 273 \qquad T_R = t_F + 460 \qquad \textit{Absolute Temperatures}$$

- The following relations apply for thermal expansion of solids:

$$\Delta L = \alpha L_0 \Delta t \qquad L = L_0 + \alpha L_0 \Delta t \qquad \textit{Linear Expansion}$$

$$\Delta A = \gamma A_0 \Delta t \qquad A = A_0 + \gamma A_0 \Delta t \qquad \gamma = 2\alpha \qquad \textit{Area Expansion}$$

$$\Delta V = \beta V_0 \Delta t \qquad V = V_0 + \beta V_0 \Delta t \qquad \beta = 3\alpha \qquad \textit{Volume Expansion}$$

- The volume expansion of a liquid uses the same relation as for a solid except, of course, that there is no linear expansion coefficient α for a liquid. Only β is needed.

Questions

13-1. Define the following terms:
  a. thermal energy
  b. temperature   **h.** Fahrenheit scale
  c. thermal equilibrium   **i.** absolute zero
  d. thermometer   **j.** kelvin scale
  e. ice point   **k.** Rankine scale
  f. steam point   **l.** coefficient of linear expansion
  g. Celsius scale   **m.** heat
13-2. Two lumps of hot iron ore are dropped into a container of water. The system is insulated and allowed to reach thermal equilibrium. Is it necessarily true that the iron ore and the water have the same thermal energy? Is it necessarily true that they have the same temperature? Discuss.
13-3. Distinguish clearly between thermal energy and temperature.
13-4. If a flame is placed underneath a mercury-in-glass thermometer, the mercury column first drops and then rises. Explain.
13-5. What factors must be considered in the design of a sensitive thermometer?
13-6. Given an unmarked thermometer, how would you proceed to graduate it in Celsius degrees?
13-7. A 6-in. ruler expands 0.0014 in. when the temperature is increased 1 C°. How much would a 6-cm ruler expand during the same temperature interval if it were made of the same material?
13-8. A brass rod connects the opposite sides of a brass ring. If the system is heated uniformly, will the ring remain circular?
13-9. A brass nut is used with a steel bolt. How is the closeness of fit affected if the bolt alone is heated? If the nut alone is heated? If they are both heated equally?
13-10. An aluminum cap is screwed tightly to the top of a pickle jar at room temperature. After the jar has been stored in the refrigerator for a day or two, the cap cannot easily be removed. Explain. Suggest a way to remove the cap with very little effort. How might the manufacturer solve this problem?
13-11. May we use Eqs. (13-3) and (13-4) to arrive at similar formulas for converting between specific temperatures on the kelvin and Rankine scales? If so, explain. If not, what equation will work for such conversions?

Problems

13-1. On a cold morning the temperature is reported as 23°F. What is the corresponding Celsius temperature? During the day, the temperature increases by 20 C°. What is the new Fahrenheit temperature?

Answer −5°C; 59°F.

13-2. A steel rod is removed from a furnace at 450°C. What is the corresponding Fahrenheit temperature? If the rod cools by 200 F°, what is the new Celsius temperature?
13-3. Convert the following Celsius temperatures to the corresponding Fahrenheit temperatures: 350, 78, 15, −60°C.

Answer 662°F; 172°F; 59°F; −76°F.

13-4. Convert the following Fahrenheit temperatures to the corresponding Celsius temperatures: 72, 39, −60, 462°F.

13-5. A neutral oxyacetylene welding flame will produce a temperature of 5600°F. What is this temperature in degrees Celsius and in kelvins?

Answer 3093°C; 3366 K.

13-6. Fill in the blank spaces with the corresponding temperatures on each scale:

	Kelvin	Celsius	Rankine	Fahrenheit
(a)	86 K			
(b)		31°C		
(c)			580°R	
(d)				−200°F

13-7. Acetone boils at 56.5°C. Liquid nitrogen boils at −196°C. Express the difference in these temperatures in Celsius degrees, in Fahrenheit degrees, and in kelvins.

Answer 252.5 C°; 454.5 F°; 454.5 R°.

13-8. A wall of fire brick has an inside temperature of 313°F and an outside temperature of 73°F. What is the difference in the surface temperatures in Celsius degrees? In kelvins?

13-9. A piece of copper alloy is removed from a furnace at 200°C and cooled to 20°C. Express the change of temperature in Fahrenheit degrees and in kelvins.

Answer 324 F°; 180 K.

13-10. Prove that the Celsius and Fahrenheit scales have the same numerical value at −40°.

13-11. A piece of copper tubing is 6 m long at 20°C. How much will it increase in length if it is heated to 80°C? What is its new length?

Answer 6.12 mm; 6.00612 m.

13-12. A silver bar is 14 in. long at 70°F. What is its length when it is placed into boiling water (212°F)?

13-13. A steel tape measures the length of an aluminum bar as 6.000 m at 20°C. What will be the reading at 80°C? Assume the tape reads correctly at 20°C, and account for the changes in length of both materials. The expansion coefficients of steel and aluminum are $1.2 \times 10^{-5}/C°$ and $2.4 \times 10^{-5}/C°$, respectively.

Answer 6.0043 m.

13-14. Steel rails 40 ft long are laid when the temperature is 38°F. If the maximum expected temperature is 128°F, what gap should be allowed between the rails? Express your answer in inches.

13-15. An automobile engine has a valve pushrod that is exactly 6 in. long at 70°F. The pushrod is made of steel, and the engine must operate at temperatures as high as 200°F. What must be the minimum clearance between the rod and the rocker arm at 70°F?

Answer 0.00515 in.

13-16. The diameter of a hole in a steel plate is 90 mm when the temperature is 20°C. What is the new diameter when the plate is heated to a temperature of 200°C?

13-17. A brass pipe is 12 ft long at 90°F. What is its new length if the pipe is cooled to 10°F?

Answer 11.9904 ft.

13-18. The laboratory apparatus for measuring the coefficient of linear expansion is illustrated in Fig. 13-10. The temperature of a metal rod is increased by passing steam through an enclosed jacket. The resulting increase in length is measured with the micrometer screw at one end. Since the original length and temperature are

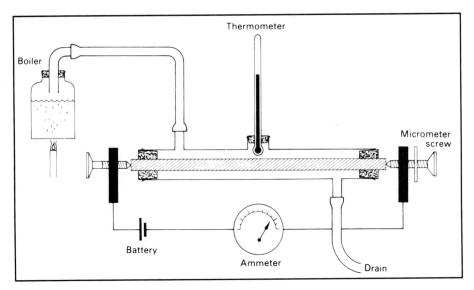

FIG. 13-10 Apparatus for measuring the coefficient of linear expansion.

known, we can calculate the expansion coefficient from Eq. (13-11). The following data were recorded during an experiment with a rod of unknown metal:

$$L_0 = 600 \text{ mm} \qquad t_0 = 23°C$$

$$\Delta L = 1.04 \text{ mm} \qquad t_f = 98°C$$

What is the coefficient of linear expansion α? What do you think the metal is, according to Table 13-1?

13-19. The area of a sheet of Pyrex glass is 120 mm² at 20°C. What is the increase in area if the temperature rises to 70°C?

Answer 0.036 mm².

13-20. A rectangular sheet of aluminum measures 18 in. by 24 in. when its temperature is 70°F. What is its new area if the temperature rises to 200°F?

13-21. A round brass plug has a diameter of 80.01 mm at 28°C. To what temperature must the plug be cooled if it is to fit snugly into a hole with a diameter of 80.00 mm?

Answer 21.2°C.

13-22. What is the increase in volume of 16 L of ethyl alcohol when its temperature is changed from 20 to 50°C?

13-23. A copper cup will hold 20.00 in.³ of a liquid when its temperature is 40°F. What volume of liquid will it hold at 200°F?

Answer 20.0902 in.³.

13-24. A 100-gal steel tank is filled with gasoline ($\beta = 0.0006/F°$) at 40°F. If the temperature rises to 120°F, how much gasoline will overflow?

13-25. A Pyrex glass beaker is filled to the top with 200 cm³ of mercury at 22°C. How much mercury overflows at a temperature of 78°C?

Answer 1.915 cm³.

13-26. If 300 cm³ of petroleum ($\beta = 0.0009/C°$) exactly fills an aluminum cup at 40°C and the system is cooled to 10°C, how much additional petroleum may be added to the cup without overflowing?

14

QUANTITY OF HEAT AND HEAT TRANSFER

OBJECTIVES

As a result of completing this chapter, you should be able to

1. Define quantity of heat in terms of the *calorie*, the *kilocalorie*, the *joule*, and the *British thermal unit* (Btu).
2. Write a formula for the *specific heat capacity* of a material and apply it to the solution of problems involving the loss or gain of heat.
3. Write formulas for calculating the *latent heats of fusion* and *vaporization* and apply them to the solution of problems in which heat produces a change in the phase of a substance.
4. Demonstrate by example and definition your understanding of heat transfer by conduction, convection, and radiation.

We define *heat transfer* as the process by which thermal energy produces some physical change, such as the melting of ice or the change in temperature of a substance. The concept of heat as a form of energy has a multitude of industrial applications. The automotive technician is concerned with the amount of heat released by a fuel mixture in a combustion chamber. The process of molding steel into useful shapes requires an understanding of how heat may be used to change the physical properties of a material. Hot-water heaters, industrial ovens, refrigeration devices, and oil furnaces are other applications concerned with the absorption, release, or transfer of heat. In this chapter, you will learn how to measure the quantity of heat, and you will study how it is transferred from one place to another.

14-1 ■ QUANTITY OF HEAT

We have defined *thermal energy* as the total internal energy of an object, that is, the sum of the potential and kinetic energies of the molecules. However, it is not possible to measure the position and velocity of every molecule in order to determine the thermal energy. We can measure the *change* in thermal energy by relating it to a change of temperature. When a flame is placed under a piece of metal, energy is transferred from the flame to the metal, causing the temperature of the metal to increase. Such transfers of energy are referred to as *heat*. Heat is not something that a body *has* but something that it *gives up* or *absorbs*. It is simply another form of energy that can be measured only in terms of the effect it produces. The unit of heat is established as the thermal energy required to produce some standard change. Three common units, which are determined from heating a known mass of water, are the *calorie,* the *kilocalorie,* and the *British thermal unit:*

> One **calorie** *(cal) is the quantity of heat required to change the temperature of one gram of water through one Celsius degree.*

> One **kilocalorie** *(kcal) is the quantity of heat required to change the temperature of one kilogram of water through one Celsius degree. (1 kcal = 1000 cal.)*

> One **British thermal unit** *(Btu) is the quantity of heat required to change the temperature of water having a mass of one standard pound ($^1/_{32}$ slug) through one Fahrenheit degree.*

Since heat is a form of energy, there is no real reason to separate heat units from mechanical units for energy. The SI unit for heat is, therefore, the *joule* (J). In modern times the mechanical equivalent of heat has been accurately determined. It is often measured by determining the electric energy necessary to raise the temperature of water through one degree. The accepted results are

$$1 \text{ cal} = 4.186 \text{ J} \qquad 1 \text{ kcal} = 4186 \text{ J}$$
$$1 \text{ Btu} = 1055 \text{ J} \qquad 1 \text{ Btu} = 778 \text{ ft·lb}$$

Other useful relationships between the heat units are

$$1 \text{ Btu} = 252 \text{ cal} \qquad 1 \text{ kcal} = 3.968 \text{ Btu}$$

The British thermal unit is being phased out, but it continues to be widely used in the United States. The pound unit (lb_m) which appears in the definition of the Btu must be considered as the *mass of the standard pound*. This is a departure from the USCS units, in which the unit *pound* was reserved for weight. Thus when we refer to 1 lb$_m$ of water, we mean the mass of water equal to about $\frac{1}{32}$ slug. This distinction is necessary because the pound of water must represent a constant mass of water independent of location. By definition, the pound-mass is related to the gram and kilogram as follows:

$$1 \text{ lb}_m = 454 \text{ g} = 0.454 \text{ kg}$$

Now that we have defined units for the measurement of heat, the distinction between quantity of heat and temperature should be clear. For example, suppose we pour 1 kg of water into one container and 4 kg of water into another container, as shown in Fig. 14-1. The initial temperature of the water in each container is 20°C. Each container is heated for the same length of time, delivering 20 kcal of heat to the water in each container. The temperature of the 1-kg sample increases by 20 C°, but the temperature of the 4-kg sample increases by only 5 C°. Yet the same quantity of heat was supplied to each sample.

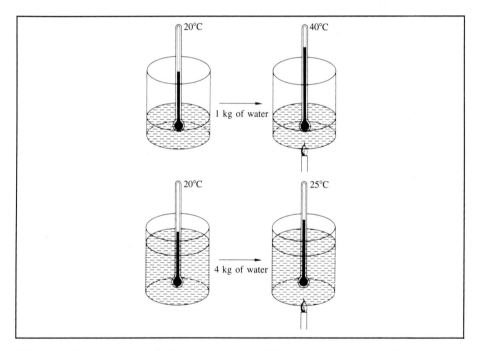

FIG. 14-1 The same quantity of heat is applied to different masses of water. The larger mass experiences a smaller rise in temperature.

14-2 ■ SPECIFIC-HEAT CAPACITY

We have defined a quantity of heat as the thermal energy required to raise the temperature of a given mass. But the amount of thermal energy needed to raise the temperature of a substance varies for different materials. For example, suppose we consider the four metal blocks of Fig. 14-2. Each block is constructed so that it has the same surface area on the bottom and the same mass (1 kg). Because of the different densities, the heights of the blocks vary, but the masses and hence the weights are identical. The quantity of heat required to raise the temperature of the aluminum block from room temperature (20°C) to 100°C is 17.6 kcal. But the heat required to accomplish the same change of temperature varies for each of the other blocks, as labeled on the figure. Aluminum is a more efficient absorber of heat than iron, copper, or lead.

Since the aluminum and iron blocks absorbed more heat than the copper and lead blocks, we might expect that they would release more heat on cooling. That this is true can be seen by placing each of the blocks (at 100°C) on a block of ice, as shown in Fig. 14-2b. The aluminum and iron blocks melt more ice and thus sink deeper than the other blocks. (Note that nothing was said about the rate at which the heat was released. The

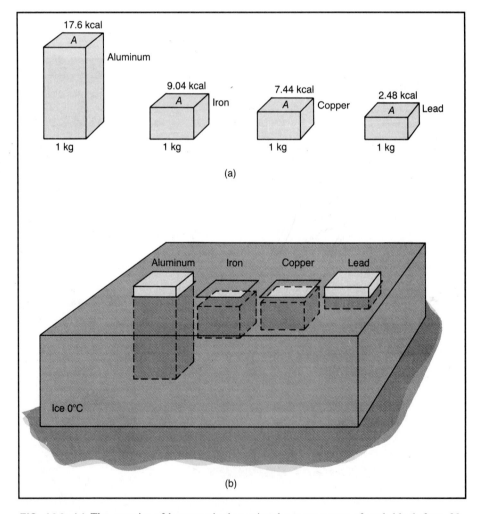

FIG. 14-2 (a) The quantity of heat required to raise the temperature of each block from 20 to 100°C varies with the material. (b) Each material of same mass, temperature, and cross section will sink to a different depth due to different specific heats.

number of calories absorbed or released in a unit of time is determined by another property of materials which will be discussed later.)

Clearly, there must be some property of materials that accounts for the differences observed in Fig. 14-2. This property must be a measure of the quantity of heat Q required to change the temperature of an object by an interval Δt. But it also should be related to the mass m of the object. We call this property the *specific-heat capacity,* denoted c:

> The **specific-heat capacity** of a material is the quantity of heat needed to raise the temperature of a unit mass by one degree.

The formula based on this definition may be written in the following useful forms:

$$c = \frac{Q}{m\,\Delta t} \qquad Q = mc\,\Delta t \qquad\qquad (14\text{-}1)$$

The SI unit of specific heat assigns the joule for heat, the kilogram for mass, and the kelvin for temperature. Thus c has units of joules per kilogram per kelvin J/(kg·K). In industry most temperature measurements are made in Celsius or Fahrenheit degrees, and the calorie and the Btu are still the dominant units for the quantity of heat. For the immediate future, this text will emphasize the following units for specific heat in the metric system and USCS: cal/g·C° and Btu/lb$_m$·F°.

We defined the calorie as the heat required to change the temperature of one gram of water by one Celsius degree. Therefore, the specific heat of water is equal to 1 cal/g·C° (or 1 kcal/kg·C°) by definition. Applying similar reasoning to the definition of a Btu will show that the specific heat of water is also 1 Btu/lb$_m$·F°. An important consequence of these definitions is that the specific heat of any object is numerically the same in either system of units. The values of c for some common substances are given in Table 14-1.

TABLE 14-1 SPECIFIC-HEAT CAPACITIES

Substance	c, cal/g·C° or Btu/lb$_m$·F°
Aluminum	0.22
Brass	0.094
Copper	0.093
Ethyl alcohol	0.60
Glass	0.20
Gold	0.03
Ice	0.50
Iron	0.113
Lead	0.031
Mercury	0.033
Silver	0.056
Steam	0.480
Steel	0.114
Turpentine	0.42
Zinc	0.092

EXAMPLE 14-1   What heat is required to raise the temperature of a 200-kg steel ingot from 25 to 450°C?

Solution   Since the values for c in Table 14-1 are in cal/g·C°, the mass must be expressed in grams. Substitution into Eq. (14-1) yields

$$Q = mc \; \Delta t = (200,000 \text{ g})(0.114 \text{ cal/g·C°})(450°C - 25°C)$$
$$= 9.69 \times 10^6 \text{ cal}$$

14-3 ■ CALORIMETERS

Heat measurements are often made by applying the law of conservation of energy to an insulated system. The principle of thermal equilibrium tells us that whenever objects are placed together in an insulated enclosure, they will eventually reach the same temperature. If energy is conserved, the heat lost by the warm bodies will be equal to the heat gained by the cool bodies. That is,

$$\boxed{\text{Heat lost} = \text{heat gained}} \qquad (14\text{-}2)$$

This equation expresses the net heat transfer within a system.

A laboratory device for measuring heat losses or gains is the calorimeter, which is illustrated in Fig. 14-3. It consists of a thin metallic cup held centrally within an outer jacket by a rubber gasket. The following example demonstrates its use.

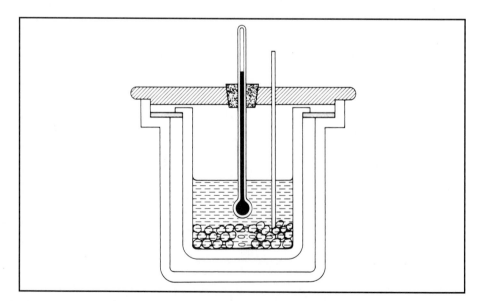

FIG. 14-3 A calorimeter is a laboratory device used to determine the specific-heat capacity of a substance.

EXAMPLE 14-2   In a laboratory experiment it is desired to use a calorimeter to find the specific heat of iron. Eighty grams of dry iron shot is placed in a cup and heated to a temperature of 95°C. The mass of the inner aluminum cup and of the aluminum stirrer is 60 g. The calorimeter is partially filled with 150 g of water at 18°C. The hot shot is quickly poured into the cup, and the calorimeter is sealed, as shown in Fig. 14-3. After the system has reached thermal equilibrium, the final temperature is 22°C. Compute the specific heat of iron.

Solution   The heat lost by the iron shot must equal the heat gained by the water plus the heat gained by the aluminum cup and stirrer. We can assume that the initial temperature of the cup is the same as that of the water and stirrer (18°C). We will calculate the heat gained by the water and by the aluminum separately:

$$Q_{\text{water}} = mc\,\Delta t = (150 \text{ g})(1 \text{ cal/g·C}°)(22°C - 18°C)$$
$$= (150 \text{ g})(1 \text{ cal/g·C}°)(4 \text{ C}°) = 600 \text{ cal}$$
$$Q_{\text{Al}} = mc\,\Delta t = (60 \text{ g})(0.22 \text{ cal/g·C}°)(22°C - 18°C)$$
$$= (60 \text{ g})(0.22 \text{ cal/g·C}°)(4 \text{ C}°) = 52.8 \text{ cal}$$

Now the total heat gained is the sum of these values:

$$\text{Heat gained} = 600 \text{ cal} + 52.8 \text{ cal} = 652.8 \text{ cal}$$

This amount must equal the heat lost by the iron shot:

$$\text{Heat lost} = Q_s = mc_s\,\Delta t = (80 \text{ g})c_s\,(95°C - 22°C)$$

Setting the heat lost equal to the heat gained gives

$$(80 \text{ g})c_s\,(73 \text{ C}°) = 652.8 \text{ cal}$$

Solving for c_s, we obtain

$$c_s = \frac{652.8 \text{ cal}}{(80 \text{ g})(73 \text{ C}°)} = 0.11 \text{ cal/g·C}°$$

In this experiment the heat gained by the thermometer was neglected. In an actual experiment, the portion of the thermometer inside the calorimeter would absorb about the same amount of heat as an extra 0.5 g of water. This quantity, called the *water equivalent* of the thermometer, should be added to the mass of water in an accurate experiment.

14-4 ■ CHANGE OF PHASE

When a substance absorbs a given amount of heat, the speed of its molecules usually increases and its temperature rises. Depending on the specific heat, the rise in temperature is proportional to the heat supplied. But a curious thing happens when a solid melts or when a liquid boils. In these cases, the temperature remains constant until all the solid melts or until all the liquid boils.

To understand what happens to the supplied energy, let us consider a simple model, as illustrated in Fig. 14-4. Under the proper conditions all substances can exist in three *phases,* solid, liquid, or gas. In the solid phase, the molecules are held together in a rigid structure giving a definite shape and volume. As heat is supplied, the energies of the particles gradually increase and the temperature rises. Eventually, the kinetic energy becomes so great that some of the particles overcome the elastic forces that hold them in fixed positions. Their increased separation gives them the freedom of motion which we associate with the liquid phase. At this point, the energy absorbed is used to separate the molecules more than in the solid phase. The temperature does not increase during such a change of phase. The change of phase from a solid to a liquid is called *fusion.* The temperature at which this occurs is called the *melting point.*

The quantity of heat required to melt a unit mass of a substance at its melting point is called the *latent heat of fusion* for that substance.

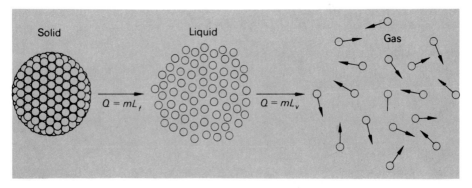

FIG. 14-4 A simplified model showing the relative molecular separation in the solid, liquid, and gaseous phases. During a change of phase, the temperature remains constant.

*The **latent heat of fusion** L_f of a substance is the heat per unit mass needed to change the substance from a solid to a liquid at its melting temperature:*

$$L_f = \frac{Q}{m} \qquad Q = mL_f \qquad\qquad (14\text{-}3)$$

The latent heat of fusion L_f is expressed in Btu per pound, calories per gram, or joules per kilogram. The heat of fusion for water is 80 cal/g, or 144 Btu/lb$_m$. This means that 1 g of ice absorbs 80 cal of heat in forming 1 g of water at 0°C.

After all the solid melts, the kinetic energy of the particles in the resulting liquid increases and the temperature rises. Eventually, the temperature levels off again as the thermal energy is used to change the substance from a liquid to a gas or vapor. The change in phase from a liquid to a vapor is called *vaporization,* and the temperature associated with this change is called the *boiling point* of the substance.

The **latent heat of vaporization** L_v *of a substance is the heat per unit mass required to change the substance from a liquid to a vapor at its boiling temperature:*

$$L_v = \frac{Q}{m} \qquad Q = mL_v \tag{14-4}$$

The heat of vaporization for water is 540 cal/g, or 970 Btu/lb$_m$. In other words, 1 g of water absorbs 540 cal of heat in forming 1 g of water vapor at 100°C. Values for L_f and L_v for many common substances are given in Table 14-2.

TABLE 14-2 HEAT OF FUSION AND HEAT OF VAPORIZATION FOR VARIOUS SUBSTANCES

Substance	Melting point, °C	Heat of fusion, cal/g	Boiling point, °C	Heat of vaporization cal/g
Alcohol, ethyl	−117.3	24.9	78.5	204
Aluminum	658	76.8	2057	
Ammonia	−75	108.1	−33.3	327
Copper	1080	42	2310	
Helium	−269.6	1.25	−268.9	5
Lead	327.3	5.86	1620	208
Mercury	−39	2.8	358	71
Oxygen	−218.8	3.3	−183	51
Silver	960.8	21	2193	558
Water	0	80	100	540
Zinc	420	24	918	475

In studying the changes of phase of a substance, it is often helpful to plot a graph showing how the temperature of the substance varies as heat is applied. Such a graph is shown in Fig. 14-5 for water. If a quantity of ice is taken from a freezer at −20°C and heated, its temperature will increase gradually until the ice begins to melt at 0°C. For each degree rise in temperature, each gram of ice will absorb 0.5 cal of heat. During the melting process, the temperature remains constant, and each gram of ice will absorb 80 cal of heat in forming 1 g of water.

Once all the ice has melted, the temperature begins to rise again at a uniform rate until the water begins to boil at 100°C. For each degree increase in temperature, each gram will absorb 1 cal of heat. During the vaporization process, the temperature remains constant. Each gram of water absorbs 540 cal of heat in forming 1 g of water vapor at 100°C. If the resulting water vapor is contained and the heating is continued until all the water is gone, the temperature of the water vapor will again start to rise. The specific heat of steam is 0.48 cal/g·C°.

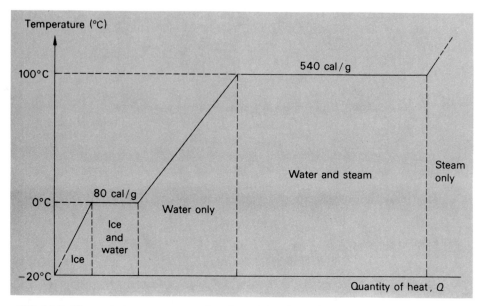

FIG. 14-5 Change in temperature as a function of the change in thermal energy for water.

EXAMPLE 14-3   What quantity of heat is needed to change 20 lb_m of ice at 12°F to steam at 212°F?

Solution   The heat needed to raise the temperature of the ice to its melting point is

$$Q_1 = mc \ \Delta t = (20 \ lb_m)(0.5 \ Btu/lb_m \cdot F°)(32°F - 12°F) = 200 \ Btu$$

The heat to melt the ice is given from Eq. (14-3):

$$Q_2 = mL_f = (20 \ lb_m)(144 \ Btu/lb_m) = 2880 \ Btu$$

The heat needed to raise the temperature of the resulting water is

$$Q_3 = mc \ \Delta t = (20 \ lb_m)(1 \ Btu/lb_m \cdot F°)(212°F - 32°F) = 3600 \ Btu$$

Finally, the heat to vaporize the water is given from Eq. (14-4):

$$Q_4 = mL_v = (20 \ lb_m)(970 \ Btu/lb_m) = 19,400 \ Btu$$

The total heat required is $Q_1 + Q_2 + Q_3 + Q_4$, or

$$Q = (200 + 2880 + 3600 + 19,400) \ Btu = 26,080 \ Btu$$

When heat is *removed* from a gas, its temperature drops until it reaches the temperature at which it boiled. As more heat is removed, the vapor returns to the liquid phase. This process is called *condensation*. In condensing, a vapor gives up an amount of heat equivalent to the heat required to vaporize it. Thus the *heat of condensation* is equivalent to L_v. The difference is only in the direction of heat transfer.

Similarly, when heat is removed from a liquid, its temperature will drop until it reaches its melting temperature. As more heat is removed, the liquid returns to its solid phase. This process is called *freezing*, or *solidification*. The heat of solidification is exactly equal to the heat of fusion. Thus the only distinction between freezing and melting lies in whether heat is being released or absorbed.

EXAMPLE 14-4   What energy (in calories) is required to raise the temperature of 3 kg of copper from 20°C to its melting point (1080°C) and completely melt the copper?

Solution   We first find the heat required to raise its temperature. Then we must add the heat needed to melt the copper.

$$Q = mc \, \Delta t + mL_f$$
$$= (3000 \text{ g})(0.093 \text{ cal/g·C°})(1080°C - 20°C) + (3000 \text{ g})(42 \text{ cal/g})$$
$$= 296,000 \text{ cal} + 126,000 \text{ cal} = 422,000 \text{ cal}$$

The heats of fusion and vaporization sometimes enter problems involving mixtures. In such cases, the quantity of heat absorbed must still equal the quantity of heat released. This principle holds even if a change of phase occurs.

14-5 ■ HEAT OF COMBUSTION

Whenever a substance is burned, it releases a definite quantity of heat. The quantity of heat per unit mass, or per unit volume, when the substance is completely burned is called the *heat of combustion*. Commonly used units are Btu per pound-mass, Btu per cubic foot, calories per gram, and joules per kilogram. For example, the heat of combustion of coal is approximately 13,000 Btu/lb$_m$. This means that each pound-mass of coal, when completely burned, should release 13,000 Btu of heat. The heat of combustion is given by

$$H_c = \frac{Q}{m} \qquad Q = mH_c \tag{14-5}$$

EXAMPLE 14-5   The heat of combustion of fuel oil is about 15,000 Btu/lb$_m$. It is determined that 55,000 Btu is needed to raise the temperature of water in a 50-gal hot-water tank to a temperature of 180°F. How much fuel oil must be burned to accomplish this task, assuming that all the energy is absorbed by the water?

Solution   The quantity of heat Q required is equal to the mass m of burned fuel oil times the heat of combustion:

$$Q = mH_c$$
$$55,000 \text{ Btu} = m(15,000 \text{ Btu/lb}_m)$$

Solving for the mass, we obtain

$$m = \frac{55,000 \text{ Btu}}{15,000 \text{ Btu/lb}_m} = 3.67 \text{ lb}_m$$

14-6 ■ HEAT TRANSFER

Heat is a form of energy in motion. Whenever there is a temperature difference between two bodies or between two portions of the same body, heat is said to *flow* in a direction from higher to lower temperature. There are three principal methods by which this transfer of heat occurs: conduction, convection, and radiation. Examples of all three are given in Fig. 14-6.

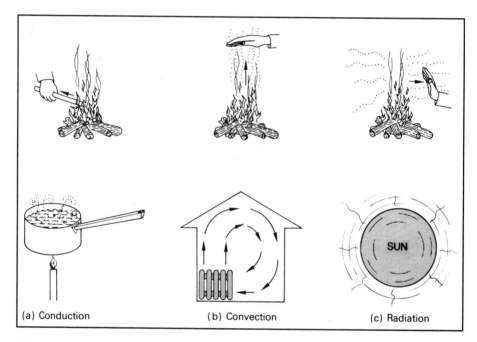

(a) Conduction   (b) Convection   (c) Radiation

FIG. 14-6

Conduction is heat transfer by molecular collisions between neighboring molecules. For example, if you hold one end of an iron rod in a fire, the heat will eventually reach your hand through the process of conduction. The increased molecular activity at the heated end is passed on from molecule to molecule until it reaches the hand. The process will continue as long as there is a difference in temperature along the rod.

> **Conduction** *is the process in which heat is transferred by molecular collisions through a material medium. The medium itself does not move.*

The rate at which heat is lost or gained by conduction is different for different materials. A metal object feels colder than a wooden object, even when both are at the same temperature. This is because the metal conducts heat from the hand faster than does the wood. Metals such as silver, copper, aluminum, and gold are excellent conductors of heat. Nonmetallic solids such as glass, wood, wallboard, asbestos, and Fiberglas are poor conductors (good insulators).

In the case of *convection,* heat is carried by the motion of a heated fluid. In Fig. 14-6*b*, heat is transferred to the hand by rising molecules of air above the fire. Convection differs from conduction in that the material medium is moving. Heat is transferred by moving masses instead of being passed along by neighboring molecules.

> **Convection** *is the process by which heat is transferred by the actual mass motion of a fluid.*

Convection currents form the basic means of heating and cooling most homes. If the motion of a fluid is caused by a difference in density which accompanies a change of temperature, the current produced is called *natural convection.* When a heated fluid is caused to move by the action of a pump or fan, the current produced is called *forced convection.* Many homes are heated by using fans to force hot air from a furnace to the rooms of the home. In the system illustrated by Fig. 14-7, forced convection currents circulate heated water and return it to the furnace. The room is heated by natural convection currents in the air. A solar heating system is constructed by heating the water in collectors attached to the roof of a house instead of using the conventional furnace.

When you hold your hand near the side of a fire or underneath a red-hot iron ball, the heat you feel is through *radiation.* Radiation is the emission or absorption of heat through electromagnetic waves originating at the atomic level. Such waves travel at the speed of light (186,000 mi/s) and require no material medium for their passage.

> **Radiation** *is the process by which heat is transferred by electromagnetic waves.*

The most obvious source of radiant energy is our sun. Neither conduction nor convection can occur in the space between the sun and the earth. The enormous amount of heat received on the earth is carried by electromagnetic waves. Where a material medium is involved, however, the transfer of heat by radiation is usually minor in comparison with other methods.

All objects are continuously emitting radiant energy. At low temperatures the rate of emission is small, but the rate increases rapidly with a rise in temperature. Experiments have shown that the rate of radiation *varies directly with the fourth power of the absolute temperature of a body.* Thus, if the *absolute* temperature of an object is doubled, the rate at which it emits thermal energy will be increased sixteenfold ($2^4 = 16$).

Another factor which affects the rate of emission or absorption of radiation is the nature of the radiant surface. Rough, dark surfaces emit or absorb radiation at a faster rate than smooth, light surfaces. For example, suppose a darkened glass ball and a silvered glass ball are heated to the same temperature and then removed from the source of heat. The temperature of the darkened ball will drop faster than will the temperature of the silvered ball. The choice of construction materials must reflect an understanding of emission and absorption of heat. A black shirt absorbs more of the sun's radiant energy than a lighter shirt. Therefore, its external temperature will be higher than your body temperature, making you uncomfortable.

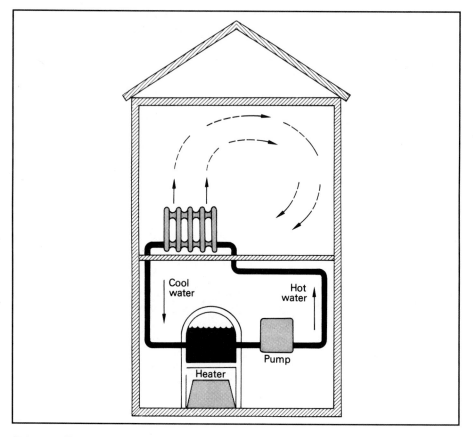

FIG. 14-7 Forced convection currents circulate the heated water and return it to the furnace. The room is heated by natural convection.

Summary

In this chapter you have studied the quantity of heat as a measurable quantity that is based on a standard change. The British thermal unit and the calorie are measures of the heat required to raise the temperature of a unit mass of water by a unit degree. By applying these standard units to experiments with a variety of materials, we have learned to predict heat losses or heat gains in a constructive fashion. The essential concepts presented in this chapter are as follows:

- The **calorie** is the heat required to raise the temperature of one gram of water by one Celsius degree.
- The **British thermal unit** (Btu) is the heat required to change the temperature of one pound-mass of water one Fahrenheit degree.
- Several conversion factors may be useful for problems involving thermal energy:

1 Btu = 252 cal = 0.252 kcal	1 cal = 4.186 J
1 Btu = 778 ft·lb	1 kcal = 4186 J

■ The **specific-heat capacity** c is used to determine the quantity of heat Q absorbed or released by a unit mass m as the temperature changes by an interval Δt:

$$c = \frac{Q}{m\,\Delta t} \qquad Q = mc\,\Delta t \qquad \textit{Specific-Heat Capacity}$$

■ Conservation of thermal energy requires that in any exchange of thermal energy the heat lost must equal the heat gained:

$$\text{Heat lost} = \text{heat gained} \qquad \Sigma(mc\,\Delta t)_{\text{loss}} = \Sigma(mc\,\Delta t)_{\text{gain}}$$

As an example, suppose body 1 transfers heat to bodies 2 and 3 as the system reaches an equilibrium temperature t_e:

$$m_1 c_1(t_1 - t_e) = m_2 c_2(t_e - t_2) + m_3 c_3(t_e - t_3)$$

■ The latent **heat of fusion** L_f and the latent **heat of vaporization** L_v are heat losses or gains by a unit mass m during a phase change. There is no change in temperature.

$$L_f = \frac{Q}{m} \qquad Q = mL_f \qquad \textit{Latent Heat of Fusion}$$

$$L_v = \frac{Q}{m} \qquad Q = mL_v \qquad \textit{Latent Heat of Vaporization}$$

If a change of phase occurs, the above relationships must be added to the calorimetry equation as appropriate.

■ The three principal methods of heat transfer are as follows:
Conduction: Heat transfer by molecular collision throughout a material medium.
Convection: Heat transfer by actual mass motion of a fluid.
Radiation: Heat transfer by electromagnetic waves.

Questions

14-1. Define the following terms:

- **a.** heat
- **b.** temperature
- **c.** calorie
- **d.** British thermal unit
- **e.** specific-heat capacity
- **f.** calorimeter
- **g.** heat of fusion
- **h.** heat of vaporization
- **i.** boiling point
- **j.** melting point
- **k.** condensation
- **l.** conduction
- **m.** convection
- **n.** radiation

14-2. Blocks of five different metals—aluminum, copper, zinc, iron, and lead—are constructed with the same mass and the same cross-sectional base area. Each block is heated to a temperature of 100°C and placed on a block of ice. Which will melt the ice to the greatest depth? List the remaining four in order of decreasing penetration depths.

14-3. Distinguish clearly between *temperature* and *heat*. Is temperature a measure of heat?

14-4. On a cold day you pick up a shovel which has been lying outdoors for a long time. Explain why the blade of the shovel feels colder to the touch than does the handle. Is the temperature of the blade likely to be the same as that of the handle?

14-5. A mechanical analogy to the concept of thermal equilibrium is given in Fig. 14-8. When the valve is opened, the water will flow until it has the same level in each container. What are the similarities to temperature and thermal energy?

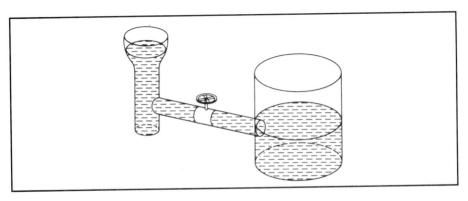

FIG. 14-8 A mechanical analogy to the equalization of temperature.

14-6. Discuss the change of phase from solid to liquid to vapor in terms of the molecular theory of matter.

14-7. In a mixture of ice and water, the temperature of both the ice and the water is 0°C. Why does the ice feel colder to the touch?

14-8. Why does steam at 100°C produce a far worse burn than water at the same temperature?

14-9. Discuss the vacuum bottle and explain how it minimizes the transfer of heat by conduction, convection, and radiation. What determines the direction of heat transfer?

14-10. A pan of water is placed over a gas burner on a kitchen stove until the water boils vigorously. Discuss the heat transfers that take place. How would you explain the fact that the bubbles forming in the water are carried to the surface in the form of an inverted cone instead of rising directly to the surface?

14-11. A physics instructor places a flame underneath a paper cup filled with water. Soon the water begins to boil without burning the paper bottom. The instructor then wraps a piece of paper tightly around a wooden stick and another piece of paper tightly around an iron rod. When a flame is placed under each, the paper around the wood burns and the other paper does not. Explain these demonstrations.

14-12. Copper conducts heat at a rate that is about twice that for aluminum under similar circumstances. Yet the specific heat of copper is a little less than half that of aluminum. A rectangular block is made of each material so that they have identical masses and the same surface area at their bases. Each block is heated to 300°C and placed on a large cube of ice. Which block will stop sinking first? Which will sink deeper into the ice?

Problems

14-1. How much heat in calories is required to raise the temperature of 200 g of lead from 20 to 100°C? How much heat is needed for the same quantity of iron?
Answer 496 cal; 4520 cal.

14-2. What quantity of heat will be released when 40 lb_m of copper cools from 78 to 32°F? How much heat would be released by the same quantity of aluminum?

14-3. A 4-lb_m copper sleeve must be heated from 70 to 250°F so that it will expand enough to slip over a shaft. How much heat is needed?
Answer 67.0 Btu.

14-4. In a heat-treating operation, a hot copper part is cooled quickly in water (quenched). If the temperature of the part drops from 400 to 30°C and loses 80 kcal of heat, what was the mass of the copper part?

14-5. A 60-g brass bolt cools to a temperature of 24°C while losing 400 cal of heat in the process. What was the initial temperature of the bolt?
Answer 94.9°C.

14-6. What mass of copper at 100°C must be added to 200 g of water at 20°C so that the equilibrium temperature will be 40°C? Neglect any other gains or losses of heat.

14-7. Three hundred grams of steel shot at 90°C is added to an unknown quantity of water initially at 20°C. What was the mass of the water if the equilibrium temperature is 30°C?
Answer 205 g.

14-8. Assume the same conditions for the calorimeter problem described in the text by Example 14-3 except that the steel shot is replaced with brass shot. What mass of brass shot at 95°C must be added to obtain the same equilibrium temperature (22°C)?

14-9. If 200 g of steel at 220°C is added to 500 g of water at 12°C, what will be the equilibrium temperature? Ignore other heat transfers.
Answer 20.1°C.

14-10. How much iron at 212°F must be mixed with 10 lb_m of water at 68°F in order to bring the equilibrium temperature to 100°F?

14-11. A worker who needs to know the temperature inside an oven removes a 2-lb_m iron bar from the oven and places it in a 1-lb_m aluminum container partially filled with 2 lb_m of water. The system is immediately insulated, and the temperature of the water and container rises from 70°F to the equilibrium temperature of 120°F. What was the oven temperature?
Answer 611°F.

14-12. A 450-g cylinder of lead is heated to 100°C and dropped into a 50-g copper calorimeter. The calorimeter contains 100 g of water initially at 10°C. Find the specific heat of lead if the equilibrium temperature of the mixture is 21.1°C.

14-13. How much heat is needed to completely melt 20 g of silver at its melting temperature?
Answer 420 cal.

14-14. A foundry has an electric furnace that can completely melt 1200 lb_m of copper. If the temperature of the copper was initially 70°F, how much total heat was required?

14-15. A heating element supplies heat at the rate of 20 kcal/min. How much time is required to completely melt a 3-kg block of aluminum?

Answer 11.5 min.

14-16. The mechanical output of a motor is 2 kW. This represents 80 percent of the input electric energy; the remainder is lost to heat. Express this heat loss in kilocalories per second.

14-17. What quantity of heat is required to convert 2 kg of ice at −25°C to steam at 100°C?

Answer 1465 kcal.

14-18. How much heat is released when 0.5 lb_m of steam at 212°F is converted to ice at 10°F?

14-19. In an experiment to determine the latent heat of vaporization of water, a student determines the mass of an aluminum calorimeter cup to be 50 g. After a quantity of water is added, the combined mass of the water and the cup is 120 g. The initial temperature of the water and the calorimeter is 18°C. A quantity of steam at 100°C is passed into the calorimeter, and the system is allowed to reach equilibrium. The equilibrium temperature is 47.4°C, and the total mass of the final mixture is 124 g. (*a*) What was the mass of water in the cup before the steam was introduced? (*b*) What mass of steam condensed? (*c*) What value will be obtained for the heat of vaporization?

Answer (*a*) 70 g; (*b*) 4 g; (*c*) 543 cal/g.

14-20. What will be the equilibrium temperature when 4 g of steam at 100°C is mixed with 20 g of ice at −5°C?

14-21. What equilibrium temperature is reached when 2 lb_m of ice at 0°F is dropped into 7.5 lb_m of water at 200°F? The water is contained in a 3-lb_m aluminum calorimeter cup.

Answer 135.4°F.

14-22. How many kilograms of coal must be burned to melt completely 100 kg of ice in a heater? Assume that all the heat is used and that the heat of combustion for coal is 7200 cal/g.

14-23. How much fuel oil (15,000 Btu/lb) is needed to raise the temperature of 120 lb_m of steel from 75 to 900°F?

Answer 0.746 lb_m.

15

GAS LAWS AND THERMODYNAMICS

As a result of completing this chapter, you should be able to

1. State and apply the relationship between the volume and the pressure of a gas at constant temperature (*Boyle's law*).
2. State and apply the relationship between the volume and the temperature of a gas under conditions of constant pressure (*Charles' law*).
3. State and apply the relationship between the temperature and the pressure of a gas under conditions of constant volume (*Gay-Lussac's law*).
4. Write and apply a general formula which considers changes in volume, pressure, and temperature for a given sample of a gas (*general gas law*).
5. State and give examples to illustrate your understanding of the first law of thermodynamics.
6. State and give examples to illustrate your understanding of the second law of thermodynamics.
7. Calculate the ideal efficiency of an engine when given input and output temperatures.
8. Describe the operation of a refrigerator and calculate the coefficient of performance.

One of the primary applications of heat is the transformation of thermal energy to work or the conversion of work to heat. Steam produced by boiling water is used to drive turbines in the generation of electricity. Gasoline-powered heat engines use the pressure from burned gases to drive pistons and to perform useful work. Refrigeration devices must remove large quantities of heat from food and other products. In each of these examples, it is necessary to understand the relationship between heat and work. Additionally, since the working element in many machines is a gas, you need to know the effects of changes in volume, temperature, and pressure. This chapter begins with a summary of useful gas laws and concludes with an introduction to thermodynamics.

15-1 ■ BOYLE'S LAW

The first experimental measurements of the thermal behavior of gases were made by Robert Boyle (1627–1691). He made an exhaustive study of the changes in the volume of gases as a result of changes in pressure. All other variables, such as mass and temperature, were kept constant. In 1660, Boyle demonstrated that the volume of a gas is inversely proportional to its pressure. In other words, doubling the volume *decreases* the pressure to one-half its original value. This finding is now known as *Boyle's law:*

> **Boyle's law:** *Provided that the mass and temperature of a sample of gas are held constant, the volume of the gas is inversely proportional to its absolute pressure.*

Another way of stating Boyle's law is to say that the product of the pressure P of a gas and its volume V will be constant as long as the temperature does not change. Consider, for example, a closed cylinder equipped with a movable piston, as shown in Fig. 15-1. In Fig. 15-1a, the initial state of the gas is described by its pressure P_1 and its volume V_1. If the piston is pressed downward until it reaches the new position shown in Fig. 15-1b, its pressure will increase to P_2 while its volume decreases to V_2. If the process occurs without a change in temperature, Boyle's law reveals that

$$P_1V_1 = P_2V_2$$

With Constant m and T (15-1)

In other words, the product of pressure and volume in the initial state is equal to the product of pressure and volume in the final state. Equation (15-1) is a mathematical statement of Boyle's law. The pressure P must be the *absolute* pressure and not *gauge* pressure. (See Chap. 12.)

EXAMPLE 15-1 What volume of hydrogen gas at atmospheric pressure is required to fill a 2-ft^3 tank under an absolute pressure of 2500 lb/in.^{2}?

Solution Recalling that atmospheric pressure is equal to 14.7 lb/in.^{2}, we apply Eq. (15-1):

$$P_1V_1 = P_2V_2$$
$$(14.7 \text{ lb/in.}^2)V_1 = (2500 \text{ lb/in.}^2)(2 \text{ ft}^3)$$
$$V_1 = 340 \text{ ft}^3$$

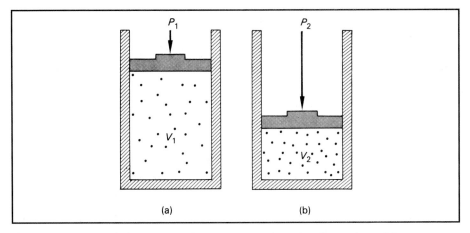

FIG. 15-1 When a gas is compressed at constant temperature, the product of its pressure and its volume is always constant, that is, $P_1V_1 = P_2V_2$.

Notice that it was not necessary to convert pressure in pounds per square inch to units of pounds per square foot in order to be consistent with volume in cubic feet. Since P and V appear on both sides of the equation, it is only necessary to use consistent units for P and consistent units for V. The unit for P does not have to be consistent with the unit chosen for V.

15-2 ■ CHARLES' LAW

In Chap. 13 we used the fact that the volume of a gas increased directly with its temperature to help us define absolute zero. We found the result ($-273°C$) by extending the line on the graph in Fig. 15-2. Of course, any real gas will become a liquid before its volume reaches zero. But the direct relationship is a valid approximation for most gases which are not subjected to extreme conditions of temperature and pressure.

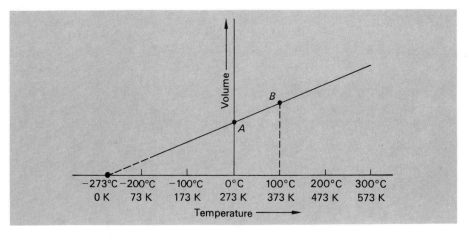

FIG. 15-2 Change in volume as a function of temperature. When the volume is extrapolated to zero, the temperature of a gas is absolute zero (0 K).

This direct proportionality between volume and temperature was first experimentally tested by Jacques Charles in 1787. *Charles' law* may be stated as follows:

Charles' law: *Provided that the mass and pressure of a gas are held constant, the volume of the gas is directly proportional to its absolute temperature.*

If we use the subscript 1 to refer to an initial state of a gas and the subscript 2 to refer to the final state, a mathematical statement of Charles' law is obtained:

$$\frac{V_1}{T_1} = \frac{V_2}{T_2} \qquad \textit{With Constant m and P} \quad (15\text{-}2)$$

In this equation V_1 refers to the volume of a gas at the *absolute* temperature T_1, and V_2 is the later volume of the same sample of gas when its absolute temperature is T_2.

EXAMPLE 15-2   A large balloon filled with air has a volume of 200 L at 0°C. What will its volume be at 57°C if the pressure is unchanged?

Solution   Since Charles' law applies only for absolute temperatures, we must first convert the given temperatures to kelvins:

$$T_1 = 273 \text{ K} \qquad T_2 = 57 + 273 = 330 \text{ K}$$

Now we may substitute and solve Eq. (15-2) for V_2:

$$\frac{V_1}{T_1} = \frac{V_2}{T_2}$$

$$\frac{200 \text{ L}}{273 \text{ K}} = \frac{V_2}{330 \text{ K}}$$

$$V_2 = \frac{(200 \text{ L})(330 \text{ K})}{273 \text{ K}} = 242 \text{ L}$$

15-3 ■ GAY-LUSSAC'S LAW

The three quantities which determine the state of a given mass of gas are its pressure, its volume, and its temperature. Boyle's law deals with changes in pressure and volume under constant temperature, and Charles' law applies for volume and temperature under constant pressure. The variation in pressure as a function of temperature is described in a law attributed to Gay-Lussac:

Gay-Lussac's law: *If the volume of a sample of gas remains constant, the absolute pressure of the gas is directly proportional to its absolute temperature.*

This means that doubling the pressure applied to a gas will cause its absolute temperature to double also. In equation form, Gay-Lussac's law may be written as

$$\frac{P_1}{T_1} = \frac{P_2}{T_2} \qquad \textit{With Constant m and V} \quad (15\text{-}3)$$

EXAMPLE 15-3   An automobile tire is inflated to a gauge pressure of 30 lb/in.^{2} at a time when the surrounding pressure is 14.4 lb/in.^{2} and the temperature is 70°F. After driving, the temperature of the air in the tire increases to 100°F. Assuming that the volume changes only slightly, what will be the new gauge pressure in the tire?

Solution   Gay-Lussac's law applies for constant volume, but we must first convert to absolute pressure and to absolute temperature:

$$P_1 = 30 \text{ lb/in.}^2 + 14.4 \text{ lb/in.}^2 = 44.4 \text{ lb/in.}^2$$
$$T_1 = 70 + 460 = 530°R \qquad T_2 = 100 + 460 = 560°R$$

Now we calculate the new pressure P_2 from Eq. (15-3):

$$\frac{P_1}{T_1} = \frac{P_2}{T_2}$$

$$\frac{44.4 \text{ lb/in.}^2}{530°R} = \frac{P_2}{560°R}$$

$$P_2 = \frac{(44.4 \text{ lb/in.}^2)(560°R)}{530°R} = 46.9 \text{ lb/in.}^2$$

Note that 46.9 lb/in.^{2} will be the absolute pressure. The gauge pressure will be 14.4 lb/in.^{2} less:

$$\text{Gauge pressure} = 46.9 \text{ lb/in.}^2 - 14.4 \text{ lb/in.}^2 = 32.5 \text{ lb/in.}^2$$

15-4 ■ GENERAL GAS LAW

Each of the laws discussed so far requires that one of the three quantities P, V, and T remain constant. A more generally applicable formula combines these laws into one general equation, called the *general gas law*. For a given mass we may write

$$\frac{P_1 V_1}{T_1} = \frac{P_2 V_2}{T_2} \qquad \textit{General Gas Law} \quad (15\text{-}4)$$

where P_1, V_1, T_1 apply to an initial state of a gas and P_2, V_2, T_2 apply to the final state of the same sample of gas. In other words, for a given mass, the ratio PV/T is constant. Equation (15-4) can be remembered by "a private (PV/T) is always a private." Notice that the general gas law reduces to Boyle's law, to Charles' law, or to Gay-Lussac's law when the appropriate quantity is constant.

EXAMPLE 15-4   An oxygen tank with internal volume of 20 L is filled with oxygen under an absolute pressure of 6000 kPa (6×10^6 N/m^2) at 20°C. The oxygen is to be used in a high-flying aircraft, where the absolute pressure is 70 kPa and the temperature is -20°C. What volume of oxygen can the tank supply under these conditions?

Solution   After converting the temperatures to the absolute kelvin scale, we apply Eq. (15-4):

$$\frac{P_1V_1}{T_1} = \frac{P_2V_2}{T_2}$$

$$\frac{(6000 \text{ kPa})(20 \text{ L})}{293 \text{ K}} = \frac{(70 \text{ kPa})(V_2)}{253 \text{ K}}$$

$$V_2 = \frac{(6000 \text{ kPa})(20 \text{ L})(253 \text{ K})}{(293 \text{ K})(70 \text{ kPa})} = 1480 \text{ L}$$

The general gas law as given in Eq. (15-4) is useful for initial and final states of a given sample of gas. However, it does not allow us to determine the pressure, volume, or temperature of a particular kind of gas in a given thermodynamic state. For example, we cannot determine the volume of a gas just from knowing its pressure, mass, and temperature. To accomplish this task requires a discussion of molecular theory that is beyond the scope of this text.

15-5 ■ HEAT AND WORK

Work, like heat, involves a transfer of energy, but there is a very important difference between the two terms. In mechanics, we define *work* as the product of a force and a displacement. Temperature plays no role in this definition. Heat, however, is energy that flows from one body to another because of a difference in temperature. A temperature difference is a necessary condition for the transfer of heat. Displacement is the necessary condition for work.

Consider, for example, the simple piston arrangement illustrated by Fig. 15-3. The movable piston allows external work to be done on the enclosed gas, causing its temperature to rise (an increase in internal energy). In this case (Fig. 15-3a) work is converted to heat. The reverse process will occur if heat is put into the system, as in Fig. 15-3b. Here input heat is being converted to mechanical energy. Any device which converts heat energy to mechanical energy (or vice versa) is called a *heat engine*. The automobile engine, the steam turbine, and even a refrigerator are examples of heat engines.

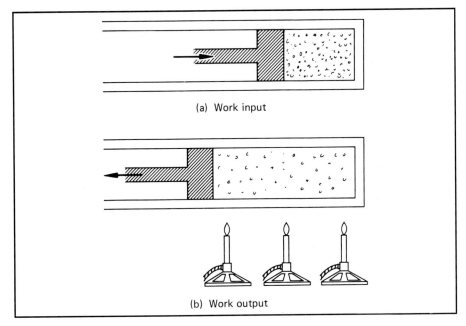

(a) Work input

(b) Work output

FIG. 15-3 (*a*) Work is done on a gas, causing a rise in the temperature of the circulated gas. (*b*) Heat is absorbed by the gas, and work is done by the piston.

Because energy can be transferred by both heat and work, we often speak of the *mechanical equivalent* of heat or of the *heat equivalent* of work. The conversion factors are given below:

$$1 \text{ cal} = 4.184 \text{ J}$$
$$1 \text{ kcal} = 4184 \text{ J}$$
$$1 \text{ Btu} = 778 \text{ ft·lb}$$

15-6 ■ FIRST LAW OF THERMODYNAMICS

In a heat engine, energy must be conserved. We can't expect more out of an engine than we put into it. The first law of thermodynamics is simply a restatement of the principle of conservation of energy:

Energy cannot be created or destroyed but can change from one form to another.

In the case of a heat engine, this means that the net heat put into a system may be converted to a combination of work and an increase in the internal energy of the system. But the sum of the increase in internal energy and the work done cannot exceed the heat energy absorbed by the engine.

First law of thermodynamics: *In any thermodynamic process, the net heat absorbed by a system is equal to the sum of the heat equivalent of the work done by the system and the change in internal energy of the system:*

$$\Delta Q = W_{\text{out}} + \Delta U \qquad (15\text{-}5)$$

In the mathematical statement, the symbol ΔQ is used to describe the net heat absorbed by a system, and ΔU is the net change in internal energy of the system. The work W_{out} in the equation is the net work done *by* the system (the output work). To understand this relationship, it is useful to consider an example. Consider the fluid in Fig. 15-4 which expands due to the heat provided by a flame. If we consider the gas as a system, there is a net heat transfer ΔQ imparted to the gas. This energy is used in two ways: (1) The internal energy ΔU of the gas is increased by a portion of the input thermal energy, and (2) the gas does an amount of work on the piston equivalent to the remainder of the available energy.

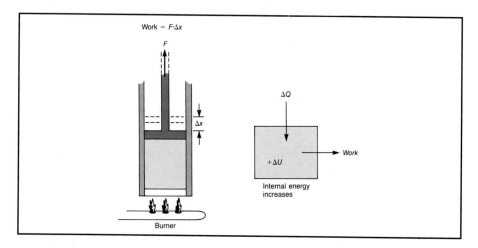

FIG. 15-4 Part of the energy supplied to the enclosed gas by the flame results in external work. The remainder increases the internal energy of the gas.

Many practical engines operate in a cycle which constantly returns the internal energy to its initial state. In such cases, $\Delta U = 0$, and the output work is equal to the difference between the heat input and the operational heat losses:

$$W_{\text{out}} = \Delta Q = Q_{\text{in}} - Q_{\text{out}}$$

It should be pointed out that if an engine is operated in reverse, the total energy put into the system is the work W_{in}, and the first law might be written as follows:

$$W_{\text{in}} = \Delta Q + \Delta U$$

A later example of this situation will be described for a refrigerator where the work done on the system by a compressor is equal to the *net* heat *discharged* by the system plus the change in internal energy. In a complete cycle, ΔU is once again zero, and the work done by the compressor is the difference between the heat discharged and the heat extracted from the contents of the refrigerator.

15-7 ■ SECOND LAW OF THERMODYNAMICS

When we rub our hands together rapidly, the work done against friction increases the internal energy and causes a rise in temperature. The surrounding air forms a large reservoir at a lower temperature, and the heat energy is transferred to the air without changing its temperature very much. When we stop rubbing, our hands return to the same state as before. According to the first law, mechanical energy has been transformed to heat with 100 percent efficiency. Such a transformation can be continued indefinitely as long as work is supplied.

Let us now consider the reverse process. Is it possible to convert heat energy to work with 100 percent efficiency? In the above example, is it possible to capture all the heat transferred to the air and return it to our hands, causing them to rub together indefinitely of their own accord? On a cold winter day, we might like to take advantage of this process. Unfortunately, such a process cannot occur even though it does not violate the first law. Neither is it possible to retrieve all the heat lost in braking a car in order to start the wheels rolling again.

We shall see that the conversion of heat energy to mechanical work is always a losing process. The first law tells us that we cannot win in such an experiment. In other words, it is impossible to get more work out of an engine than the heat put into the engine. It does not, however, prevent us from breaking even. Clearly, we need another rule which states that the 100 percent conversion of heat to useful work is not possible. This rule forms the basis for the second law of thermodynamics.

> **Second law of thermodynamics:** *It is impossible to construct an engine which, operating continuously, produces no effect other than the extraction of heat from a reservoir and the performance of an equivalent amount of work.*

To give more insight and application to this principle, suppose we study the operation and efficiency of heat engines. A particular system might be a gasoline engine, a jet engine, a steam engine, or even the human body. The operation of a heat engine is best described by a diagram similar to that shown in Fig. 15-5. During the operation of such a general engine, three processes occur:

1. A quantity of heat Q_{in} is supplied to the engine from a reservoir at a high temperature T_{in}.
2. Mechanical work W_{out} is done by the engine through the use of a portion of the heat input.
3. A quantity of heat Q_{out} is released to a reservoir at a low temperature T_{out}.

Since the system is periodically returned to its initial state, the net change in internal energy is zero. Thus, the first law tells us that

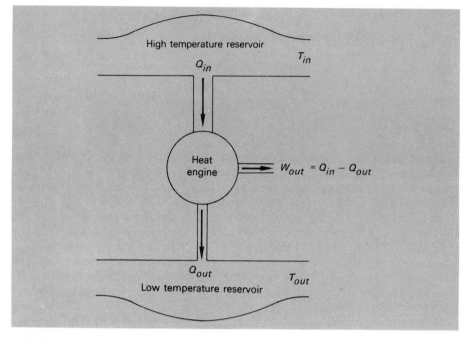

FIG. 15-5 Schematic diagram for a heat engine.

Work output = heat input − heat output

$$W_{out} = Q_{in} - Q_{out} \tag{15-6}$$

The *efficiency* of a heat engine is defined as the ratio of the useful work done by the engine to the heat put into the engine, and it is usually expressed as a percentage:

$$\text{Efficiency} = \frac{\text{work output}}{\text{heat input}}$$

$$E = \frac{Q_{in} - Q_{out}}{Q_{in}} \tag{15-7}$$

For example, an engine that is 25 percent efficient ($E = 0.25$) might absorb 1000 Btu, perform 250 Btu of work, and reject 750 Btu as wasted heat. A 100 percent efficient engine is one in which all the input heat is converted to useful work. In this case, no heat would be rejected to the environment ($Q_{out} = 0$). Although such a process would conserve energy, it violates the second law of thermodynamics. The most efficient engine is the one which rejects the *least* possible heat to the environment.

As an example of a real heat engine, consider the reciprocating steam engine illustrated in Fig. 15-6. This device is called an *external-combustion engine* because the fuel is burned outside the engine to produce heat. Superheated steam enters the steam chamber from the boiler and passes through the opening A into the cylinder containing a piston. The expanding steam pushes the piston downward. The used steam from the upstroke is exhausted at the same time through opening B and out the exhaust pipe. The sliding valve is operated by a cam so that it opens and closes at exactly the right times.

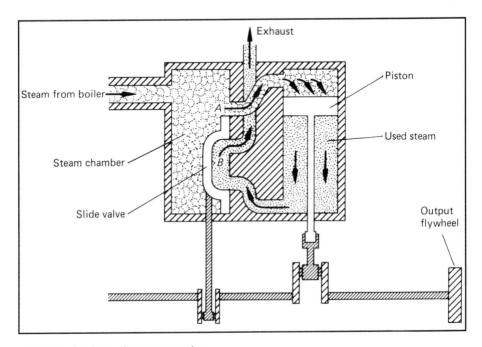

FIG. 15-6 Reciprocating steam engine.

15-8 ■ *EFFICIENCY OF AN IDEAL ENGINE*

The efficiency of a real engine is difficult to predict from Eq. (15-7) because the quantities Q_{in} and Q_{out} are difficult to calculate. Frictional and heat losses through the cylinder walls and around the piston, incomplete burning of the fuel, and even the physical properties of different fuels all frustrate our attempts to measure the efficiency of such engines. However, we can imagine an *ideal engine,* one which is not restricted by these practical difficulties. The efficiency of such an engine depends only on the quantities of heat absorbed and rejected between two well-defined heat reservoirs. It does not depend on the thermal properties of the working fuel. In other words, regardless of the internal changes in pressure, volume, length, or other factors, all ideal engines have the same efficiency when they are operating between the same two temperatures (T_{in} and T_{out}).

> An **ideal engine** is one which has the highest possible efficiency for the temperature limits within which it operates.

If we can define the efficiency of an engine in terms of input and output temperatures instead of in terms of the input and output heat, we will have a more useful formula. For an ideal engine it can be shown that the ratio of Q_{in}/Q_{out} is the same as the ratio of T_{in}/T_{out}. The actual proof is beyond the scope of this text. The efficiency of an ideal engine can, therefore, be expressed as a function of the absolute temperatures of the input and output reservoirs. Equation (15-7), for an ideal engine, becomes

$$E = \frac{T_{in} - T_{out}}{T_{in}}$$

(15-8)

It can be shown that no engine operating between the same two temperatures can be more efficient than would be indicated by Eq. (15-8). This ideal efficiency thus represents an upper limit to the efficiency of any practical engine. The greater the difference in temperature between two reservoirs, the greater the efficiency of any engine.

EXAMPLE 15-5   (a) What is the efficiency of an ideal engine operating between two heat reservoirs at 400 and 300 K? (b) How much work is done by the engine in one complete cycle if 800 cal of heat is absorbed from the high-temperature reservoir? (c) How much heat is delivered to the low-temperature reservoir?

Solution (a)   The ideal efficiency is found from Eq. (15-8):

$$E = \frac{T_{in} - T_{out}}{T_{in}} = \frac{400 \text{ K} - 300 \text{ K}}{300 \text{ K}} = 0.25$$

Thus, the ideal efficiency is 25 percent.

Solution (b)   The efficiency is the ratio of W_{out}/Q_{in}, so that

$$\frac{W_{out}}{Q_{in}} = 0.25 \qquad \text{or} \qquad W_{out} = 0.25 Q_{in}$$

$$W_{out} = (0.25)(800 \text{ cal}) = 200 \text{ cal}$$

A 25 percent efficient engine delivers one-fourth of the input heat to useful work. The rest must be lost (Q_{out}).

Solution (c)   The first law of thermodynamics requires that

$$W_{out} = Q_{in} - Q_{out}$$

Solving for Q_{out}, we obtain

$$Q_{out} = Q_{in} - W_{out} = 800 \text{ cal} - 200 \text{ cal} = 600 \text{ cal}$$

The work output is usually expressed in joules. Conversion to these units gives

$$W_{out} = (200 \text{ cal})(4.186 \text{ J/cal}) = 837 \text{ J}$$

15-9 ■ *INTERNAL-COMBUSTION ENGINES*

An internal-combustion engine generates the input heat within the engine itself. The most common engine of this variety is the four-stroke gasoline engine, in which a mixture of gasoline and air is ignited by a spark plug in each cylinder. The thermal energy released is converted to useful work by the pressure exerted on a piston by the expanding gases. The four-stroke process is illustrated in Fig. 15-7. During the *intake stroke* (Fig. 15-7*a*), a mixture of air and gasoline vapor enters the cylinder through the intake valve. Both valves are closed during the *compression stroke* (Fig. 15-7*b*) as the piston moves upward, causing a rise in pressure. Just before the piston reaches the top, the mixture is ignited, causing a sharp increase in temperature and pressure. In the *power stroke* (Fig. 15-7*c*), the expanding gases force the piston downward, performing external work. The fourth stroke (Fig. 15-7*d*) pushes the burned gases out of the cylinder through the exhaust valve. The entire cycle is then repeated for as long as the combustible fuel is supplied to the cylinder. The ideal efficiency of such an engine is around 57 percent. Actual efficiencies are around 30 percent due to uncontrolled heat losses and friction.

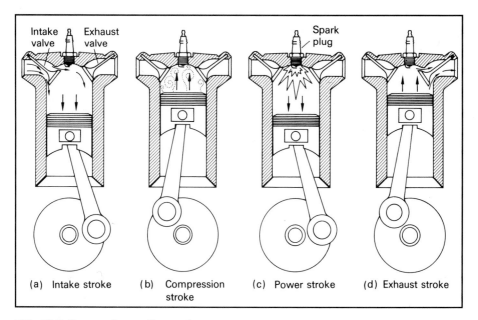

FIG. 15-7 Four-stroke gasoline engine.

Another type of internal-combustion engine is the diesel engine. The four strokes of a diesel engine are shown in Fig. 15-8. Only air is drawn into the cylinder on the intake stroke. Then the air is compressed to a high temperature and pressure near the top of the cylinder. Diesel fuel, injected into the cylinder at this point, ignites and pushes the piston downward. The waste gases are expelled during the exhaust stroke.

The efficiency of an internal-combustion engine can be improved by increasing the volume of gas in the cylinder at the end of the intake stroke as compared with the volume at the end of the compression stroke. The ratio of these volumes, V_i/V_c, is

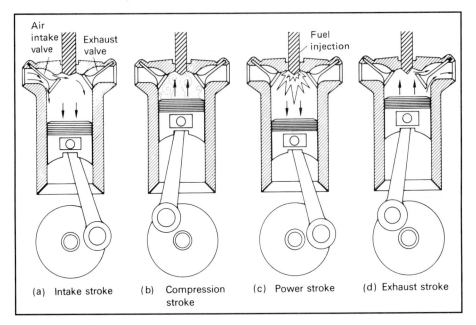

FIG. 15-8 Diesel engine.

called the *compression ratio*. It can be seen that when this ratio is larger, the gases expand farther and do more work. However, the temperature of the gases in the cylinder rises as the gases are compressed. In fact, if the ratio is made too high, the mixture will explode prematurely. For this reason, the compression ratio in modern gasoline engines is limited to around 10 to 1.

The diesel engine is not limited in this way, since only air is drawn into the cylinder during the intake stroke. Compression ratios as high as 16 to 1 are not uncommon for diesel engines. This explains why greater efficiencies are realized with diesel engines. However, the construction must be such that very high pressures can occur in the cylinders without structural damage.

15-10 ■ REFRIGERATION

A refrigerator can be thought of as a heat engine operated in reverse. A schematic diagram of a refrigerator is shown in Fig. 15-9. During every cycle, a compressor or similar device supplies mechanical work W_{in} to the system. A quantity of heat Q_{cold} is extracted from a cold reservoir, and a quantity of heat Q_{hot} is deposited in a hot reservoir. According to the first law, the input work is given by

$$W_{in} = Q_{hot} - Q_{cold} \qquad (15\text{-}9)$$

The effectiveness of any refrigerator is determined by the amount of heat Q_{cold} extracted for the least expenditure of work W_{in}. The ratio W_{cold}/W_{in} is, therefore, a measure of the cooling efficiency of a refrigerator. This ratio is called the *coefficient of performance* η and is given by

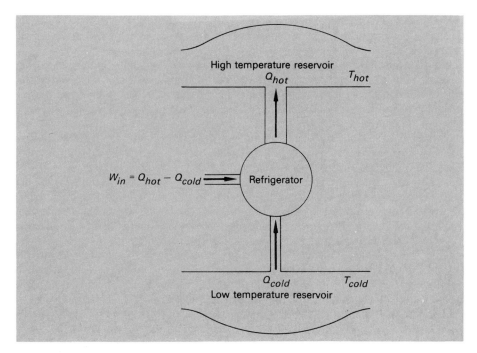

FIG. 15-9 Schematic diagram for a refrigerator.

$$\eta = \frac{Q_{\text{cold}}}{W_{\text{in}}} = \frac{Q_{\text{cold}}}{Q_{\text{hot}} - Q_{\text{cold}}}$$

Coefficient of Performance (15-10)

Just as we can imagine an ideal heat engine, we can also consider an ideal refrigerator. The efficiency of an ideal refrigerator is expressed in terms of absolute temperatures:

$$\eta = \frac{T_{\text{cold}}}{T_{\text{hot}} - T_{\text{cold}}}$$

(15-11)

To understand the refrigeration process better, let us consider the general schematic in Fig. 15-10. This diagram may refer to a number of refrigeration devices, from a commercial plant to a household refrigerator. The working substance, called the *refrigerant,* is a gas which is easily liquefied by an increase in pressure or a drop in temperature. In the liquid phase it can be vaporized easily by passing it through a narrow constriction, called the *throttling valve.* When the liquid vaporizes, it absorbs the heat of condensation from the surrounding area. Common refrigerants are ammonia, Freon-12, methyl chloride, and sulfur dioxide. Ammonia, the most common industrial refrigerant, boils at −28°F under ambient pressure (1 atm). Freon-12, the most common household refrigerant, boils at −22°F at atmospheric pressure. Variation in pressure radically affects the condensation and evaporation temperatures of all refrigerants.

As shown in the schematic, a typical refrigerator consists of a *compressor,* a *condenser,* a *liquid storage tank,* a *throttling valve,* and an *evaporator.* The compressor provides the necessary input work to move the refrigerant through the system. As the piston moves to the right, it sucks the refrigerant through the intake valve at a little above atmospheric pressure and near room temperature. During the power stroke, the intake valve closes, and the discharge valve opens. The emergent refrigerant, at high temperature and pressure, passes into the condenser, where it is cooled until it lique-fies. The condenser may be cooled by running water or by an electric fan. It is during this phase that a quantity of heat Q_{hot} is rejected from the system. The condensed liquid refrigerant, still at high pressure and temperature, is collected in a liquid reservoir. Then the liquid refrigerant is drawn from the storage tank through a throttling valve, causing a sudden drop in temperature and pressure. As the cold liquid refrigerant flows through the evaporator coils, it absorbs a quantity of heat Q_{cold} from the space and products being cooled. This heat boils the liquid refrigerant and is carried away by the resulting gaseous refrigerant as latent heat of vaporization. This phase is the "payoff" for the entire operation, and all components just contribute to the effective transfer of heat to the evaporator. Finally, the refrigerant vapor leaves the evaporator and is sucked into the compressor to begin another cycle.

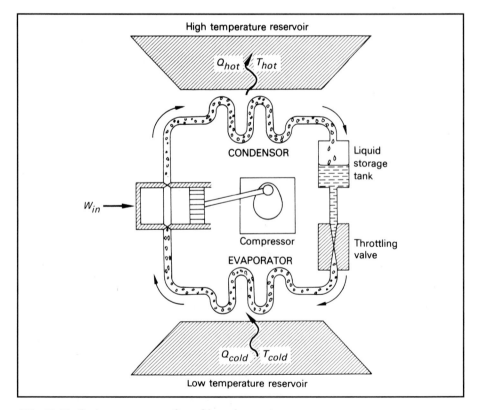

FIG. 15-10 Basic components of a refrigeration system.

In refrigerators and air conditioners, the removal of Q_{cold} is used to lower temperatures, freeze liquids, etc. But Q_{hot} may also be used to increase temperatures, melt solids, etc. The heat pump functions as an air conditioner in the summer, removing Q_{cold} from the building. In the winter, the condenser coil becomes the evaporator and vice versa, so that the heat pump adds Q_{hot} to the building.

Summary

We began this chapter by discussing the expansion of gases and the relationships among pressure, volume, and temperature. The expansion of gases is the working element in heat engines, which are devices to convert thermal energy to mechanical energy (or vice versa). The essential concepts presented in this chapter are as follows:

- **Boyle's law** states that the product of the pressure and the volume of a sample of gas will be constant as long as the temperature is unchanged:

$$P_1 V_1 = P_1 V_2 \qquad \text{\textit{Boyle's Law}}$$

- **Charles' law** states that the volume of a sample of gas is directly proportional to its absolute temperature, under conditions of constant pressure:

$$\frac{V_1}{T_1} = \frac{V_2}{T_2} \qquad \text{\textit{Charles' Law}}$$

- **Gay-Lussac's law** states that the absolute pressure of a sample of gas is directly proportional to its absolute temperature, if the volume is constant:

$$\frac{P_1}{T_1} = \frac{P_2}{T_2} \qquad \text{\textit{Gay-Lussac's Law}}$$

- The **general gas law** is a combination of all three laws and applies as long as the mass does not change:

$$\frac{P_1 V_1}{T_1} = \frac{P_2 V_2}{T_2} \qquad \text{\textit{General Gas Law}}$$

- When you are applying the gas laws, remember that the pressure is *absolute pressure* and the temperature is *absolute temperature:*

Absolute pressure = gauge pressure + atmospheric pressure
$$1 \text{ atm} = 101.3 \text{ kPa} = 1.013 \times 10^5 \text{ N/m}^2 = 14.7 \text{ lb/in.}^2$$
$$T_K = t_C + 273 \qquad T_R = t_F + 460$$

■ The **first law of thermodynamics** is a restatement of the conservation of energy. It points out that the net heat ΔQ put into a system is equal to the work done *by* the system plus the change in internal energy ΔU of the system:

$$Q = W_{\text{out}} + \Delta U$$

For a heat engine, where the net change in internal energy ΔU is zero, $\Delta Q = W_{\text{out}}$ and the first law becomes

$$W_{\text{out}} = Q_{\text{in}} - Q_{\text{out}}$$

■ The **second law of thermodynamics** places restrictions on the possibility of satisfying the first. It points out that a 100 percent efficient engine, one which converts all input heat to useful work, is not possible.
■ In general, a heat engine operates between a high-temperature reservoir, which is the source on input heat, and a low-temperature reservoir, for output heat. The actual efficiency of such an engine is the ratio of the work output to the heat input:

$$E = \frac{\text{work output}}{\text{heat input}} \qquad\qquad E = \frac{Q_{\text{in}} - Q_{\text{out}}}{Q_{\text{in}}}$$

For an ideal engine, $Q_{\text{out}}/Q_{\text{in}}$ is equal to the corresponding ratio of absolute temperatures. Thus, the ideal efficiency is

$$E = \frac{T_{\text{in}} - T_{\text{out}}}{T_{\text{in}}} \qquad\qquad \textit{Ideal Efficiency}$$

■ A **refrigerator** is a heat engine operated in reverse. A measure of the performance is the amount of cooling obtained compared with the input work. The **coefficient of performance** η is given by

$$\eta = \frac{\text{heat extracted}}{\text{work input}} \qquad\qquad \eta = \frac{Q_{\text{cold}}}{Q_{\text{hot}} - Q_{\text{cold}}}$$

For an ideal refrigerator, the absolute temperatures of the reservoirs may be used to calculate the ideal coefficient:

$$\eta = \frac{T_{cold}}{T_{hot} - T_{cold}}$$

Questions

15-1. Define the following terms:

a. Boyle's law	**g.** refrigerator
b. Charles' law	**h.** refrigerant
c. Gay-Lussac's law	**i.** efficiency
d. general gas law	**j.** coefficient of performance
e. heat engine	**k.** first law of thermodynamics
f. ideal engine	**l.** second law of thermodynamics

15-2. Explain Boyle's law and Charles' law in terms of the molecular theory of matter.

15-3. Why is it necessary to use absolute temperature in the general gas law?

15-4. A closed steel tank is filled with a quantity of gas, and the system is heated uniformly. What happens to the volume, density, and pressure of the enclosed gas?

15-5. If both heat and work are expressed in the same units, why is it necessary to distinguish between them?

15-6. Describe what happens to the volume, temperature, and pressure of the enclosed gases during the compression stroke of an internal-combustion engine. What happens during the power stroke?

15-7. What determines the efficiency of a heat engine? Why is it generally so low? Why is a diesel engine more efficient than a gasoline engine?

15-8. What happens to the temperature of a gas when (*a*) work is done *on* the gas, (*b*) work is done *by* the gas?

15-9. In the text, only one statement was given for the second law of thermodynamics. Discuss each of the following statements, showing it to be equivalent to that given in the text:

a. It is impossible to construct a refrigerator which, operating in a complete cycle, will extract heat from a cold body and exhaust it to a hot body without the performance of work on the system.

b. The natural direction of heat flow is from a body at a high temperature to a body at a low temperature, regardless of the size of each reservoir.

c. All natural spontaneous processes are irreversible.

d. Natural events always proceed in the direction from order to disorder.

15-10. It is energetically possible to extract the heat energy contained in the ocean and use it to power a steamship across the sea. What objections can you offer?

15-11. In an electric refrigerator, heat is transferred from the cool interior to warmer surroundings. Why is this not a violation of the second law?

15-12. If natural processes tend to decrease order in the universe, how can you explain the evolution of biological systems to a highly organized state? Does this violate the second law of thermodynamics?

15-13. Will keeping the door of an electric refrigerator open warm or cool a room? Explain.

15-14. What temperature must the cold reservoir have if an ideal engine is to have an efficiency of 100 percent? Can this ever happen? If it is not possible for an ideal engine to have an efficiency of 100 percent, why is it called an *ideal* engine?

Problems

15-1. The gauge on an oxygen tank reads 50 kPa at 40°C. What are the absolute pressure and absolute temperature of the gas?

Answer 601.3 kPa; 313 K.

15-2. The absolute pressure inside an automobile tire is 48 lb/in.² when its temperature is 90°F. What are the gauge pressure and absolute temperature of the air inside the tire?

15-3. If the temperature and mass of a 15-L sample of gas are held constant while its absolute pressure increases from 200 to 600 kPa, what will be its new volume? What will be the final gauge pressure?

Answer 5 L; 498.7 kPa.

15-4. What volume of air at atmospheric pressure can be stored in a 12-ft³ tank which can withstand an absolute pressure of 120 lb/in.²?

15-5. Four hundred cubic centimeters of gas at 27°C is heated at constant pressure. If the volume of the gas increases to 600 cm³, what was the final temperature in kelvins? In degrees Celsius?

Answer 450 K; 177°C.

15-6. A 6-L container holds a sample of gas under an absolute pressure of 600 kPa and at a temperature of 57°C. What will be the new pressure if the same sample of gas is placed into a 3-L container at 7°C?

15-7. An air compressor takes in 2 m³ of air at 20°C and atmospheric pressure (101.3 kPa). If the compressor discharges into a 0.3-m³ tank at an absolute pressure of 1500 kPa, what is the temperature of the discharged air?

Answer 651 K.

15-8. If 0.8 L of a gas at 10°C is heated to 90°C at constant pressure, what will the new volume be?

15-9. A tank with a capacity of 14 L contains helium gas at 76°F under a gauge pressure of 400 lb/in.². (*a*) What will be the volume of a balloon filled with this gas if the helium expands to an internal absolute pressure of 1 atm and if its temperature drops to −35°F? (*b*) Eventually the whole system returns to its original temperature. What is the final volume of the balloon?

Answer (*a*) 302 L; (*b*) 381 L.

15-10. A 6-ft³ container is filled with a gas at an absolute pressure of 300 lb/in.² and a temperature of 75°F. What is the new pressure if the temperature is raised to 260°F?

15-11. At 70°F the gauge pressure of a gas in a steel container is indicated as 80 lb/in.². What will the pressure gauge read when the tank is heated uniformly to 150°F?

Answer 94.3 lb/in.².

15-12. Five liters of a gas at an absolute pressure of 200 kPa and at a temperature of 25°C is heated uniformly to 60°C, and the absolute pressure is reduced to 120 kPa. What volume will the gas occupy under these conditions?

15-13. Four hundred calories of heat is absorbed by a gas in a cylinder, causing 450 J of work to be done by a piston. (*a*) What is the heat equivalent of the work done by the piston? (*b*) What is the mechanical equivalent of the heat absorbed by the system? (*c*) According to the first law of thermodynamics, what is the maximum change in internal energy of the gas?

Answer (*a*) 1674 J; (*b*) 108 cal; (*c*) 1224 J.

15-14. An *adiabatic process* is one in which the net heat exchange is zero ($\Delta Q = 0$). If 800 cal of heat is absorbed during such a process, how much work is done in joules?

15-15. An *isothermal process* is an ideal process in which there is no change in temperature. If the temperature does not change and if the operating gas does not condense, what is the change in internal energy? If 400 J of work is done in an isothermal process, how many calories of heat were absorbed?

Answer 0; 95.6 cal.

15-16. A piston does 90 J of work on a gas enclosed in a cylinder, causing the internal energy to increase by 60 J. Calculate the total heat loss during this process.

15-17. A steam engine takes heat from a boiler at 200°C and exhausts directly into the air at 100°C. What is the ideal efficiency of this engine?

Answer 50.

15-18. Assume that the actual efficiency of the steam engine in Prob. 15-17 is only one-fourth of its ideal efficiency. How much work is done in one cycle by this engine if it takes in 800 cal of heat?

15-19. An engine absorbs 600 J of energy while exhausting 400 J to a cool reservoir. How much work was done, and what was the efficiency of the engine?

Answer 200 J; 33.3%.

15-20. A heat engine operates between 300 and 110°F. What is the ideal efficiency?

15-21. An ideal engine takes 1200 cal of heat from a reservoir at 500 K and rejects a certain amount of heat to another reservoir at 300 K. (*a*) What is the ideal efficiency? (*b*) How much useful work (in joules) is done? (*c*) How much heat is exhausted during each cycle?

Answer (*a*) 40%; (*b*) 2009 J; (*c*) 720 cal.

15-22. If the engine of Prob. 15-21 is operated in reverse as a refrigerator and if it extracts 1200 cal from the cold reservoir, what is the coefficient of performance? How much heat is delivered to the high-temperature reservoir? How much mechanical work is required to accomplish one operational cycle?

15-23. A heat pump takes heat from a water reservoir at 41°F and delivers it to a system of pipes in a house at 78°F. The energy required to operate the pump is about twice that required for an ideal pump. How much mechanical work must be supplied by the pump in order to deliver 1,000,000 Btu of heat energy to the house?

Answer 1.38×10^5 Btu.

15-24. How many joules of work must be done by the compressor in a refrigerator in order to change 1 kg of water at 20°C to ice at −10°C? The coefficient of performance is 3.5.

15-25. The ideal efficiency of a heat engine is 25 percent. How much work is done if 120 Btu of heat is lost in each cycle?

Answer 40 Btu.

15-26. The coefficient of performance of an ideal refrigerator is 4.2. If the temperature of the cold reservoir is to be held at 250 K, what is the temperature of the hot reservoir?

15-27. An engine operating between 300 and 80°C has an actual efficiency that is only one-third its ideal efficiency. How much work is done in each cycle if 800 cal is absorbed from the high-temperature reservoir?

Answer 102 cal.

16

ELECTROSTATICS

OBJECTIVES

As a result of completing this chapter, you should be able to

1. Demonstrate the existence of two kinds of electric charge and verify and explain the *first law of electrostatics*, using appropriate laboratory materials.
2. Demonstrate or explain with diagrams the processes of charging by *contact* or by *induction*.
3. State or write *Coulomb's law* for electrostatic forces and apply it to the solution of problems.
4. Define and illustrate your understanding of the concepts of *electric field*, *electric field intensity*, and *electric field lines*.
5. Write and apply a relationship for calculating the *electric field intensity* as a function of the *force* on a given *charge*.
6. Distinguish by definition and example between *electric potential energy* and *electric potential difference*.
7. Write and apply a relationship among *potential difference*, *electric field intensity*, and *plate separation* for two parallel conductors.
8. Use your knowledge of potential difference to calculate the work required to move a known charge from one point to another in an electric field.
9. Define *capacitance* and apply a relationship among *capacitance*, *applied voltage*, and *total charge*.
10. Discuss how the capacitance is affected by plate *area*, plate *separation*, and the insertion of a *dielectric*.

For an understanding of the nature of electricity, we begin with a discussion of electric charges at rest. This branch of physics is referred to as *electrostatics*. In this chapter, you will learn what is meant by an *electric charge*. You will see how charges interact to produce an electric force, an electric field, and an electric potential. Applications of these concepts are numerous in industry. Electrostatic filters are placed in factory smokestacks to remove dust and soot that would otherwise pollute the air. Valuable metals are sometimes separated electrostatically from a fluid which passes between charged plates. Even the construction of certain abrasive materials, such as sandpaper, is accomplished by the attraction of charged particles to an adhesive surface.

16-1 ■ ELECTRIC CHARGE

The best way to begin a study of electrostatics is to experiment with objects which become electrified or "charged" by rubbing. The materials commonly found in a physics laboratory are a hard-rubber rod resting on a piece of cat's fur, a glass rod resting on a piece of silk, and a pith-ball electroscope. A pith ball is a very light sphere of wood pith painted with metallic paint and usually suspended from a silk thread. An *electroscope* is a sensitive laboratory instrument used to detect the presence of an electric charge.

The pith-ball electroscope can be used to study the effects of electrification. Two metallic-coated pith balls are suspended by silk threads from a common point. We begin by vigorously rubbing the rubber rod with cat's fur (or wool). Then if the rubber rod is brought near the electroscope, the suspended pith balls will be attracted to the rod, as shown in Fig. 16-1a. After remaining in contact with the rod for an instant, the balls will be repelled from the rod and from each other. When the rod is removed, the pith balls remain separated, as shown in the figure. The initial attraction (due to a redistribution of charge on the neutral pith balls) will be explained later. The repulsion must be due to some property acquired by the pith balls as a result of their contact with the charged rod. We may reasonably assume that some of the charge has been transferred from the rod to the pith balls and that all three objects become similarly charged. From these observations, we can state that

> A force of repulsion exists between two substances which are charged in the same way.

Let us continue our experiment by picking up the glass rod and rubbing it vigorously on a silk cloth. When the charged rod is brought near the pith balls, the same sequence of events occurs as with the rubber rod. (See Fig. 16-1b.) Does this mean that the nature of the charge is the same on both rods? Our experiment neither proves nor disproves this assumption. In each case, the rod and balls were electrified in the same way, and so repulsion occurs in each case.

To test whether the two processes are identical, let us charge one pith ball with a glass rod and the other with a rubber rod. As shown in Fig. 16-2, a force of attraction exists between the balls charged in this manner. We conclude that the charges produced on the glass and rubber rods are different.

Similar experimentation with many different materials demonstrates that all electrified objects can be divided into two groups: (1) those which have a charge like that produced on glass and (2) those which have a charge like that produced on rubber.

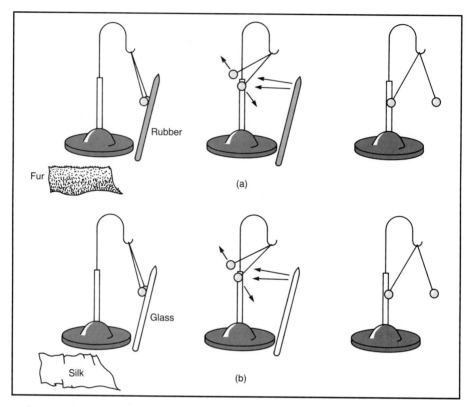

FIG. 16-1 (*a*) Charging the pith-ball electroscope with a rubber rod. (*b*) Charging the pith balls with a glass rod.

According to a convention established by Benjamin Franklin, objects in the former group are said to have a *positive* (+) *charge,* and objects belonging to the latter group are said to have a *negative* (−) *charge*. These terms have no mathematical significance; they simply denote the two opposite kinds of electric charge.

We are now in a position to state the *first law of electrostatics:*

Like charges repel and unlike charges attract.

Two negatively charged objects or two positively charged objects repel each other, as demonstrated by Fig. 16-1*a* and *b*, respectively. Figure 16-2 demonstrates that a positively charged object attracts a negatively charged object and vice versa.

16-2 ■ ELECTRONS

What actually occurs during the rubbing process that causes the electrification of objects? Benjamin Franklin thought that all bodies contained a specified amount of electric fluid which served to keep them in an uncharged state. When two different substances were rubbed together, he postulated that one accumulated an excess of fluid and became positively charged, whereas the other lost fluid and became negatively charged. It is now known that the substance transferred is not a fluid but very small amounts of negative electricity called *electrons*.

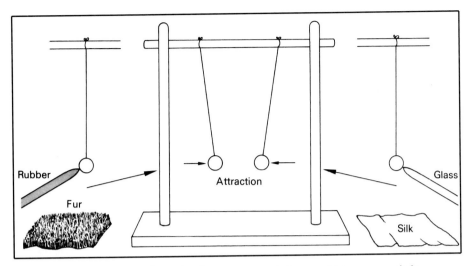

FIG. 16-2 A force of attraction exists between two substances which are oppositely charged.

The modern atomic theory of matter holds that all substances are made up of atoms and molecules. Each atom has a positively charged central core, called the *nucleus,* which is surrounded by a cloud of negatively charged electrons. The nucleus consists of a number of *protons,* each with a single unit of positive charge, and (except for hydrogen) one or more *neutrons.* As the name suggests, a neutron is a neutral particle. Normally, an atom of matter is in a *neutral* or *uncharged* state because it contains the same number of protons in its nucleus as there are electrons surrounding the nucleus. A schematic diagram of the neon atom is shown in Fig. 16-3. If, for some reason, a

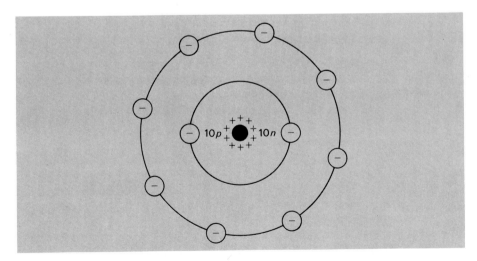

FIG. 16-3 The neon atom consists of a tightly packed nucleus containing 10 protons (p) and 10 neutrons (n). The atom is electrically neutral because it is surrounded by 10 electrons.

neutral atom loses one or more of its outer electrons, the atom has a net positive charge and is referred to as a *positive ion*. A *negative ion* is an atom which has gained one or more additional electrons.

When two particular materials are brought in close contact, some of the loosely held electrons may be transferred from one material to the other. For example, when the hard-rubber rod is rubbed against fur, electrons are transferred from the fur to the rod, leaving an excess of electrons on the rod and a deficiency of electrons on the fur. Similarly, when a glass rod is rubbed on a silk cloth, electrons are transferred from the glass to the silk. We can now state the following:

> *An object which has an excess of electrons is negatively charged, and an object which has a deficiency of electrons is positively charged.*

A laboratory demonstration of the transfer of charge is illustrated in Fig. 16-4. A hard-rubber rod is rubbed on a piece of fur. One pith ball is charged negatively with the rod, and the other is touched with the fur. The resulting attraction shows that the fur is oppositely charged. The process of rubbing has left a deficiency of electrons on the fur.

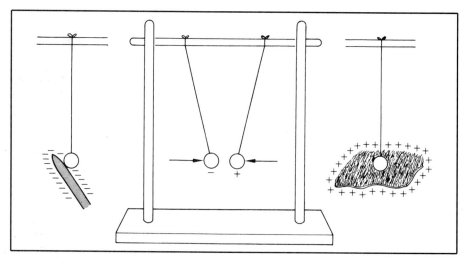

FIG. 16-4 Rubbing a hard-rubber rod on a piece of fur transfers electrons from the fur to the rod.

16-3 ■ INSULATORS AND CONDUCTORS

A solid piece of matter is composed of many atoms arranged in a manner peculiar to that material. Some materials, primarily metals, have a large number of *free electrons*, which can move about through the material. These materials have the ability to transfer charge from one object to another, and they are called *conductors*:

> A **conductor** is a material through which charge can easily be transferred.

Most metals are good conductors. In Fig. 16-5 a copper rod is supported by a glass stand. The pith balls can be charged by touching the right end of the copper with a

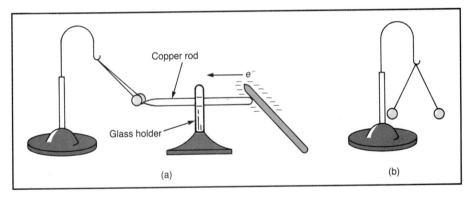

FIG. 16-5 Electrons are conducted by the copper rod to charge the pith balls.

charged rubber rod. The electrons are transferred or *conducted* through the rod to the pith balls. Note that none of the charge is transferred to the glass support or to the silk thread. These materials are poor conductors and are referred to as *insulators:*

> An **insulator** *is a material which resists the flow of charge.*

Other examples of good insulators are rubber, plastic, mica, Bakelite, sulfur, and air.

> A **semiconductor** *is a material intermediate in its ability to carry charge.*

Examples are silicon, germanium, and gallium arsenide. The ease with which a semiconductor carries charge can be greatly varied by the addition of impurities or by a change in temperature.

16-4 ■ CHARGING BY CONTACT AND BY INDUCTION

When a negatively charged rod is brought close to an uncharged pith ball, there is an initial attraction, as shown in Fig. 16-6. The attraction of the uncharged object is due to the separation of positive and negative electricity within the neutral body. The proximity of the negatively charged rod repels loosely held electrons to the opposite side of the uncharged object, leaving a deficiency (positive charge) on the near side and an excess (negative charge) on the far side. Since the unlike charge is nearer to the rod, the force of attraction will exceed the force of repulsion and the electrically neutral object will be attracted to the rod. No charge is gained or lost during this process; the charge on the neutral body is simply redistributed. When the pith ball actually makes contact with the rod, electrons are transferred to the neutral ball. The ball then has an excess of electrons and is said to have been charged by *contact.*

The redistribution of charge due to the presence of a nearby charged object can be useful in charging objects without contact. This process, called charging by *induction,* can be accomplished without any loss of charge from the charging body. For example, consider two neutral metal spheres placed in contact as shown in Fig. 16-7. When a negatively charged rod is brought near the left sphere (without touching it), a redistribution of charge occurs. Electrons are forced from the left sphere to the right sphere through the point of contact. Now, if the spheres are separated in the presence of the

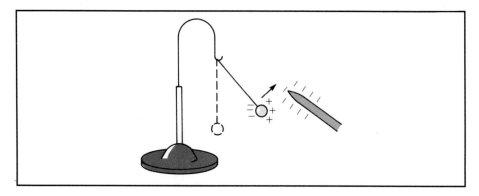

FIG. 16-6 Attraction of a neutral body due to a redistribution of charge.

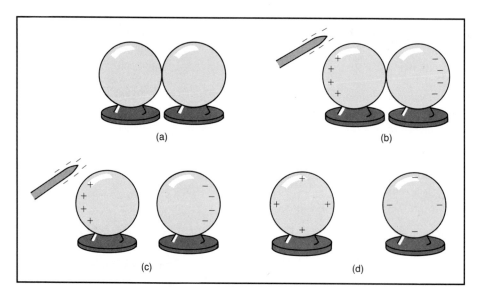

FIG. 16-7 Charging two metal spheres by induction.

charging rod, the electrons cannot return to the left sphere. Thus the left sphere will have a deficiency of electrons (a positive charge), and the right sphere will have an excess of electrons (a negative charge).

A charge can also be induced on a single sphere. This process is illustrated in Fig. 16-8. A negatively charged rod is placed near the metal sphere, causing a redistribution of charge. The repelled electrons move to the right side of the sphere. Touching the right side with a finger or connecting a wire from the sphere to the earth provides a path by which the repelled electrons can leave the sphere. This is sometimes referred to as *grounding* the sphere. The body or the ground will acquire a negative charge equal to the positive charge (deficiency) left on the sphere. When the charging rod is removed, the sphere is left with a residual positive charge which is opposite to that of the charging body.

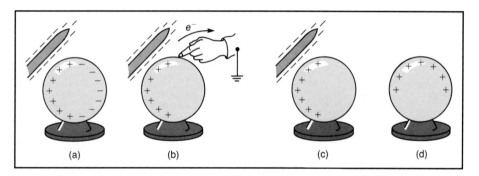

FIG. 16-8 Charging a single metal sphere by induction.

16-5 ■ COULOMB'S LAW

As usual, the task of the physicist is to measure the interactions between charged objects in some quantitative fashion. It is not sufficient to state that an electric force exists; we must be able to predict its magnitude.

The first theoretical investigation of the electric forces between charged bodies was accomplished by Charles Augustin de Coulomb in 1784. His studies were made with a torsion balance to measure the variation in force with separation and quantity of charge. The separation r of two charged objects is the straight-line distance between their centers. The quantity of charge q can be thought of as the excess of electrons or protons in the body.

Coulomb found that the force of attraction or repulsion between two objects is inversely proportional to the square of their separation distance. In other words, if the distance between two charged objects is reduced by one-half, the force of attraction or repulsion between them will be increased fourfold.

The concept of a quantity of charge was not clearly understood in Coulomb's time. There was no established unit of charge and no means for measuring it, but his experiments clearly showed that the electric force between two charged objects is directly proportional to the product of the quantity of charge on each object. Today, his conclusions are stated as a physical law.

Coulomb's law: *The force of attraction or repulsion between two point charges is directly proportional to the product of the two charges and inversely proportional to the square of the distance between them.*

$$F = \frac{kqq'}{r_2} \qquad (16\text{-}1)$$

The formula resulting from Coulomb's law can be understood from Fig. 16-9 for two charges q and q' separated by a distance r. The force F is an attraction for unlike charges and a repulsion for like charges, and k is the constant of proportionality. When the electric force is measured in *newtons* (N) and the separation is measured in *meters* (m), the SI unit of electric charge must be the *coulomb* (C). A formal definition of this new unit requires a discussion of current electricity. In terms of electrons, we note that a charge of -1 C corresponds to the charge of 6.25×10^{18} electrons.

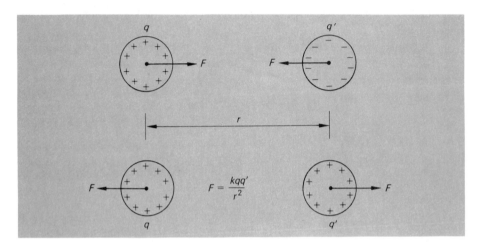

FIG. 16-9 Illustrating Coulomb's law.

The coulomb is an enormously large unit for most electrostatic applications. The charge of one electron in coulombs is

$$1 \; e^- = -1.6 \times 10^{-19} \; C$$

A more convenient unit for electrostatics is the *microcoulomb* (μC), defined by

$$1 \; \mu C = 1 \times 10^{-6} \; C$$

The constant k in Eq. (16-1) has been determined experimentally and is approximately equal to 9×10^9 N·m²/C². When one applies Coulomb's law in SI units, Eq. (16-1) becomes

$$F = \frac{(9 \times 10^9 \text{ N·m}^2/\text{C}^2)qq'}{r^2} \tag{16-2}$$

It must be remembered that F represents the force on a charged particle and is, therefore, a vector quantity. The direction of the force is determined solely by the nature of the charges q and q'. Therefore, it is easier to use the absolute values of q and q' in Coulomb's law, Eq. (16-2). Then remember that like charges repel and opposite charges attract to get the direction of the force. When more than one force acts on a charge, the resultant force is the vector sum of the separate forces.

EXAMPLE 16-1   A $-3\text{-}\mu C$ charge is placed 100 mm from a $+3\text{-}\mu C$ charge. Calculate the force between the two charges.

Solution   First we convert to appropriate units:

$$3 \ \mu C = 3 \times 10^{-6} \ C \qquad 100 \ mm = 100 \times 10^{-3} \ m = 0.1 \ m$$

Then we use the absolute values, so that both q and q' are equal to 3×10^{-6} C. Applying Coulomb's law, we obtain

$$F = \frac{kqq'}{r^2} = \frac{(9 \times 10^9 \ \text{N·m}^2/\text{C}^2)(3 \times 10^{-6} \ \text{C})(3 \times 10^{-6} \ \text{C})}{(0.1 \ \text{m})^2}$$

$$= 8.1 \ N \qquad \text{attraction}$$

This is a force of *attraction* because the charges had opposite signs.

*EXAMPLE 16-2   Two charges $q_1 = -8 \ \mu C$ and $q_2 = +12 \ \mu C$ are placed 120 mm apart in the air. What is the resultant force on a third charge $q_3 = -4 \ \mu C$ placed midway between the other two charges?

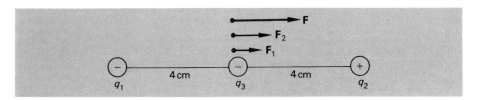

FIG. 16-10 Computing the resultant force on a charge placed midway between two other charges.

Solution   We convert the charges to coulombs ($1 \ \mu C = 1 \times 10^{-6}$ C), use their absolute values, and convert the distance to meters (120 mm = 0.12 m). One-half of 0.12 m is 0.06 m. A sketch is drawn in Fig. 16-10 to help you visualize the forces. The force on q_3 due to q_1 is directed to the right and is calculated from Coulomb's law:

$$F_1 = \frac{kq_1 q_3}{r^2} = \frac{(9 \times 10^9 \ \text{N·m}^2/\text{C}^2)(8 \times 10^{-6} \ \text{C})(4 \times 10^{-6} \ \text{C})}{(0.06 \ \text{m})^2}$$

$$= \frac{0.288 \ \text{N·m}^2}{0.0036 \ \text{m}^2} = 80 \ N \qquad \text{repulsion, to the right}$$

Similarly, the force F_2 on q_3 due to q_2 is equal to

$$F_2 = \frac{kq_2 q_3}{r^2} = \frac{(9 \times 10^9 \ \text{N·m}^2/\text{C}^2)(12 \times 10^{-6} \ \text{C})(4 \times 10^{-6} \ \text{C})}{(0.06 \ \text{m})^2}$$

$$= 120 \ N \qquad \text{attraction, also to the right}$$

The resultant force **F** is the vector sum of $\mathbf{F}_1$ and $\mathbf{F}_2$. Thus

$$\mathbf{F} = 80 \text{ N} + 120 \text{ N}$$
$$= 200 \text{ N} \qquad \text{directed to the right}$$

Note that the signs of the charges were used only to determine the direction of the forces; they were not substituted into the calculations.

16-6 ■ ELECTRIC FIELD

Both the electric force and the gravitational force are examples of *action-at-a-distance forces,* which are difficult to visualize. To aid in visualization, physicists have postulated the existence of electric and gravitational fields which surround charges and masses. We will consider that the mere presence of an electric charge alters the space surrounding it in such a way as to produce an electric force on another nearby charge. We say that an *electric field* exists in this space.

> An **electric field** *is said to exist in a region of space in which an electric charge will experience an electric force.*

This definition provides a test for the existence of an electric field. Simply place a charge at the point in question. If an electric force is observed, an electric field exists at that point.

We define the *electric field intensity* **E** at a point in terms of the force **F** experienced by a small positive charge $+q$ when it is placed at that point. See Fig. 16-11. The magnitude of the electric field intensity is the force per unit charge at a given point; it is found from

$$E = \frac{F}{q} \qquad \textit{Electric Field Intensity} \quad (16\text{-}3)$$

In the metric system, a unit of electric field intensity is the *newton per coulomb* (N/C). The usefulness of this definition rests with the fact that if we know the field intensity at any point, we can predict the force on any charge placed at that point.

Since the electric field intensity is defined in terms of a positive charge, its direction at any point is the same as the electric force on a positive charge placed at that point.

> *The direction of the electric field intensity* **E** *at a point in space is the same as the direction in which a positive charge would move if it were placed at that point.*

On this basis, the electric field in the vicinity of a positive charge $+Q$ would be outward or away from the charge. The field direction near a negative charge $-Q$ would be toward the charge.

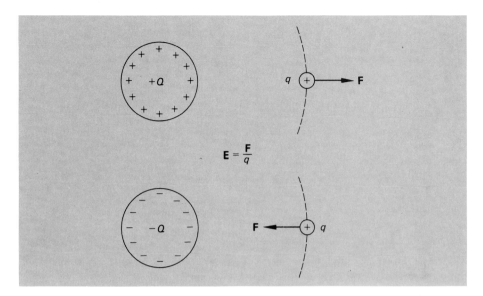

FIG. 16-11 The direction of the electric field intensity at a point is the direction in which a positive charge $+q$ placed at that point would move. Its magnitude is the force per unit charge (F/q).

An excellent method of visualizing both the strength and the direction of an electric field is with imaginary lines called *electric field lines*.

> **Electric field lines** *are imaginary lines drawn in such a manner that their direction at any point is the same as the direction of the electric field at that point.*

Such electric field lines would be drawn radially outward from a positive charge and radially inward toward a negative charge. The field intensity would be strong where the lines are close together and weak where they are far apart. A graphical description of the field for two common cases is given in Fig. 16-12. As a consequence of how these lines are drawn, they will always leave positive charges and enter negative charges.

Remember that the electric field intensity is a property assigned to the *space* which surrounds charged objects. A gravitational field exists above the earth whether or not a mass is positioned above the earth. Similarly, an electric field exists in the neighborhood of charged bodies whether or not another charge is placed in the field. If an external charge is positioned in an electric field at a point where the resultant field intensity is **E**, it will experience a force **F** given by

$$\boxed{\mathbf{F} = q\mathbf{E}} \tag{16-4}$$

as determined from Eq. (16-3). If q is positive, **E** and **F** will have the same direction; if q is negative, the force **F** will be directed opposite to the field **E**.

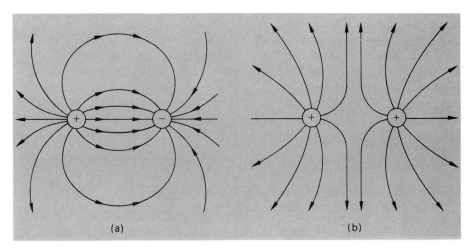

(a)   (b)

FIG. 16-12 (*a*) Graphical illustration of the electric field lines in the region surrounding two opposite charges. (*b*) Field lines between two positive charges.

EXAMPLE 16-3   The electric field intensity between the two plates in a cathode-ray tube (CRT) is constant and directed downward. (See Fig. 16-13.) The magnitude of the electric field intensity is 60,000 N/C. What are the magnitude and direction of the electric force exerted on an electron projected horizontally between the plates?

Solution   Since the direction of the field intensity **E** is defined in terms of a positive charge, the force on an electron will be upward, or opposite to **E.** The charge on an electron is -1.6×10^{-19} C. Thus the electric force will be

$$F = qE = (1.6 \times 10^{-19} \text{ C})(60,000 \text{ N/C})$$
$$= 9.6 \times 10^{-15} \text{ N upward}$$

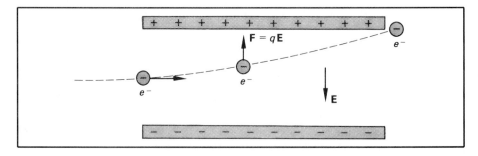

FIG. 16-13 An electron projected into an electric field of constant intensity **E.**

Remember that the absolute value of the charge is used. The directions of **F** and **E** are the same for positive charges and opposite for negative charges.

In the above example, we neglected the downward force due to the weight of the electron (approximately 9×10^{-30} N). The electric force is larger than the gravitational force in this instance by a factor of 1×10^{15}.

16-7 ■ ELECTRIC POTENTIAL ENERGY

One of the best ways to understand electric potential energy is to compare it with gravitational potential energy. In the gravitational case, consider that a mass m is moved from point A in Fig. 16-14 to point B. An external force $\mathbf{F}$ equal to the weight $m\mathbf{g}$ must be applied to move the mass against gravity. The work done by this force is the product of mg and h. When the mass m reaches point B, it has a *potential* for doing work relative to point A. The system has a *potential energy E_p* which is equal to the work done against gravity:

$$E_p = mg \cdot h$$

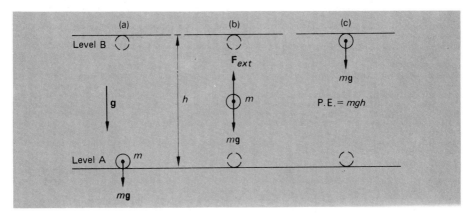

FIG. 16-14 Lifting a mass m against a gravitational field results in a potential energy of mgh at level B. The symbol $\mathbf{F}_{ext}$ represents the external force needed to move the mass against the field.

This expression represents the *potential* for doing work after the mass m is released at point B. Therefore, its magnitude does not depend on the path taken to reach point B.

Now, let us consider a positive charge $+q$ resting at point A in a uniform electric field $\mathbf{E}$ between two oppositely charged plates. (See Fig. 16-15.) An electric force $q\mathbf{E}$ acts downward on the charge. The work done against the electric field in moving the charge from A to B is equal to the product of the force $q\mathbf{E}$ and the distance d. Hence, the electric potential energy at point B relative to point A is

$$E_p = qE \cdot d \qquad (16\text{-}5)$$

Before we proceed, we should point out an important difference between gravitational potential energy and electric potential energy. In the case of gravity, there is only one kind of mass, and the forces involved are always forces of attraction. Therefore, a mass at higher elevations always has greater potential energy relative to the earth. This is not true in the electrical case because of the existence of negative charge. In the example, a positive charge, as in Fig. 16-15, has a greater potential energy at point B than at point A. This is true regardless of the reference point chosen for measuring the energy because work has been done *against* the field. (Refer to Fig. 16-16.) However,

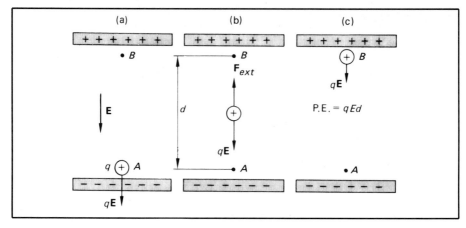

FIG. 16-15 A positive charge $+q$ moved against an electric field **E** results in a potential energy at qEd at point B relative to A. The external force $\mathbf{F}_{ext}$ is needed to move the positive charge against the field.

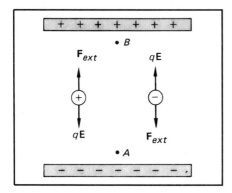

FIG. 16-16 A positive charge increases its potential energy when it is moved from A to B; a negative charge loses its potential energy.

if a negative charge were moved from point A to point B, work would be done *by* the field. A negative charge would have a *lower* potential energy at B, which is exactly opposite to the situation for a positive charge.

> *Whenever a positive charge is moved against an electric field, the potential energy increases; whenever a negative charge moves against an electric field, the potential energy decreases.*

The above rule is a direct consequence of the fact that the direction of the electric field is defined in terms of a positive charge.

16-8 ■ POTENTIAL DIFFERENCE

In practical electricity, we want to know the work required to move a unit charge from one point to another. The work done against electric forces in moving a charge from point A to point B would be equal to the difference in the potential energy at the two locations. This leads us to the concept of *potential difference*.

The **potential difference** V *between two points is the work per unit charge done against electric forces in moving a small test positive charge from one point to another point.*

The unit of potential difference is the *joule per coulomb,* defined as a *volt* (V). For two points A and B, we may denote the potential difference by V_{AB} and write

$$V_{AB} = \frac{\text{work}_{A \to B}}{q} \qquad (16\text{-}6)$$

Thus a potential difference of 6 V between two points A and B indicates that 6 J of work is required to move each coulomb (1 C) of charge from point A to point B. The potential difference is sometimes loosely referred to as the *voltage.*

Now let us return to the example of a uniform electric field between two oppositely charged plates, as in Fig. 16-17. Assume that the plates are separated by a distance d. A charge q placed in the region between plates A and B will experience a force given by $\mathbf{F} = q\mathbf{E}$. The work done by this force in moving the charge q from A to B is

$$\text{Work}_{A \to B} = qE \cdot d$$

The potential difference V_{AB} is found by dividing both sides of this equality by the charge q:

$$V_{AB} = \frac{\text{work}_{A \to B}}{q} = \frac{qEd}{q}$$

This yields a very useful relationship between V and E:

$$\boxed{V = Ed} \qquad (16\text{-}7)$$

The potential difference between two oppositely charged plates is equal to the product of the field intensity and the plate separation.

EXAMPLE 16-4   The potential difference between two plates 5 mm apart is 6 kV. (a) How much work is done against the electric field in moving a 2-μC charge from the negative plate to the positive plate? (b) What is the electric field intensity between the two plates?

Solution (a)   Solving Eq. (16-6) for work$_{A \to B}$, we obtain

$$\text{Work}_{A \to B} = qV_{AB}$$
$$= (2 \times 10^{-6} \text{ C})(6000 \text{ V}) = 0.012 \text{ J}$$

The unit *joule* comes from the fact that 1 V = 1 J/C.

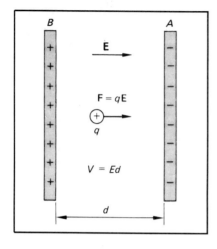

FIG. 16-17 Potential difference between two oppositely charged plates.

Solution (b)   To find the electric field intensity, we solve Eq. (16-7) for E, giving

$$E = \frac{V}{d} = \frac{6000 \text{ V}}{0.005 \text{ m}} = 1{,}200{,}000 \text{ V/m}$$

As an extra exercise, you should demonstrate that the *volt per meter* (V/m) is equivalent to the *newton per coulomb* (N/C). The SI unit for electric intensity is the volt per meter, which is sometimes referred to as the *potential gradient*.

16-9 ■ CAPACITORS AND CAPACITANCE

Any charged conductor can be looked on as a storehouse or source of electric charge. If another conductor is placed close to it, some of the charge on the initial conductor will flow to the second conductor. It is sometimes desirable to store large quantities of charge on conductors so that they may be used as sources of electric charge. Two such conductors placed close together constitute what is called a *capacitor*.

A **capacitor** *consists of two closely spaced conductors carrying equal and opposite charges.*

The simplest capacitor is the *parallel-plate capacitor* illustrated in Fig. 16-18. A potential difference between two such plates can be realized by connecting a battery to them, as shown in the figure. Electrons are transferred from plate B to plate A, producing an equal and opposite charge on the plates. The transfer of charge will continue until a maximum charge is obtained. This maximum quantity of charge for a given potential difference is called the *capacitance* of the arrangement.

The **capacitance** *between two conductors having equal and opposite charges* $(+Q$ *and* $-Q)$ *is the ratio of the magnitude of the charge on either conductor to the potential difference V between the two conductors:*

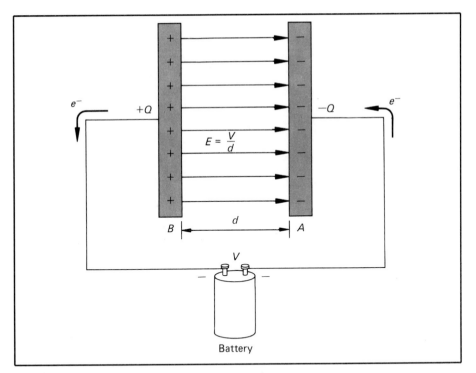

FIG. 16-18 A capacitor consists of two closely spaced conductors. It is charged by transferring charge from one plate to the other.

$$C = \frac{Q}{V}$$

Capacitance (16-8)

The unit for capacitance is the *coulomb per volt* (C/V), which is redefined as a *farad* (F). Thus, a capacitance of 1 F indicates that a charge of 1 C can be stored by a potential difference of 1 V. Because of the large size of the coulomb as a unit of charge, the farad is usually too large a unit for most applications. The following metric multiples are frequently used:

$$1 \ microfarad \ (1 \ \mu F) = 1 \times 10^{-6} \ F$$
$$1 \ picofarad \ (1 \ pF) = 1 \times 10^{-12} \ F$$

Capacitance as low as a few picofarads are not uncommon in some electrical communications applications.

EXAMPLE 16-5   A capacitor having a capacitance of 4 μF is connected to a 60-V battery. What is the charge on the capacitor?

Solution The charge on a capacitor refers to the magnitude of the charge on either plate. From Eq. (16-8), we obtain

$$Q = CV = (4 \ \mu\text{F})(60 \ \text{V}) = 240 \ \mu\text{C}$$

Since the charge is expressed in microcoulombs, it is not necessary to convert microfarads to farads in this example.

The capacitance of a given parallel-plate capacitor is determined by a number of physical characteristics. It can be shown that the area of the plates, their separation distance, and the medium between the plates all have an effect on the capacitance. In general, a larger conductor can hold more charge, and a capacitor can store more charge than a single conductor because of the induced charge resulting from two closely spaced conductors. The following observation applies:

> *The capacitance of a parallel-plate capacitor is directly proportional to the area of the plates and inversely proportional to their separation.*

In other words, if we double the surface area of the plates, the capacitance will also be doubled. Also, we can double the capacitance by decreasing the separation to one-half its original distance.

Parallel-plate capacitors are frequently made with a stack of several plates, as shown in Fig. 16-19. By making one of the sets of plates movable, a variable capacitor can be constructed. Rotating one set of plates relative to the other set varies the effective area of the capacitor plates, causing a variation in the capacitance. Variable capacitors are often used in the tuning circuits of radios.

Sometimes a capacitor will *break down* due to the fact that the material between the plates becomes a conductor and leaks charge from one plate to the other. To prevent this occurrence, and to increase the capacity for storing charge, a nonconducting material is often positioned between the plates. Such a material is called a *dielectric*. The following advantages are realized by the insertion of a dielectric:

1. A dielectric material provides for a small plate separation without contact.
2. A dielectric increases the capacitance.
3. Higher voltages may be used without danger of dielectric breakdown.
4. A dielectric often provides greater mechanical strength.

Common dielectric materials are mica, paraffined paper, ceramics, and plastics. Alternating thin sheets of metal foil and paraffin-coated paper can be rolled up to provide a compact capacitor with a capacitance of several microfarads.

The number of times a dielectric is as effective as air is called the *dielectric constant* K of that substance. If we denote the capacitance of a capacitor with a vacuum or air between its plates by C_0 and the capacitance with the dielectric by C, the dielectric constant K is given by

$$K = \frac{C}{C_0} \qquad \textit{Dielectric Constant} \quad (16\text{-}9)$$

Thus air has a dielectric constant of 1, mica has a constant of about 5, and waxed paper may have a K value of about 3. If the capacitance were 4 μF in air, the insertion of a piece of mica would increase the capacitance to 5 × 4 μF, or 20 μF.

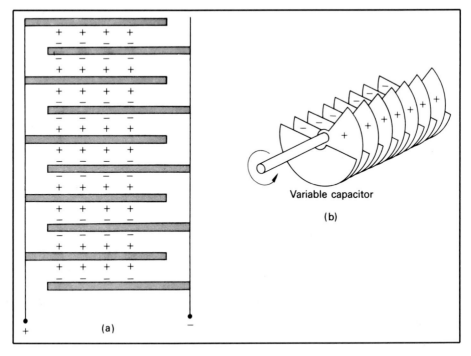

FIG. 16-19 (*a*) A capacitor consisting of a number of stacked plates with alternating positive and negative charges. (*b*) A variable capacitor allows one set of plates to move relative to the other, causing a variation in the effective area.

Summary

Electrostatics has been presented as the science which treats charges at rest. The electric force, electric field intensity, electric potential difference, and capacitance are concepts which have been defined and discussed at length in this chapter. The main ideas are summarized below:

- An object which has an excess of electrons is said to be *negatively charged,*and an object which has a deficiency of electrons is said to be *positively charged.*
- The first law of electrostatics states that *like charges repel each other and unlike charges attract each other.*
- Coulomb's law states that *the force of attraction or repulsion between two point charges is directly proportional to the product of the two charges and inversely proportional to the square of their separation distance.*

$$F = \frac{kqq'}{r^2} \qquad k = 9 \times 10^9 \ \text{N·m}^2/\text{C}^2$$

■ An *electric field* is said to exist in a region of space in which an electric charge will experience an electric force. The *magnitude* of the electric field intensity **E** is given by the force **F** per unit charge q.

$$\mathbf{E} = \frac{\mathbf{F}}{q}$$

Electric Field Intensity

■ The *direction* of **E** at a given point is the direction in which a positive charge would move if it were placed at that point. The *electric field lines* which describe the field surrounding charges always lead *away* from positive charges and *toward* negative charges.

■ *Electric potential energy* is defined as the work done against electric forces in moving a charge from one point to another. When a distance d is moved against a constant **E**,

$$E_p = qEd$$

Potential Energy

■ Due to the existence of two kinds of charge, it must be remembered that the potential energy *increases* as a *positive* charge is moved *against* the field **E** and the potential energy *decreases* as a *negative* charge is moved *against* the same field **E**.

■ The *potential difference* V_{AB} between two points A and B is the work per unit charge done against electric forces in moving a small positive test charge from the point of lower potential energy to the point of higher potential energy.

$$V_{AB} = \frac{\text{work}_{A \to B}}{q}$$

Potential Difference

■ The potential difference between two oppositely charged plates is equal to the product of E and d:

$$V = Ed$$

Plate Voltage

■ The *capacitance* C of a parallel-plate capacitor is the ratio of the charge Q on either plate to the applied voltage V across the plates. One *farad* (F) is equal to one *coulomb* (C) per *volt* (V).

$$C = \frac{Q}{V}$$

Capacitance

- The capacitance is *directly* proportional to the plate area A and *inversely* proportional to the plate separation d.
- The material between the plates of a capacitor is called a *dielectric*. The *dielectric constant K* is the ratio of the capacitance with the dielectric to the capacitance with only air or a vacuum between the plates.

Questions

16-1. Define the following terms:

a. electrostatics	**i.** electric potential energy
b. ion	**j.** potential difference
c. positive charge	**k.** volt
d. negative charge	**l.** potential gradient
e. induced charge	**m.** capacitor
f. Coulomb's law	**n.** capacitance
g. electric field	**o.** dielectric
h. electric field intensity	**p.** dielectric constant

16-2. In the process of rubbing a glass rod with a silk cloth, is electric charge *created?* Explain what happens. What is the nature of the charge on the silk cloth?

16-3. An insulated stand supports a charged metal ball in the laboratory. How can you determine the nature of its charge?

16-4. During a laboratory experiment, two bodies are seen to attract each other. Is this conclusive proof that they are *both* charged? Explain. If the bodies are seen to repel each other, is this proof that they are *both* charged? Why?

16-5. When you are charging a metal sphere by induction, should your finger be taken away before the charging rod is removed? Why?

16-6. Two identical charges experience a repulsive force of 20 N when they are separated by a distance of 10 mm. What will be the new force if their separation is increased to 20 mm?

16-7. Can an electric field exist in a region of space in which an electric charge would *not* experience a force? Is it necessary that a charge be placed at a point in order to have an electric field at that point? Explain.

16-8. Discuss the similarities between electric fields and gravitational fields. How does gravitational potential energy differ from electric potential energy?

16-9. Can electric field lines ever intersect? Can a line begin and end on the same conductor? Discuss.

16-10. Why is the electric field intensity constant in the region between two parallel plates of equal and opposite charge? Draw vector diagrams of the field due to each plate at various locations between the plates.

16-11. Distinguish clearly between positive and negative work. Distinguish between positive and negative potential energy.

16-12. The direction of electric field intensity is from a point A to a point B. Will a positive charge have a higher potential energy at A or at B? What about a negative charge?

16-13. Distinguish between potential difference and a difference in potential energy.

16-14. Is it possible to increase the potential energy of a mass m by moving it to a lower elevation? Is it possible for a charged object to increase its potential energy by moving in the same direction as the electric field?

16-15. Air is pumped from one metal tank to another, creating a partial vacuum in one tank and a high pressure in the other. When the pump is removed, potential energy is stored. The energy is released if the two tanks are reconnected and the pressure becomes equal in the tanks. In what ways is this mechanical example similar to charging and discharging a capacitor?

16-16. The Leyden jar is a capacitor which consists of a glass jar coated inside and out with tinfoil, as shown in Fig. 16-20. Contact with the inside coating is made with a metal chain connected to the central metal rod. From the figure, explain how the capacitor becomes charged. What is the function of the ground wire?

16-17. What effect do the following actions have on the capacitance, assuming that other parameters remain constant? (*a*) Doubling the area of the plates, (*b*) doubling the separation of the plates, (*c*) inserting a dielectric whose K value is 4.

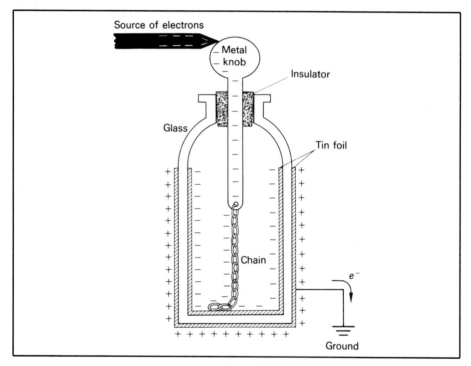

FIG. 16-20 Leyden jar.

Problems

16-1. In an experiment 40,000 electrons are transferred from body A to body B. How many coulombs of charge are on each body, and which is the positively charged body? (Remember that the charge of one electron is -1.6×10^{-19} C.)

Answer 6.4×10^{-15} C; body A.

16-2. If it were possible to place 1 C of charge on an object, how many electrons would have to be transferred? How many electrons are needed for 1 μC?

16-3. A small metal sphere is given a charge of $+40$ μC, and a second sphere located 100 mm away is given a charge of -10 μC. (*a*) What is the force between them? Is it attraction or repulsion? (*b*) The two spheres are now allowed to touch for a minute. Consider what happens as a result of the contact. Now the spheres are again separated a distance of 100 mm. What is the force between them? Is it attraction or repulsion?

> *Answer* (*a*) 360 N, attraction; (*b*) 202 N, repulsion.

16-4. An alpha particle consists of two protons and two neutrons. What is the force of repulsion between two alpha particles separated by a distance of 2×10^{-9} m?

16-5. Two balls, each having a charge of $+6 \ \mu C$, are separated by 2 cm. What is the force between them? Is it a repulsion or attraction?

> *Answer* 810 N; repulsion.

16-6. A charge of $-5 \ \mu C$ is placed 8 cm from a charge of $+12 \ \mu C$. What is the force between them? Is it attraction or repulsion?

16-7. A $+300\text{-}\mu C$ charge is placed 80 mm to the left of a $+200\text{-}\mu C$ charge. What is the resultant force on a $-40\text{-}\mu C$ charge placed midway between the first two charges?

> *Answer* -1.44×10^4 N, left.

16-8. Two identical charges of $+50 \ \mu C$ are 0.006 m apart. What is the resultant force on a $+20\text{-}\mu C$ charge placed on a line joining the two charges at a distance of 0.002 m from the charge on the right side?

16-9. A charge of $+2 \ \mu C$ placed in an electric field experiences a force of 8×10^{-4} N. What is the magnitude of the electric field intensity?

> *Answer* 400 N/C.

16-10. A $-12\text{-}\mu C$ charge experiences an upward electric force of 0.5 N. What are the magnitude and direction of the electric field intensity?

16-11. The electric field intensity between two horizontal plates is a constant 40,000 N/C directed vertically upward. What are the magnitude and direction of the electric force on a $+6\text{-}\mu C$ charge that is projected horizontally between the two plates? What would be the magnitude and direction of the electric force on a $-3\text{-}\mu C$ charge?

> *Answer* 0.24 N, upward; 0.12 N, downward.

16-12. A $-8\text{-}\mu C$ charge is projected upward between two vertical plates. The charge experiences an electric force of 6 N directed to the right. What are the magnitude and direction of the field intensity? What would be the force on a $+4\text{-}\mu C$ charge placed between these plates?

16-13. The potential difference between two plates 6 mm apart is 800 V. (*a*) How much work is required to move a 4×10^{-9} C charge from the positive plate to the negative plate? (*b*) What is the electric field intensity between the plates?

> *Answer* (*a*) 3.2×10^6 J; (*b*) 1.33×10^5 V/m.

16-14. What is the potential difference between two metal spheres if 4 J of work is done against electric forces in moving a $-3\text{-}\mu C$ charge from point A to point B? Is the motion in the same direction as the electric field?

16-15. Two metal plates are separated by 40 mm and are oppositely charged, so that the electric field between them is 4×10^4 N/C. How much work must be done *against* the electric field in order to move a $-2\text{-}\mu C$ charge from the negative plate to the positive plate? How much work is done *by* the electric field? What is the potential energy after the charge is moved? What is the potential difference between the plates?

> *Answer* -0.0032 J; $+0.0032$ J; -0.0032 J; 1600 V.

16-16. How much work is done *by* the electric field in moving a $-3\text{-}\mu C$ charge between the plates of Prob. 16-15? What is the potential energy after the charge is moved? Is this energy negative? Explain.

16-17. A pair of large parallel plates 50 mm apart in air are charged until they have a potential difference of 1600 V. Determine the potential gradient between the plates.

Answer 32,000 V/m.

16-18. Two large parallel plates are 80 mm apart and have a potential difference of 800 V. What is the magnitude of the force which would act on an electron placed between the two plates? What would be the kinetic energy of the electron if it moved from one plate to the other?

16-19. Find the capacitance of a parallel-plate capacitor if 1200 μC of charge is held on its plates when 40 V is applied across the capacitor.

Answer 30 μF.

16-20. How much charge will be stored by a 40-μF capacitor when it is connected to a 120-V potential difference?

16-21. A capacitor has a capacitance of 12 μF when the dielectric is air. It is charged to 800 V by means of a battery. (*a*) What is the charge on each plate? (*b*) If the dielectric is replaced by mica ($K = 5$), what will the new capacitance be? (*c*) What will be the new charge stored on each plate, assuming the same potential difference?

Answer (*a*) 9600 μC; (*b*) 60 μF; (*c*) 48,000 μC.

16-22. What potential difference is needed to store 1200 μC if the capacitance is 6 μF and the dielectric is air? What voltage is required to store the same charge if the dielectric is changed to porcelain ($K = 6$)?

16-23. The breakdown voltage of a particular capacitor is 500 V. What is the maximum charge on each plate if the capacitance is 7 μF?

Answer 3500 μC.

17

DIRECT ELECTRIC CURRENT

As a result of completing this chapter, you should be able to

1. Demonstrate by definition and examples your understanding of the concepts of *electric current* and *electromotive force,* including their units.
2. Write *Ohm's law* and apply it to the solution of electrical problems involving resistance.
3. Discuss the effects of *material, length, area,* and *temperature* on electric resistance.
4. Compute the power losses in an electric circuit when any two of the following quantities are known: voltage, current, and resistance.
5. Illustrate and write statements concerning the voltage, current, and effective resistance for resistors connected in *series* and in *parallel.*
6. Calculate the effective resistance for a group of resistors connected in series or in parallel.
7. Determine the current and voltage for each resistor in a circuit containing known resistors connected in series and in parallel with a given source of emf.
8. Understand and apply the relationship among the terminal voltage, the emf, the internal resistance, and the load resistance for a given direct-current (dc) circuit.

Almost every technician or other worker in industry is expected to check the operation of equipment by making measurements. Usually such checks require the worker to understand electric currents well enough to identify certain electric components of equipment and replace them when they are defective. In this chapter, you will learn about the relationship among voltage, current, and resistance. Simple series and parallel circuits will be analyzed in order to familiarize you with different ways of controlling the flow of electric current. The concepts introduced in this section form the very foundation of basic electricity. You cannot study the operation of ammeters, voltmeters, generators, motors, and many other electric devices without first mastering the ideas contained in this unit.

17-1 ■ MOTION OF ELECTRIC CHARGE

We now leave electrostatics and enter a discussion of charges in motion. We have been concerned with forces, electric fields, and potential energies as they relate to charged conductors. For example, excess electrons evenly distributed over an insulated spherical surface will remain at rest. However, if a wire is connected from the sphere to ground, the electrons will flow through the wire to the ground. The flow of charge constitutes an *electric current*.

Let us begin our discussion of moving charges by considering the discharge of a capacitor. The potential difference V between the two capacitor plates in Fig. 17-1a is a function of the total charge Q, given by

$$Q = CV$$

where C is the capacitance. If a path is provided, electrons on one plate will travel to the other, decreasing the net charge on each plate. Any conductor used to connect the plates of a capacitor will cause it to discharge. However, the rate of discharge varies considerably with the size, shape, material, and temperature of the conductor.

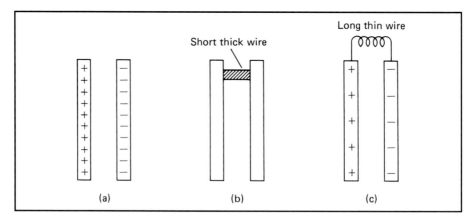

FIG. 17-1 (a) A charged capacitor is a source of current. (b) If the plates are joined by a short, thick wire, the capacitor will discharge quickly. (c) A long, thin wire allows for a gradual discharge.

If a short, thick wire is connected to the plates, as shown in Fig. 17-1b, the capacitor discharges instantly, indicating a very rapid transfer of charge. This current, which exists for a very short time, is called a *transient current*. If we replace the short, thick wire with a long, thin wire of the same material, we observe a gradual discharge, as shown in Fig. 17-1c. Such opposition to the flow of electricity is called *electrical resistance*. A quantitative description of resistance will be given later. It is introduced here to illustrate that the rate at which charge flows through a conductor varies. This rate is referred to as the *electric current*.

The **electric current** *I is the rate of flow of charge Q past a given point P on an electric conductor:*

$$I = \frac{Q}{t}$$

Electric Current (17-1)

The unit of electric current is the ampere. One *ampere* (A) represents a flow of charge at the rate of one coulomb per second past any point:

$$\text{One ampere (A)} = \frac{\text{one coulomb (C)}}{\text{one second (s)}}$$

In the example of a discharging capacitor, the current arises from the motion of electrons, as illustrated in Fig. 17-2. The positive charges in a wire are tightly bound and cannot move. The electric field created in the wire because of the potential difference between the plates causes the free electrons in the wire to experience a drift toward the positive plate. The electrons are repeatedly deflected or stopped by processes relating to impurities and thermal motions of the atoms. Consequently, the motion of the electrons is not an accelerated one but a drifting or diffusion process. The average drift velocity of electrons is typically of the order of a few meters per hour. This velocity of charge, which is a *distance* per unit of time, should not be confused with current, which is a *quantity* of charge per unit of time.

An analogy to water flowing through a pipe is useful in understanding current flow. The rate of flow of the water in gallons per minute is analogous to the rate of flow of charge in coulombs per second. For a current of 1 A, 6.25×10^{18} electrons (1 C) flow past a given point every second. Just as the size and length of a pipe affect the flow of water, the size and length of a conductor affect the flow of electrons.

EXAMPLE 17-1 How many electrons pass a point in 5 s if a constant current of 8 A is maintained in a conductor?

Solution From Eq. (17-1),

$$Q = It = (8\text{ A})(5\text{ s})$$
$$= (8\text{ C/s})(5\text{ s}) = 40\text{ C}$$
$$= (40\text{ C})(6.25 \times 10^{18}\text{ electrons/C})$$
$$= 2.50 \times 10^{20}\text{ electrons}$$

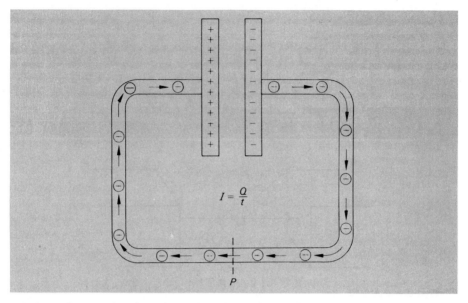

$$I = \frac{Q}{t}$$

FIG. 17-2 Current arises from the motion of electrons and is a measure of the quantity of charge passing a given point in a unit of time.

17-2 ■ DIRECTION OF ELECTRIC CURRENT

Thus far, we have discussed only the magnitude of electric current. The choice of direction is purely arbitrary as long as we apply our definition consistently. The flow of charge caused by an electric field in a gas or a liquid consists of a flow of positive ions in the direction of the field or a flow of electrons opposite to the field direction. As we have seen, the current in a metallic material consists of electrons flowing against the field direction. However, a current that consists of negative particles moving in one direction is electrically the same as a current consisting of positive charges moving in the opposite direction.

There are a number of reasons for preferring the motion of positive charge as an indicator of direction. First, all the concepts introduced in electrostatics were defined in terms of positive charges, e.g., the electric field, potential energy, and potential difference. An electron flows contrary to the electric field and "up a potential hill" from the negative plate to the positive plate. If we define current as a flow of *positive charge,* the loss in energy as charge encounters resistance will be from + to −, or "down a potential hill." By convention, we consider all currents as consisting of a flow of positive charge.

> *The direction of conventional current is always the same as the direction in which positive charges would move, even if the actual current consists of a flow of electrons.*

For a metallic conducting wire, both electron flow and conventional current are indicated in Fig. 17-3. The zigzag line is used to indicate the electrical resistance R. Note that the conventional current flows from the positive plate of the capacitor, neutralizing negative charge on the other plate. Conventional current is in the same direction as the electric field **E** producing the current.

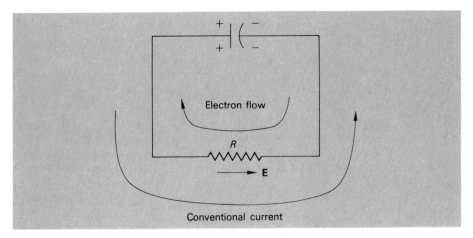

FIG. 17-3 In a metallic conductor, the direction of conventional current is opposite to the flow of electrons.

17-3 ■ ELECTROMOTIVE FORCE

The currents discussed in the preceding sections are called *transient currents* because they exist only for a short time. Once the capacitor has been completely discharged, there will no longer be a potential difference to promote the flow of additional charge. If some means were available to keep the capacitor continually charged, a continuous current could be maintained. This would require that electrons be continuously supplied to the negative plate to replace those leaving. In other words, energy must be supplied to replace the energy lost by the charge in the external circuit. In this manner, the potential difference between the plates could be maintained, allowing for a continuous flow of charge. A device with the ability to maintain a potential difference between two points is called a *source of electromotive force* (emf).

The most familiar sources of emf are batteries and generators. Batteries convert chemical energy to electric energy, and generators transform mechanical energy into electric energy. The detailed nature and operation of these devices will be discussed in a later chapter.

> A **source of emf** is a device which converts chemical, mechanical, or other forms of energy to the electric energy necessary to maintain a continuous flow of electric charge.

In an electric circuit, the source of emf is usually represented by the symbol ε.

The function of a source of emf in an electric circuit is similar to the function of a water pump in maintaining the continuous flow of water through a system of pipes. In Fig. 17-4a the water pump must perform the work on each unit volume of water necessary to replace the energy lost by each unit volume flowing through the pipes. In Fig. 17-4b the source of emf must do work on each unit of charge which passes through it in order to raise it to a higher potential. This work must be supplied at a rate equal to the rate at which energy is lost in flowing through the circuit.

By convention, we have assumed that the current consists of a flow of positive charge even though in most cases it is negative electrons. Therefore, the charge loses energy in passing through the resistor from a high potential to a low potential. In the

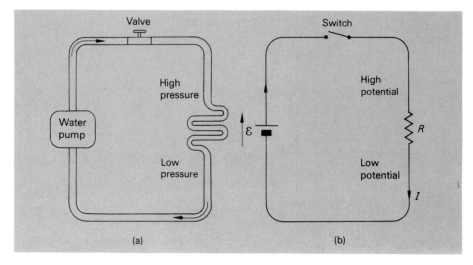

FIG. 17-4 The mechanical analogy of a water pump can be used to explain the function of a source of emf in an electric circuit.

hydraulic analogy, water passes from high pressure to low pressure. When the shutoff valve is closed, pressure exists but there is no water flow. Similarly, when the electric switch is *open*, there is voltage but no current.

Since emf is work per unit charge, it is expressed in the same unit as potential difference, i.e., the *joule per coulomb,* or *volt:*

> *A source of emf of one volt will perform one joule of work on each coulomb of charge which passes through it.*

For example, a 12-V battery performs 12 J of work on each coulomb of charge transferred from the low-potential side ($-$ terminal) to the high-potential side ($+$ terminal). An arrow ($\uparrow$) is usually drawn next to the symbol ε for an emf to indicate the direction in which the source, acting alone, would cause a positive charge to move through the external circuit. The conventional current is directed away from the positive terminal of a battery, and the hypothetical positive charge flows ''downhill'' through external resistance to the negative terminal of the battery.

In the following sections, we will frequently use circuit diagrams, like that in Fig. 17-4b, to describe electric systems. Many of the symbols we shall use are defined in Fig. 17-5.

17-4 ■ OHM'S LAW; RESISTANCE

Resistance R is defined as opposition to the flow of electric charge. Although most metals are good conductors of electricity, all offer some opposition to the surge of electric charge through them. This electrical resistance is fixed for many specific materials of known size, shape, and temperature. It is independent of the applied emf and the current passing through it.

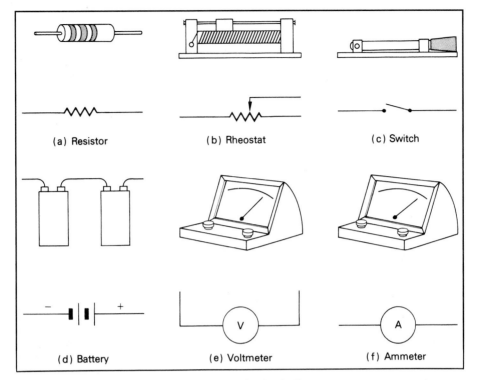

FIG. 17-5 Conventional symbols used in electric circuit diagrams.

The effects of resistance in limiting the flow of charge were first studied quantitatively by Georg Simon Ohm in 1826. He discovered that, *for a given resistor at a particular temperature, the current is directly proportional to the applied voltage.* Just as the rate of flow of water between two points depends on the difference in height between them, the rate of flow of electric charge between two points depends on the difference in potential between them. This proportionality is usually stated as *Ohm's law:*

> *The current produced in a given conductor is directly proportional to the difference of potential between its endpoints.*

The current I which is observed for a given voltage V is therefore an indication of resistance. Mathematically, the resistance R of a given conductor can be calculated from

$$R = \frac{V}{I} \qquad V = IR \qquad (17\text{-}2)$$

The greater the resistance R, the smaller the current I for a given voltage V. The unit of measurement of resistance is the *ohm,* for which the symbol is the Greek capital letter *omega* (Ω). From Eq. (17-2), it is seen that

$$1 \, \Omega = \frac{1 \, V}{1 \, A}$$

A resistance of *one ohm* will allow a current of *one ampere* when a potential difference of *one volt* is impressed across its terminals.

EXAMPLE 17-2   The difference of potential between the terminals of an electric heater is 80 V when there is a current of 6 A in the heater. What will the current be if the voltage is increased to 120 V?

Solution   According to Ohm's law, the resistance of the coils in the heater is

$$R = \frac{V}{I} = \frac{80 \, V}{6 \, A} = 13.3 \, \Omega$$

Therefore, if the voltage is increased to 120 V, the new current will be

$$I = \frac{V}{R} = \frac{120 \, V}{13.3 \, \Omega} = 9 \, A$$

Here we have neglected any change in resistance due to a rise in the temperature of the heating coils.

Four devices commonly used in the laboratory to study Ohm's law are the battery, the voltmeter, the ammeter, and the rheostat. As their names imply, the voltmeter and the ammeter measure voltage and current. The rheostat is simply a variable resistor. A sliding contact changes the number of resistance coils through which charge can flow. A laboratory collection of these electric devices is illustrated in Fig. 17-6. You should study the circuit diagram in Fig. 17-6a and justify the electric connections shown pictorially in Fig. 17-6b. Note that the voltmeter is connected in parallel with the battery, whereas the ammeter is connected in series. In general, the positive terminals are color-coded red, and the negative terminals are black.

EXAMPLE 17-3   The apparatus shown in Fig. 17-6 is used to study Ohm's law in the laboratory. The voltage V is determined by the source of emf and remains at 6 V. (*a*) What is the resistance when the rheostat is varied to indicate a current of 0.4 A? (*b*) If this resistance is doubled, what will the new current be?

Solution (a)   From Ohm's law,

$$R = \frac{V}{I} = \frac{6 \, V}{0.4 \, A} = 15 \, \Omega$$

Solution (b)   Doubling the resistance to 30 Ω would result in a current given by

$$I = \frac{6 \, V}{30 \, \Omega} = 0.2 \, A$$

Note that doubling the resistance in a circuit reduces the current by one-half.

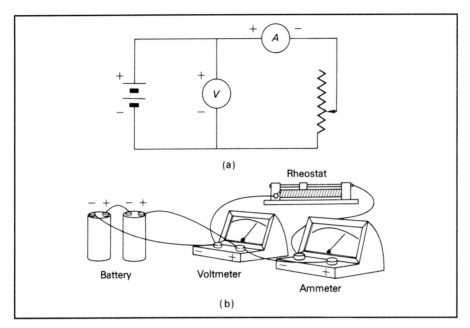

FIG. 17-6 (*a*) Circuit diagram for studying Ohm's law. (*b*) Pictorial diagram showing how the various elements in the circuit are connected in the laboratory.

17-5 ■ *ELECTRIC POWER AND HEAT LOSS*

We have seen that electric charge gains energy within a generating source of emf and loses energy in passing through external resistance. Inside the source of emf, work is done *by the source* in raising the potential energy of charge. As the charge passes through the external circuit, work is done *by* the charge on the components of the circuit. In the case of a pure resistor, the energy is dissipated in the form of heat. If a motor is attached to the circuit, the energy loss is divided between heat and useful work. In any case, the energy gained in the source of emf must equal the energy lost in the entire circuit.

The rate at which heat is dissipated in an electric circuit is referred to as the *power loss*. When charge is flowing continuously through a circuit, this power loss is given by

$$P = \frac{\text{work}}{t} = VI \qquad\qquad \textit{Electric Power} \quad (17\text{-}3)$$

When V is in volts and I is in amperes, the power loss is measured in watts. That the product of voltage and current will give a unit of power is shown as follows:

$$(\text{V})(\text{A}) = \left(\frac{\text{J}}{\text{C}}\right)\left(\frac{\text{C}}{\text{s}}\right) = \frac{\text{J}}{\text{s}} = \text{W}$$

Equation (17-3) can be expressed in alternative forms by using Ohm's law ($V = IR$). Substituting for V, we obtain

$$P = VI = I^2R \tag{17-4}$$

Substitution of $I = V/R$ into Eq. (17-3) gives another form:

$$P = VI = \frac{V^2}{R} \tag{17-5}$$

The relation given as Eq. (17-4) is so often used in electrical work that heat loss in electric wiring is often referred to as an "*I*-squared *R*" loss. In summary, the following relations are used in problems involving electric power:

$$P = VI \qquad P = I^2R \qquad P = \frac{V^2}{R}$$

EXAMPLE 17-4   A current of 6 A flows through a resistance of 300 Ω for 1 h. What is the power loss? How much heat is generated in joules?

Solution   From Eq. (17-4),

$$P = I^2R = (6 \text{ A})^2(300 \text{ Ω}) = 10{,}800 \text{ W}$$

Since the power represents the heat lost (work) per unit of time, we obtain

$$\text{Work} = Pt = (10{,}800 \text{ W})(3600 \text{ s})$$
$$\text{Heat loss} = 3.89 \times 10^7 \text{ J}$$

This represents about 36,900 Btu.

17-6 ■ FACTORS AFFECTING RESISTANCE

Just as capacitance is independent of the voltage and quantity of charge, the resistance of a conductor is independent of current and voltage. Both capacitance and resistance are inherent properties of a conductor. The resistance of a wire of uniform cross-sectional area, like the one shown in Fig. 17-7, is determined by the following four factors:

1. *The kind of material:* Different materials offer varying amounts of resistance to the flow of electrons. Iron wire, for example, offers about 7 times the resistance as copper wire of the same diameter and length. Wires of given diameters (gauges) are often rated in *ohms per foot* or in *ohms per meter.*
2. *The length:* The resistance of a conductor is directly proportional to its length.

3. *The cross-sectional area:* The resistance of a wire is inversely proportional to its cross-sectional area. A wire with twice the cross-sectional area will offer only one-half the resistance. Since area varies with the square of the diameter, a wire with 3 times the diameter will offer one-ninth as much resistance.

4. *The temperature:* For most metallic conductors the resistance increases directly with temperature. This fact forms the basis for many thermometers which use electrical resistance as an indicator of temperature. However, the resistance of most liquid and nonmetallic conductors (carbon) *decreases* with a rise in temperature.

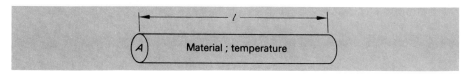

FIG. 17-7 The resistance of a wire depends on the kind of material, length, cross-sectional area, and temperature of the wire.

17-7 ■ SIMPLE CIRCUITS; RESISTORS IN SERIES

An electric circuit consists of any number of branches joined so that at least one closed path is provided for current. The simplest circuit consists of a single source of emf joined to a single external resistance, as shown in Fig. 17-8. If ε represents the emf and R indicates the total resistance, Ohm's law yields

$$\varepsilon = IR \qquad (17\text{-}6)$$

where I is the current through the circuit. All the energy gained by charge in passing through the source of emf is lost in flowing through the resistance.

Let us consider the addition of a number of elements to a circuit. Two or more elements are said to be in *series* if they have only *one* point in common that is not connected to some third element. Current can follow only a single path through elements in series. Resistors R_1 and R_2 of Fig. 17-9a are in series because point A is common to both resistors. However, the resistors in Fig. 17-9b are not in series because point B is common to three current branches. Electric current entering such a junction may follow two separate paths.

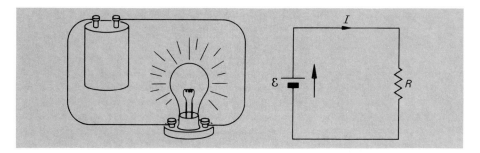

FIG. 17-8 A simple electric circuit and its diagram.

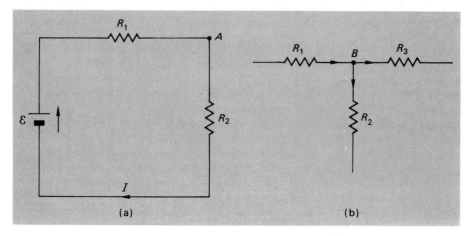

FIG. 17-9 (*a*) Resistors connected in series. (*b*) Resistors not connected in series.

Suppose that three resistors R_1, R_2, and R_3 are connected in series and enclosed in a box, indicated by the shaded portion of Fig. 17-10. The effective resistance R of the three resistors can be determined from the external voltage V and current I, as recorded by the meters. From Ohm's law

$$R = \frac{V}{I} \tag{17-7}$$

But what is the relationship of R to the three individual resistances? The current through each resistor must be identical, since a single path is provided. Thus

$$I = I_1 = I_2 = I_3 \tag{17-8}$$

Utilizing this fact and noting that Ohm's law applies equally well to any part of a circuit, we write

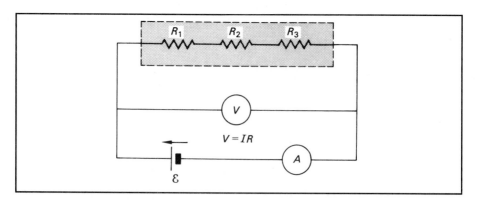

FIG. 17-10 Voltmeter-ammeter method of measuring the effective resistance of a number of resistors connected in series.

$$V = IR \qquad V_1 = IR_1 \qquad V_2 = IR_2 \qquad V_3 = IR_3 \qquad\qquad (17\text{-}9)$$

The external voltage V represents the sum of the energies lost per unit of charge in passing through each resistance. Therefore,

$$V = V_1 + V_2 + V_3$$

Finally, if we substitute from Eq. (17-9) and divide out the current, we obtain

$$IR = IR_1 + IR_2 + IR_3$$

$$\boxed{R = R_1 + R_2 + R_3 \qquad \text{series}} \qquad\qquad (17\text{-}10)$$

To summarize what has been learned about resistors connected in *series:*
1. The current in all parts of a series circuit is the same.
2. The voltage across a number of resistances in series is equal to the sum of the voltages across the individual resistors.
3. The effective resistance of a number of resistors in series is equivalent to the sum of the individual resistances.

EXAMPLE 17-5 The resistances R_1 and R_2 in Fig. 17-9a are 2 and 4 Ω, respectively. If the source of emf maintains a constant potential difference of 12V, what is the current delivered to the external circuit? What is the potential drop across each resistor?

Solution The effective resistance is

$$R = R_1 + R_2 = 2\ \Omega + 4\ \Omega = 6\ \Omega$$

The current is then found from Ohm's law:

$$I = \frac{V}{R} = \frac{12\ \text{V}}{6\ \Omega} = 2\ \text{A}$$

The voltage drops are therefore

$$V_1 = IR_1 = (2\ \text{A})(2\ \Omega) = 4\ \text{V}$$
$$V_2 = IR_2 = (2\ \text{A})(4\ \Omega) = 8\ \text{V}$$

Note that the sum of the voltage drops $(V_1 + V_2)$ is equal to the applied 12 V.

17-8 ■ RESISTORS IN PARALLEL

There are several limitations in the operation of series circuits. If a single element in a series circuit fails to provide a conducting path, the entire circuit is opened and current ceases. It would be quite annoying if all electric devices in a shop were to cease functioning whenever one lamp burned out. Moreover, each element in a series circuit adds to the total resistance of the circuit, thereby limiting the total current which can be supplied. These objections can be overcome by providing alternative paths for electric current. Such a connection, in which current can be divided between two or more elements, is called a *parallel connection*.

A *parallel circuit* is one in which two or more components are connected to two common points in the circuit. For example, in Fig. 17-11 resistors R_2 and R_3 are in parallel because they both have points A and B in common. Note that the current I, provided by the source of emf, is divided between resistors R_2 and R_3.

To arrive at an expression for the equivalent resistance R of a number of resistances connected in parallel, we follow a procedure similar to that discussed for a series connection. Assume that three resistors R_1, R_2, and R_3 are placed inside a box, as shown in Fig. 17-12. The total current I delivered to the box is determined by its effective resistance and the applied voltage:

$$I = \frac{V}{R} \qquad (17\text{-}11)$$

In a parallel connection, the voltage drop across each resistor is the same and equivalent to the total drop in voltage:

$$V = V_1 = V_2 = V_3 \qquad (17\text{-}12)$$

That this is true can be realized when we consider that the same energy must be lost by a unit of charge regardless of the path it travels in the circuit. In this example, charge may flow through any one of the three resistors. Thus the total current delivered is divided among the resistors:

$$I = I_1 + I_2 + I_3 \qquad (17\text{-}13)$$

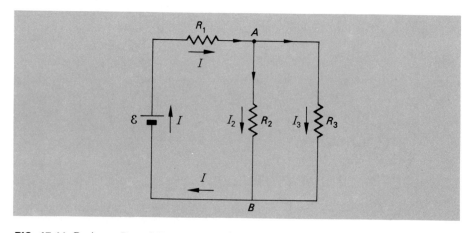

FIG. 17-11 Resistors R_2 and R_3 are connected in parallel.

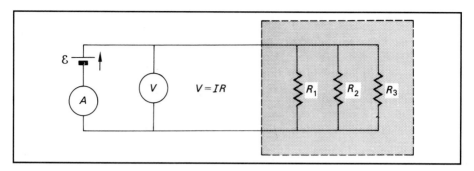

FIG. 17-12 Computing the equivalent resistance of a number of resistors connected in parallel.

Applying Ohm's law to Eq. (17-13) yields

$$\frac{V}{R} = \frac{V_1}{R_1} + \frac{V_2}{R_2} + \frac{V_3}{R_3}$$

But the voltages are equal, and we can divide them out:

$$\frac{1}{R} = \frac{1}{R_1} + \frac{1}{R_2} + \frac{1}{R_3} \qquad \text{parallel} \qquad (17\text{-}14)$$

In summary, for parallel resistors:
  1. The total current in a parallel circuit is equal to the sum of the currents in the individual branches.
  2. The voltage drops across all branches in a parallel circuit must be of equal magnitude.
  3. The reciprocal of the equivalent resistance is equal to the sum of the reciprocals of the individual resistances connected in parallel.

In the case of only two resistors in parallel,

$$\frac{1}{R} = \frac{1}{R_1} + \frac{1}{R_2}$$

Solving this equation algebraically for R, we obtain a simplified formula for computing the equivalent resistance:

$$R = \frac{R_1 R_2}{R_1 + R_2} \qquad (17\text{-}15)$$

The equivalent resistance of two resistors connected in parallel is equal to their product divided by their sum.

EXAMPLE 17-6   The total applied voltage to the circuit in Fig. 17-11 is 12 V, and resistances R_1, R_2, and R_3 are 4, 3, and 6 Ω, respectively. (a) Determine the equivalent resistance of the circuit. (b) What is the current through each resistor?

Solution (a)   The best approach to a problem which contains both series and parallel resistors is to reduce the circuit by steps to its simplest form, as shown in Fig. 17-13. We first find the equivalent resistance R' of the pair of resistors R_2 and R_3:

$$R' = \frac{R_2 R_3}{R_2 + R_3} = \frac{(3\ \Omega)(6\ \Omega)}{(3 + 6)\ \Omega} = 2\ \Omega$$

Since the equivalent resistance R' is in series with R_1, the total equivalent resistance is

$$R = R' + R_1 = 2\ \Omega + 4\ \Omega = 6\ \Omega$$

Solution (b)   The total current can be found from Ohm's law:

$$I = \frac{V}{R} = \frac{12\ V}{6\ \Omega} = 2\ A$$

The current through R_1 and R' is therefore 2 A, since they are in series. To find the currents I_2 and I_3, we must know the voltage drop V' across the equivalent resistance R':

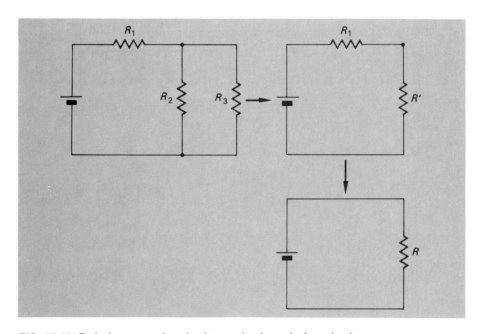

FIG. 17-13 Reducing a complex circuit to a simple equivalent circuit.

$$V' = IR' = (2 \text{ A})(2 \text{ } \Omega) = 4 \text{ V}$$

Thus the potential must drop by 4 V through each of the resistors R_2 and R_3. The currents are found from Ohm's law:

$$I_2 = \frac{V'}{R_2} = \frac{4 \text{ V}}{3 \text{ } \Omega} = 1.33 \text{ A}$$

$$I_3 = \frac{V'}{R_3} = \frac{4 \text{ V}}{6 \text{ } \Omega} = 0.67 \text{ A}$$

Note that $I_2 + I_3 = 2$ A, which is the total current delivered to the circuit.

17-9 ■ EMF AND TERMINAL POTENTIAL DIFFERENCE

In all the preceding problems, we have assumed that all resistance to current flow is due to elements of a circuit which are external to the source of emf. This is not strictly true, however, because there is an inherent resistance within sources of emf. This *internal resistance* is represented by the symbol r and is shown schematically as a small resistance in series with the source of emf. (See Fig. 17-14.) When a current I is flowing through the circuit, there is a loss of energy through the external load R_L and there is a heat loss due to the internal resistance. Thus the actual terminal voltage V_T across a source of emf ε with an internal resistance r is given by

$$\boxed{V_T = \varepsilon - Ir} \tag{17-16}$$

The voltage applied to the external load is therefore less than the emf by an amount equal to the internal potential drop. Since $V_T = IR_L$, Eq. (17-16) can be rewritten

$$V_T = IR_L = \varepsilon - Ir \tag{17-17}$$

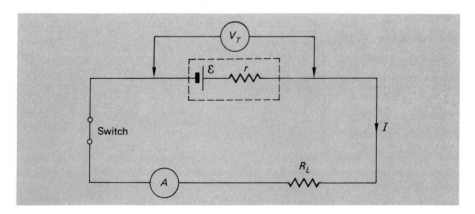

FIG. 17-14 Measuring internal resistance.

Solving Eq. (17-17) for the current I, we have

$$I = \frac{\varepsilon}{R_L + r} \tag{17-18}$$

The current in a simple circuit containing a single source of emf is equal to the emf ε divided by the total resistance in the circuit (including internal resistance).

EXAMPLE 17-7   A load resistance of 8 Ω is connected to a battery whose internal resistance is 0.2 Ω. (*a*) If the emf of the battery is 12 V, what current is delivered to the load? (*b*) What is the terminal voltage of the battery?

Solution (a)   The current delivered is found from Eq. (17-18):

$$I = \frac{\varepsilon}{R_L + r} = \frac{12 \text{ V}}{8 \text{ } \Omega + 0.2 \text{ } \Omega} = 1.46 \text{ A}$$

Solution (b)   The terminal voltage is

$$V_T = \varepsilon - Ir = 12 \text{ V} - (1.46 \text{ A})(0.2 \text{ } \Omega)$$
$$= 12 \text{ V} - 0.292 \text{ V} = 11.7 \text{ V}$$

As a check, we can find the voltage drop across the load R_L:

$$V_T = IR_L = (1.46 \text{ A})(8 \text{ } \Omega) = 11.7 \text{ V}$$

17-10 ■ MEASURING INTERNAL RESISTANCE

The internal resistance of a battery can be measured in the laboratory by using a voltmeter, an ammeter, and a known resistance. A voltmeter is an instrument which has an extremely high resistance. When a voltmeter is attached directly to the terminals of a battery, negligible current is drawn from the battery. We can see from Eq. (17-16) that for zero current this terminal voltage is equal to the emf ($V_T = \varepsilon$). In fact, the emf of a battery is sometimes referred to as its ''open-circuit'' potential difference. Thus the emf of a battery can be measured with a voltmeter. By connecting a known resistance to the circuit, we can determine the internal resistance by measuring the current delivered to the circuit.

EXAMPLE 17-8   A dry cell gives an open-circuit reading of 1.5 V when a voltmeter is connected to its terminals. When the voltmeter is removed and a load of 3.5 Ω is placed across the terminals of the battery, a current of 0.4 A is measured. What is the internal resistance of the battery?

Solution   Solving for r in Eq. (17-17), we obtain

$$r = \frac{\varepsilon - IR_L}{I} = \frac{1.5 \text{ V} - (0.4 \text{ A})(3.5 \text{ } \Omega)}{0.4 \text{ A}}$$
$$= \frac{0.10 \text{ V}}{0.4 \text{ A}} = 0.25 \text{ } \Omega$$

As a dry cell ages, its internal resistance increases while its emf remains relatively constant. The increased internal resistance causes a reduction in the current delivered. This fact accounts for the difference in intensity of light between a flashlight using old batteries and one using fresh batteries.

17-11 ■ REVERSING THE CURRENT THROUGH A SOURCE OF EMF

In a battery chemical energy is converted to electric energy in order to maintain current flow in an electric circuit. A generator performs a similar function by converting mechanical energy to electric energy. In either case, the process is reversible. If a source of higher emf is connected in direct opposition to a source of lower emf, the current will pass through the latter from its positive terminal to its negative terminal. Reversing the flow of charge in this manner results in a loss of energy as electric energy is converted to chemical or mechanical energy.

Let us consider the process of charging a battery, as illustrated in Fig. 17-15. As charge flows through the higher source of emf ε_1, it gains energy. The terminal voltage for ε_1 is given by

$$V_1 = \varepsilon_1 - Ir_1$$

in accordance with Eq. (17-17). The output voltage is reduced because of the internal resistance r_1.

Energy is lost in two ways as charge is forced through the battery against its normal output direction:

1. Electric energy in an amount equal to ε_2 is stored as chemical energy in the battery.
2. Energy is lost to the internal resistance of the battery.

Therefore, the terminal voltage V_2, which represents the total drop in potential across the battery, is given by

$$V_2 = \varepsilon_2 + Ir_2 \tag{17-19}$$

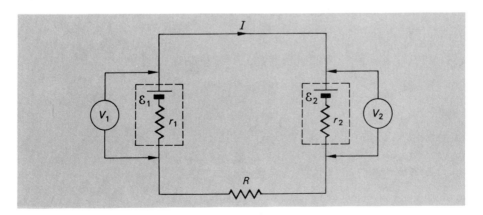

FIG. 17-15 Reversing the current through a source of emf.

where r_2 is the internal resistance. Note that in this case the terminal voltage is *greater* than the emf of the battery. The remainder of the potential supplied by the higher source of emf is lost through the external resistance R.

Throughout the entire circuit, the energy lost must equal the energy gained. Thus we can write

Energy gained per unit charge = energy lost per unit charge

$$\mathcal{E}_1 = \mathcal{E}_2 + Ir_1 + Ir_2 + IR$$

Solving for the current I yields

$$I = \frac{\mathcal{E}_1 - \mathcal{E}_2}{r_1 + r_2 + R}$$

The current supplied to a continuous electric circuit is equal to the net emf divided by the total resistance of the circuit, including internal resistance:

$$I = \frac{\Sigma \mathcal{E}}{\Sigma R} \qquad (17\text{-}20)$$

For the purposes of applying Eq. (17-20), an emf is considered negative when the current flows against its normal output direction.

EXAMPLE 17-9 Assume the following values for the parameters of the circuit in Fig. 17-15: $\mathcal{E}_1 = 12$ V, $\mathcal{E}_2 = 6$ V, $r_1 = 0.2$ Ω, $r_2 = 0.1$ Ω, and $R = 4$ Ω. (*a*) What is the current in the circuit? (*b*) What is the terminal voltage across the 6-V battery?

Solution (a) From Eq. (17-20), the current is

$$I = \frac{\mathcal{E}_1 - \mathcal{E}_2}{r_1 + r_2 + R} = \frac{12 \text{ V} - 6 \text{ V}}{0.2 \text{ Ω} + 0.1 \text{ Ω} + 4 \text{ Ω}}$$

$$= \frac{6 \text{ V}}{4.3 \text{ Ω}} = 1.40 \text{ A}$$

Solution (b) The terminal voltage of the battery being charged is, from Eq. (17-19),

$$V_2 = \mathcal{E}_2 + Ir_2$$
$$= 6 \text{ V} + (1.4 \text{ A})(0.1 \text{ Ω})$$
$$= 6.14 \text{ V}$$

Summary

In this chapter, we have introduced the concepts of current, resistance, and voltage. It is an extremely important chapter both for initial understanding of current electricity and for later concepts. You should spend a lot of time on each topic summarized below. Working problems is an excellent way to enhance your understanding.

■ *Electric current* is the rate of flow of charge Q past a given point on a conductor:

$$I = \frac{Q}{t} \qquad 1 \text{ ampere (A)} = \frac{1 \text{ coulomb (C)}}{1 \text{ second (s)}}$$

■ *Ohm's law* states that the current produced in a given conductor is directly proportional to the difference of potential between the endpoints:

$$R = \frac{V}{I} \qquad V = IR \qquad\qquad \textit{Ohm's Law}$$

The symbol R represents the *resistance* defined in *ohms* (Ω), where one ohm (Ω) is one *volt* (V) per *ampere* (A).

■ The *electric power* in watts (W) is given by

$$P = VI \qquad P = I^2R \qquad P = \frac{V^2}{R} \qquad\qquad \textit{Power}$$

■ The resistance of a wire depends on four factors: the kind of *material*, the *length*, the cross-sectional *area*, and the *temperature*. In general, the resistance is *directly* proportional to length and temperature, but *inversely* proportional to the cross-sectional area.

■ In dc circuits, resistors may be connected in *series* or in *parallel* as defined by Fig. 17-16.

■ For *series connections,* the current in all parts of the circuit is the same, the total voltage drop is the sum of the individual drops across each resistor, and the effective resistance is the sum of the individual resistances:

$$I_T = I_1 = I_2 = I_3 \qquad V_T = V_1 + V_2 + V_3$$

$$R_e = R_1 + R_2 + R_3 \qquad\qquad \textit{Series Connections}$$

■ For *parallel connections,* the total current is the sum of the individual currents, the voltage drops are all equal, and the effective resistance is given by

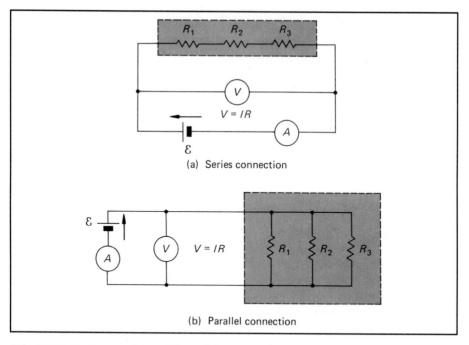

FIG. 17-16 (*a*) Series resistors; (*b*) parallel resistors.

$$I_T = I_1 + I_2 + I_3 \qquad V_T = V_1 = V_2 = V_3$$

$$\frac{1}{R_e} = \frac{1}{R_1} + \frac{1}{R_2} + \frac{1}{R_3}$$

Parallel Connections

■ For two resistors in parallel, a simpler form for R_e is

$$R_e = \frac{R_1 R_2}{R_1 + R_2}$$

Two Parallel Resistors

■ The current supplied to an electric circuit is equal to the *net* emf divided by the total resistance of the circuit, including any internal resistances:

$$I = \frac{\Sigma \, \mathcal{E}}{\Sigma \, R}$$

for example,

$$I = \frac{\varepsilon_1 - \varepsilon_2}{r_1 + r_2 + R_L}$$

The example is for two opposing batteries of internal resistances r_1 and r_2 when the external load resistance is R_L.

Questions

17-1. Define the following terms:

    a. current
    b. ampere
    c. resistance
    d. ohm
    e. Ohm's law
    f. electric power

    g. dc circuit
    h. series connection
    i. parallel connection
    j. terminal potential difference
    k. internal resistance
    l. load resistance

17-2. Distinguish clearly between electron flow and conventional current. What are the advantages of using conventional current? What is the major disadvantage?

17-3. Use the mechanical analogy of water flowing through pipes to describe the flow of charge through conductors of various lengths and cross-sectional areas.

17-4. A rheostat is connected across the terminals of a battery. What determines the positive and negative terminals of the rheostat?

17-5. Is the electromotive force really a force? What is the function of a source of emf?

17-6. Defend the following statement: The effective resistance of a group of resistors connected in parallel will be less than any of the individual resistances.

17-7. Discuss the advantages and disadvantages of connecting Christmas-tree lights (*a*) in series, (*b*) in parallel.

17-8. What is meant by the "open-circuit" potential difference of a battery?

17-9. Distinguish clearly between terminal potential difference and emf.

17-10. Many electric devices and appliances are designed to operate at the same voltage. How should such devices be connected in an electric circuit?

17-11. Should elements connected in series be designed to function at a constant current or at a constant voltage?

17-12. In an electric circuit, it is desired to decrease the effective resistance by adding resistors. Should these resistors be connected in series or in parallel?

17-13. Describe a method for measuring the resistance of a spool of wire, using a voltmeter, an ammeter, a rheostat, and a source of emf. Draw the circuit diagram. (The rheostat is used to adjust the current to the range required for the ammeter.)

17-14. Given the emf of a battery, describe a laboratory procedure for determining its internal resistance.

17-15. Can the terminal voltage of a battery be greater than its emf? Explain.

Problems

17-1. How many electrons pass a point every second in a wire carrying a current of 20 A?

                                                              Answer 1.25×10^{20}.

17-2. A current of 4 A flows through a wire whose ends are at a potential difference of 24 V. How much charge flows through the wire in 1 min? What is the resistance?

17-3. Find the current in amperes in 850 C of charge passed through a wire in 1 min.

Answer 14.2 A.

17-4. If a current of 50 A exists for 50 s, how many coulombs of charge have passed through the wire?

17-5. (*a*) What is the potential drop across a 4-Ω resistor with a current of 10 A? (*b*) What is the resistance of a rheostat if the drop in potential is 40 V and the current through it is 6 A? (*c*) Determine the current through a 5-Ω resistor that has a potential drop of 30 V across it.

Answer (*a*) 40 V; (*b*) 6.67 Ω; (*c*) 6 A.

17-6. A 1-A fuse is placed in a circuit with a battery that has a terminal voltage of 6 V. What is the minimum resistance for a circuit containing this fuse? If a 20-A fuse is inserted, what voltage will blow the circuit for the same resistance?

17-7. What emf is required to pass 50 mA through a resistance of 20,000 Ω? If the emf is reduced to 200 V, what current will flow in the same circuit?

Answer 1000 V; 0.01 A.

17-8. A soldering iron draws 0.75 A at 120 V. What is its resistance?

17-9. An electric lamp has an 80-Ω filament connected to a 110-V dc line. (*a*) What is the lamp current? (*b*) What is the power loss in watts?

Answer (*a*) 1.38 A; (*b*) 151 W.

17-10. Assume that the cost of electric energy in a home is 11¢/kWh. What is the cost of operating a 60-W lamp for 1 yr?

17-11. A 120-V generator delivers 24 kW to an electric furnace. What current is supplied? What is the resistance?

Answer 200 A; 0.6 Ω.

17-12. A water turbine delivers 2000 hp to an electric generator (1 hp = 746 W). The generator is only 80 percent efficient and has an output terminal voltage of 1200 V. What current is delivered?

17-13. A 110-V radiant heater draws a direct current of 6 A. How much heat energy is given off in 1 h?

Answer 2.38×10^6 J.

17-14. A 20-kΩ resistor has a rating of 100 W. Determine the maximum current and voltage that may be applied to the resistor.

17-15. What is the resistance of 40 m of copper wire if it has a resistance of 0.02 Ω/m? What is the current through this wire for a voltage of 60 V?

Answer 0.8 Ω; 75 A.

17-16. A spool of copper wire has a resistance of 60 Ω. What is the new resistance if (*a*) the length is doubled, (*b*) the cross-sectional area is doubled?

17-17. An 18-Ω resistor R_1 and a 9-Ω resistor R_2 are connected in parallel and then in series with a 24-V battery. What is the effective resistance for each connection? What is the current drawn from the battery in each case?

Answer 6 Ω, 27 Ω; 4 A, 0.889 A.

17-18. Answer the questions of Prob. 17-17 for 4- and 8-Ω resistors connected to a 12-V battery.

17-19. An 8-Ω resistor R_1 and a 14-Ω resistor R_2 are connected in series. What is the effective resistance? If they are reconnected in parallel, what will be the effective resistance?

Answer 22 Ω; 5.09 Ω.

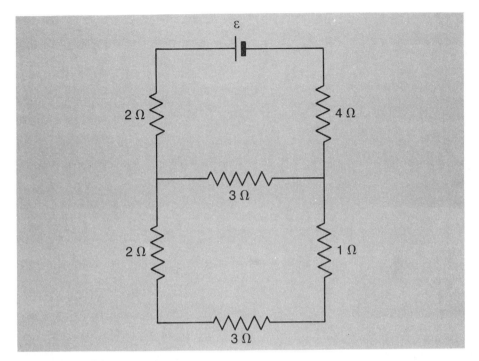

FIG. 17-17

17-20. Given three resistors $R_1 = 100 \ \Omega$, $R_2 = 150 \ \Omega$, and $R_3 = 80 \ \Omega$, find their equivalent resistance when they are connected (a) in series, (b) in parallel.

17-21. Assume that the battery of the circuit in Fig. 17-17 has zero internal resistance. What is the effective resistance of the circuit?

Answer 8 Ω.

17-22. Neglecting internal resistance, what is the effective resistance of the circuit shown in Fig. 17-18?

*17-23. Consider the circuit described by Fig. 17-17 which contains a source of emf equal to 24 V with zero internal resistance. (a) What current is delivered to the circuit? (b) What are the current and voltage drop across the central 3-Ω resistor? (c) What are the current and voltage drop across the 1-Ω resistor?

Answer (a) 3 A; (b) 2 A, 6 V; (c) 1 A, 1 V.

*17-24. A 36-V source of emf with negligible internal resistance is connected as shown in Fig. 17-18. What are the current and voltage across the 2-Ω resistor?

17-25. A resistance of 6 Ω is placed across a 12-V battery whose internal resistance is 0.3 Ω. What is the current delivered to the circuit? What is the terminal voltage?

Answer 1.9 A; 11.43 V.

17-26. In an experiment to determine the internal resistance of a battery, its open-circuit potential difference is measured to be 6 V. The battery is then connected to a 4-Ω resistor, and the current is found to be 1.4 A. What is the internal resistance?

17-27. Determine the total current delivered by the source of emf to the circuit in Fig. 17-19. What is the current through each resistor? Assume that $R_1 = 6 \ \Omega$, $R_2 = 3 \ \Omega$, $R_3 = 1 \ \Omega$, $R_4 = 2 \ \Omega$, $r = 0.4 \ \Omega$, and $\varepsilon = 24$ V.

Answer 15 A; 2 A; 4 A; 6 A; 9 A.

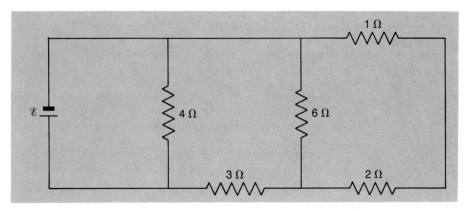

FIG. 17-18

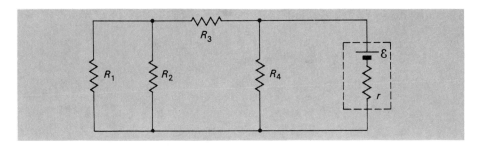

FIG. 17-19

17-28. The circuit shown in Fig. 17-14 consists of a 12-V battery, a 4-Ω resistor, and a switch. The internal resistance of the battery is 0.4 Ω. Assume that the switch is left open. (*a*) What would a voltmeter read when placed across the terminals of the battery? (*b*) If the switch is closed, what will the voltmeter read? (*c*) With the switch closed, what would the voltmeter read when placed across the 4-Ω resistor?

17-29. A 6-Ω resistor R_1 and a 4-Ω resistor R_2 are connected in parallel across a 6-V generator whose internal resistance is 0.3 Ω. (*a*) Draw a circuit diagram. (*b*) What is the total current? (*c*) What is the power *developed* by the generator? (*d*) What is the power *delivered* by the generator? (*e*) At what rate is energy lost through R_1? (*f*) At what rate is energy lost through R_2?

> *Answer* (*b*) 2.22 A; (*c*) 13.32 W; (*d*) 11.85 W; (*e*) 4.74 W; (*f*) 7.11 W.

17-30. The generator in Fig. 17-20 develops an emf ε_1 of 24 V and has an internal resistance of 0.2 Ω. The generator is used to charge a 12-V battery whose internal resistance is 0.3 Ω. The series resistors are $R_1 = 4$ Ω and $R_2 = 6$ Ω. (*a*) What current is delivered to the circuit? (*b*) What is the terminal potential difference across the generator? (*c*) What is the terminal voltage across the battery? (*d*) Show that the total drop in voltage in the circuit external to the generator is equal to the terminal voltage of the generator.

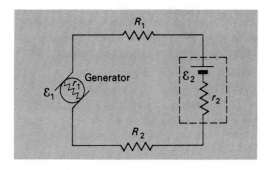

FIG. 17-20 Charging a battery with a dc generator.

18

MAGNETISM AND ELECTRIC INSTRUMENTS

As a result of completing this chapter, you should be able to

1. Demonstrate by definitions, examples, and drawings your understanding of magnetic forces, magnetic field lines, and the modern theory of magnetism.
2. Apply the *right-hand rules* to determine the direction of magnetic forces and magnetic fields of current-carrying conductors.
3. Explain the operation of a dc ammeter and calculate the shunt resistance necessary to increase the range of the ammeter.
4. Explain the operation of a dc voltmeter and calculate the multiplier resistance necessary to increase its range.
5. Disassemble a laboratory dc motor and explain the function of each of its parts, with particular emphasis on the *split-ring commutator;* if a motor is not available, explain with drawings.

An understanding of many electric devices and measuring instruments requires a knowledge of magnetic effects and magnetic forces. In this chapter you will begin by studying magnetic forces and magnetic fields. You will recognize that an electric current can produce a magnetic force and that a moving magnetic field can generate an electric current. The relationship between magnetic and electric fields provides the basis for the operation of many devices. Electric motors, generators, transformers, circuit breakers, telephones, televisions, radios, and most electric meters are only a few examples. The concepts and applications of basic physics principles in this chapter are carefully selected to help you understand the operation of many electromagnetic devices.

18-1 ■ MAGNETISM

The first magnetic effects to be observed were associated with rough fragments of lodestone (an oxide of iron) found near the ancient city of Magnesia some 2000 years ago. These natural magnets were observed to attract bits and pieces of unmagnetized iron. This force of attraction is referred to as *magnetism,* and the device which exerts a magnetic force is called a *magnet.*

If a bar magnet is dipped into a pan of iron filings and removed, the tiny pieces of iron are observed to cling most strongly to small areas near the ends, as in Fig. 18-1. These regions where the magnet's strength appears to be concentrated are called *magnetic poles.*

When any magnetic material is suspended from a string, it turns about a vertical axis. As illustrated in Fig. 18-2, the magnet aligns itself in a north-south direction. The end pointing toward the north is called the *north-seeking* pole or the north (N) pole of the magnet. The opposite, *south-seeking* end is called the south (S) pole. It is the polarization of magnetic material that accounts for its usefulness as a compass for navigation. A compass consists of a light magnetized needle pivoted on a low-friction support.

That the north and south poles of a magnet are different can be demonstrated easily. When another bar magnet is brought near a suspended magnet, as in Fig. 18-3, two north poles repel each other, whereas the north pole of one will attract the south pole of another. Such attraction and repulsion are the basis for the law of magnetic forces:

Law of magnetic forces: *Like magnetic poles repel each other, and unlike magnetic poles attract each other.*

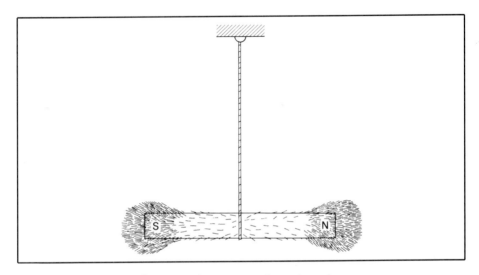

FIG. 18-1 The strength of a magnet is concentrated near its ends.

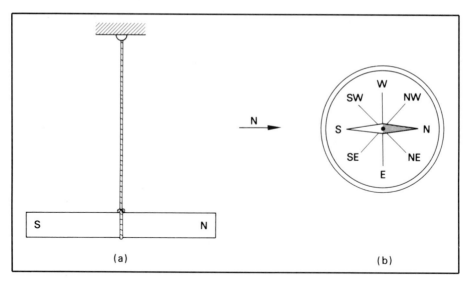

FIG. 18-2 (*a*) A suspended bar magnet will come to rest in a north-south direction. (*b*) Top view of a magnetic compass.

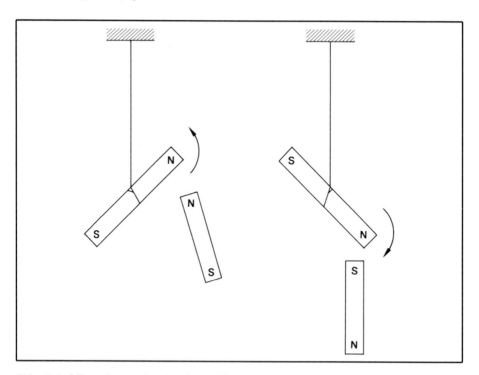

FIG. 18-3 Like poles repel each other; unlike poles attract each other.

Isolated poles have not been found in nature. No matter how many times a magnet is broken in half, each piece will become a magnet, having both a north and a south pole.

Additionally, the attraction of a magnet for unmagnetized iron and the interacting forces between magnetic poles act through all substances. In industry, ferrous (iron-containing) materials are often separated by magnets for recycling.

18-2 ■ MAGNETIC FIELDS

Every magnet is surrounded by a space in which its magnetic effects are present. Such regions are called *magnetic fields*. Just as electric field lines were useful in describing electric fields, magnetic field lines, called *magnetic flux lines*, are useful for visualizing magnetic fields. The direction of a flux line at any point is the same as the direction of the magnetic force on an imaginary isolated north pole positioned at that point. (See Fig. 18-4a.) Accordingly, magnetic flux lines leave the north pole of a magnet and enter the south pole. Unlike electric field lines, magnetic flux lines do not have origins or terminating points. They form continuous loops, passing through the metallic bar, as shown in Fig. 18-4b. The flux lines in the region between two like or unlike poles are shown in Fig. 18-5.

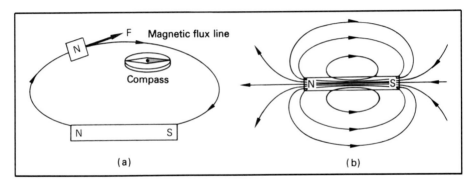

FIG. 18-4 (a) Magnetic flux lines are in the direction of the force exerted on an independent north pole. (b) Flux lines in the vicinity of a bar magnet.

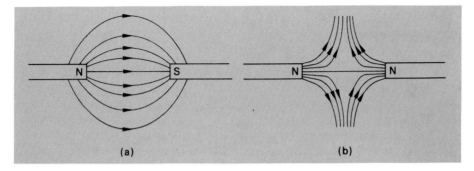

FIG. 18-5 (a) Magnetic flux lines between two unlike magnetic poles. (b) Flux lines in the space between two like magnetic poles.

18-3 ■ MODERN THEORY OF MAGNETISM

Magnetism in matter is currently believed to result from the movements of electrons in the atoms of substances. If this is true, magnetism is a property of *charge in motion* and is closely related to electrical phenomena. Individual atoms of a magnetic substance are, in effect, tiny magnets with north and south poles. The magnetic polarity of atoms stems primarily from the spin of the electrons and is due only partially to their orbital motions around the nucleus. Figure 18-6 illustrates the two kinds of electron motions.

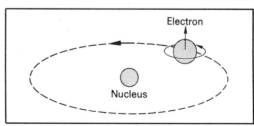

FIG. 18-6 Two kinds of electron motion responsible for magnetic properties.

The atoms in a magnetic material are grouped into microscopic magnetic regions called *domains*. All the atoms within a domain are believed to be magnetically polarized in a single direction. In an unmagnetized material, these domains are oriented in random directions, as indicated by the arrows in Fig. 18-7a. A dot is used to indicate an arrow directed out of the paper, and a cross is used to denote an arrow directed into the paper. If a large number of the domains become oriented in the same direction, as in Fig. 18-7b, the material will exhibit strong magnetic properties.

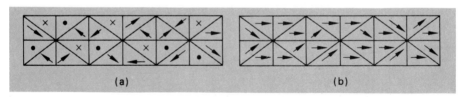

(a)   (b)

FIG. 18-7 (*a*) Magnetic domains are randomly oriented in an unmagnetized material. (*b*) Orientation of domains in a magnetized material.

This theory of magnetism is highly plausible in that it offers an explanation for many of the observed magnetic effects of matter. For example, an unmagnetized iron bar can be made into a magnet simply by holding another magnet near it or in contact with it. This process, called *magnetic induction,* is illustrated in Fig. 18-8. The tacks become temporary magnets by induction. Note that the tacks on the right become magnetized even though they do not actually touch the magnet. Magnetic induction is explained by the domain theory. The introduction of a magnetic field aligns the domains, resulting in magnetization.

Induced magnetism is often only temporary, and when the field is removed, the domains become disoriented. If the domains remain aligned to some degree after the field has been removed, the material is said to be *permanently magnetized.* The ability to retain magnetism is referred to as *retentivity.*

Another property of magnetic materials which is easily explained by the domain theory is *magnetic saturation.* There appears to be a limit to a material's degree of magnetization. Once this limit has been reached, no greater strength of an external magnet can increase the magnetization. It is believed that all its domains have been aligned.

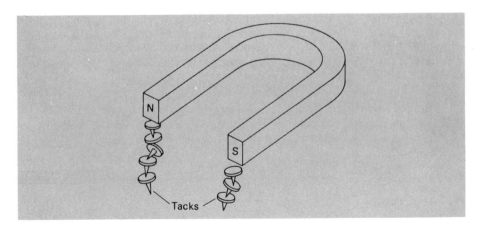

FIG. 18-8 Magnetic induction.

18-4 ■ MAGNETIC FIELD AND ELECTRIC CURRENT

Although the modern theory of magnetism holds that a magnetic field results from the motion of charges, science has not always accepted this proposition. It is fairly easy to show that a powerful magnet exerts no force on a static charge. In the course of a lecture-demonstration in 1820, Hans Oersted set up an experiment to show his students that *moving* charges and magnets also do not interact. He placed the magnetic needle of a compass near a conductor, as illustrated in Fig. 18-9. To his surprise, when a current was sent through the wire, a twisting force was exerted on the compass needle until it pointed almost perpendicular to the wire. Further, the magnitude of the force depended upon the relative orientation of the compass needle and the current direction. The maximum twisting force occurred when the wire and the compass needle were parallel before the current was established. If they were initially perpendicular, no force was experienced. Evidently, a magnetic field is set up by the charge in motion through the conductor.

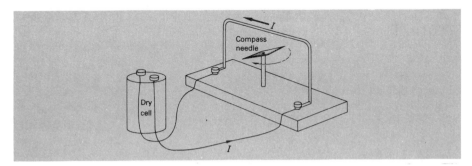

FIG. 18-9 Oersted's experiment.

The fact that a magnetic force is exerted on the compass needle in Oersted's experiment demonstrates that a magnetic field surrounds a current-carrying conductor. A laboratory method of studying such a field is illustrated by Fig. 18-10. If iron filings are sprinkled on the paper surrounding the current-carrying wire, they will become aligned in concentric circles around the wire. Similar investigation of the area surrounding the wire with a magnetic compass will confirm that the magnetic field is circular and directed in a clockwise fashion, as viewed along the direction of the conventional (positive) current. A convenient method devised by Ampère to determine the direction of the flux lines surrounding a straight wire is called the *right-hand-thumb rule* (refer to Fig. 18-10).

> **Right-hand-thumb rule:** *If a wire is grasped with the right hand so that the thumb points in the direction of conventional current, the curled fingers of that hand will point in the direction of the magnetic field.*

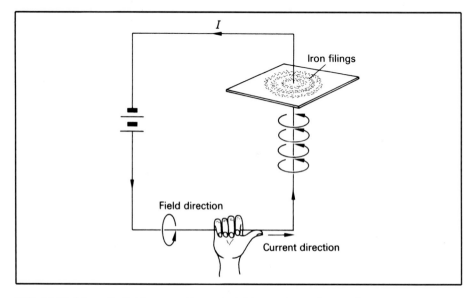

FIG. 18-10 Magnetic field surrounding a straight current-carrying conductor.

If a current-carrying wire is bent into a loop, as shown in Fig. 18-11*a*, the flux lines around the wire add, giving a magnetic field like a thin slice of a bar magnet. It will have a north and a south pole, as determined by the right-hand-thumb rule. If a wire is wound into several loops so as to form a coil, the coil is referred to as a *solenoid*. (See Fig. 18-11*b*.) When an electric current flows through such a coil, a magnetic field is produced that is very similar to that of a bar magnet. The strength of the magnetic field produced by a solenoid is increased by increasing the current through the coil or by increasing the number of turns of wire. The diameter of the solenoid is not significant if it is small compared with the length of the solenoid.

The magnetic field produced by a solenoid can also be increased by inserting a material inside the coil. Some materials, such as iron, cobalt, and nickel, are magnetized very easily. Such materials are said to be highly *permeable,* and the ease with which they establish flux lines is referred to as their *permeability.* A simple *electromagnet,* shown in Fig. 18-12, is constructed by winding an insulated copper wire around a soft-iron core. Very strong electromagnets that can lift heavy metallic materials can be designed in this fashion. Smaller electromagnets are used in the construction of electric meters, bells, relays, telephone receivers, electric motors, and generators.

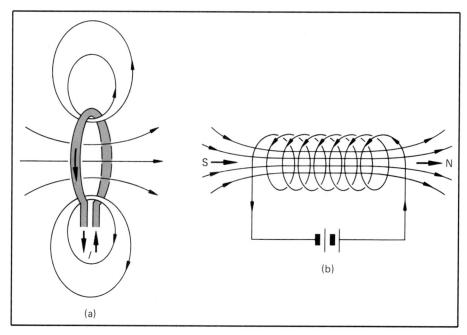

FIG. 18-11 (*a*) Magnetic field of a current loop. (*b*) Magnetic field for a solenoid.

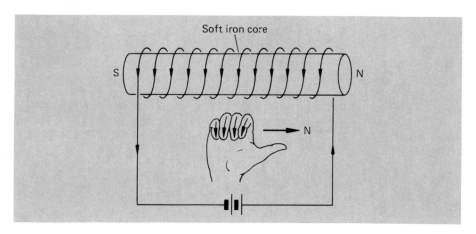

FIG. 18-12 Electromagnet.

A slight modification of the right-hand-thumb rule can be used to determine the magnetic poles of an electromagnet.

Right-hand rule for determining poles of a solenoid: *Grasp the coil with the right hand, wrapping the fingers around the coil in the direction of conventional current; the thumb will then point toward the north pole of the solenoid.*

If a current-carrying solenoid is positioned between two magnets, as shown in Fig. 18-13, the magnetic field of the solenoid will interact with the magnetic field of the stationary magnets. A torque will be produced on the solenoid as like poles repel and unlike poles attract. The twisting effect will be greatest when the angle θ that the axis of the coil makes with horizontal flux is 90°. The torque will become zero as the north and south poles of the solenoid line up with the stationary field. Such a twisting effect that results from an applied current is the basic principle behind the operation of motors and many electric instruments.

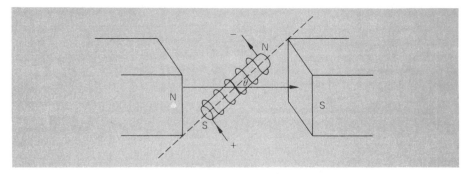

FIG. 18-13 Torque produced on a solenoid.

18-5 ■ GALVANOMETER

Any device used to detect an electric current is called a *galvanometer*. The operating principle for the majority of such instruments is based on the torque exerted on a coil in a magnetic field. The essential parts are shown in Fig. 18-14. A coil, wrapped around a soft-iron core, is pivoted on jeweled bearings between the poles of a permanent magnet. Its rotational motion is restrained by a pair of spiral springs which also serve as current leads to the coil. Depending upon the direction of the current being measured, the coil and pointer will rotate in either a clockwise or a counterclockwise direction.

In Fig. 18-15, the coil and pointer of the laboratory galvanometer are shown in the equilibrium position. Note that the permanent magnets are shaped to provide a uniform *radial magnetic field*. This ensures that the pointer deflection will be directly proportional to the current in the coil. If the current through the coil passes into the page on the right and out of the page on the left, magnetic forces in the radial field will produce a clockwise torque. The pointer and coil will move clockwise, until the resisting counterclockwise spring torque equals the magnetic torque produced by the current. Thus the position of the pointer on the marked scale is a measure of the current magnitude. Reversing the current direction will cause an equal pointer deflection in a counterclockwise direction.

The sensitivity of a galvanometer of the type shown in Fig. 18-15 is determined by the spring torque, the friction in the bearings, and the strength of the magnetic field.

EXAMPLE 18-1 The galvanometer in Fig. 18-15 has a sensitivity of 50 μA (50×10^{-6} A) per scale division. What current is required to give full deflection of the pointer through 25 scale divisions to the right or left of the equilibrium position?

Solution The current required is equal to the product of the sensitivity and the number of divisions (div):

$$I = (50 \ \mu\text{A/div})(25 \ \text{div}) = 1250 \ \mu\text{A}$$

Galvanometers of this type may be designed to read currents as low as 1 μA (10^{-6} A). Greater sensitivities require a modified design. In one such design, a coil is suspended in a magnetic field by a thin thread. The motion of the coil is then observed by mounting a mirror to the thread, utilizing the principle of an optical lever.

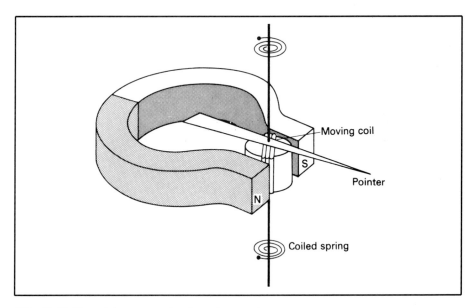

FIG. 18-14 Essential components of a galvanometer.

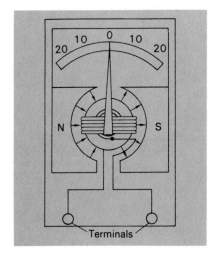

FIG. 18-15 Laboratory galvanometer.

18-6 ■ DC VOLTMETER

Many important dc measuring instruments utilize a galvanometer as an indicating element. Two of the most common are the *voltmeter* and the *ammeter*. The first will be discussed in this section as an example of how the galvanometer can be used to measure voltages.

The potential difference across a galvanometer is extremely small even when a large-scale deflection occurs. Thus, if a galvanometer is to be used to measure voltage, it must be converted to a high-resistance instrument. Suppose, for example, we want to measure the drop in voltage across the battery in Fig. 18-16. We must measure this voltage without appreciably disturbing the current through the circuit. In other words, the voltmeter must draw negligible current. To accomplish this, a large multiplier resistance R_m is placed in series with the galvanometer as an integral part of a dc voltmeter.

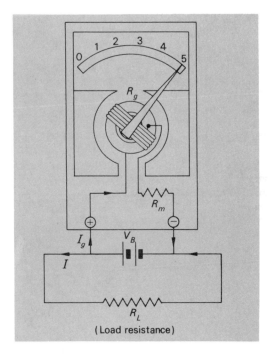

FIG. 18-16 The dc voltmeter.

Note that the galvanometer used in a voltmeter is adjusted so that its equilibrium position is to the extreme left on the scale. This allows for a greater range of measurement but unfortunately requires that the current pass through the coils in one direction only. The sensitivity of the galvanometer is determined by the current I_g required for *full-scale deflection* (maximum pointer deflection), as indicated in Fig. 18-16.

Suppose that the galvanometer coil has a resistance R_g and that the meter is designed to yield full-scale deflection for the current I_g. Such a galvanometer, acting alone, could be calibrated to record voltages from zero up to a maximum value given by

$$V_g = I_g R_g \qquad (18\text{-}1)$$

By properly choosing the multiplier resistance R_m, we can calibrate the meter to read any desired voltage.

Suppose, for example, we want full-scale deflection of the voltmeter for the voltage V_B in Fig. 18-16. The multiplier resistance R_m must be chosen so that only the small current I_g passes through the galvanometer. Under these conditions,

$$V_B = I_g R_g + I_g R_m$$

Solving for R_m, we obtain

$$R_m = \frac{V_B}{I_g} - R_g \qquad (18\text{-}2)$$

Thus we see that the multiplier resistance R_m is equal to the total resistance V_B/I_g less the galvanometer resistance R_g.

EXAMPLE 18-2 A certain galvanometer has an internal resistance of 30 Ω and gives a full-scale deflection for a current of 1 mA. Calculate the multiplier resistance necessary to convert this galvanometer to a voltmeter whose maximum range is 50 V.

Solution The multiplier resistance R_m must be such that the total drop in voltage through R_g and R_m is 50 V. From Eq. (18-2),

$$R_m = \frac{50\text{ V}}{1 \times 10^{-3}\text{ A}} - 30\text{ Ω} = 50{,}000\text{ Ω} - 30\text{ Ω}$$
$$= 49{,}970\text{ Ω}$$

Note that the total resistance of the voltmeter $(R_m + R_g)$ is 50 kΩ.

A voltmeter must be connected in *parallel* with the part of the circuit whose potential difference is to be measured. This is necessary so that the large resistance of the voltmeter will not greatly alter the circuit.

18-7 ■ DC AMMETER

An ammeter is a device which, through calibrated scales, gives an indication of the electric current without appreciably altering it. A galvanometer is an ammeter, but its range is limited by the extreme sensitivity of the moving coil. The range of the galvanometer can be extended simply by adding a very low resistance, called a *shunt*, in parallel with the galvanometer coil (refer to Fig. 18-17). Placing the shunt in parallel ensures that the ammeter as a whole will have a very low resistance, which is necessary if the current is to be essentially unaltered. The major portion of the current will pass through the shunt. Only the small current I_g required for galvanometer deflection will be drawn from the circuit. For example, if 10 A goes through an ammeter, 9.99 A may go through the shunt and only 0.01 A through the coil itself.

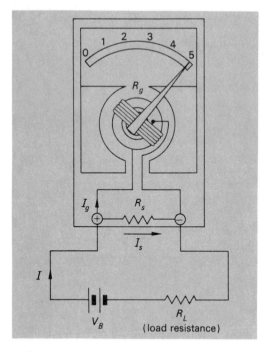

FIG. 18-17 The dc ammeter.

Suppose the range of a galvanometer is to be extended to measure a maximum current I of the circuit in Fig. 18-17. A shunt resistance R_s must be chosen such that only the current I_g required for full-scale deflection passes through the galvanometer coil. The remainder of the current I_s must pass through the shunt. Since R_g and R_s are in parallel, the IR drop across each resistance must be identical:

$$I_s R_s = I_g R_g \tag{18-3}$$

The shunt current I_s is the difference between the circuit current I and the galvanometer current I_g. Thus Eq. (18-3) becomes

$$(I - I_g)R_s = I_g R_g$$

Solving for the shunt resistance R_s, we obtain the following useful relation:

$$R_s = \frac{I_g R_g}{I - I_g} \tag{18-4}$$

EXAMPLE 18-3 A certain galvanometer has an internal coil resistance of 46 Ω, and a current of 200 mA is required for full-scale deflection. What shunt resistance must be used to convert the galvanometer to an ammeter whose maximum range is 10 A?

Solution Equation (18-4) gives

$$R_s = \frac{(0.2 \text{ A})(46 \text{ } \Omega)}{10 \text{ A} - 0.2 \text{ A}} = \frac{9.2 \text{ V}}{9.8 \text{ A}}$$

$$= 0.939 \text{ } \Omega$$

It is important to remember that an ammeter must be connected in *series* with the portion of a circuit through which the current is to be measured. The circuit must be opened at some convenient point and the ammeter inserted. If by mistake the ammeter were placed in parallel, the circuit would be short-circuited across the ammeter because of its extremely low resistance.

18-8 ■ DC MOTOR

An electric motor is a device which transforms electric energy to mechanical energy. The dc motor, like the moving coil of a galvanometer, consists of a current-carrying coil in a magnetic field. However, the motion of the coil in a motor is unrestrained by springs. The design is such that the coil will rotate continuously under the influence of magnetic torque.

A very simple dc motor, consisting of a single current-carrying loop suspended between two magnetic poles, is illustrated in Fig. 18-18. Normally, the torque exerted on a current-carrying loop would reverse its direction every half revolution. In order to provide for continuous rotation of the loop, the current in the loop must be automatically reversed each time it turns through 180°.

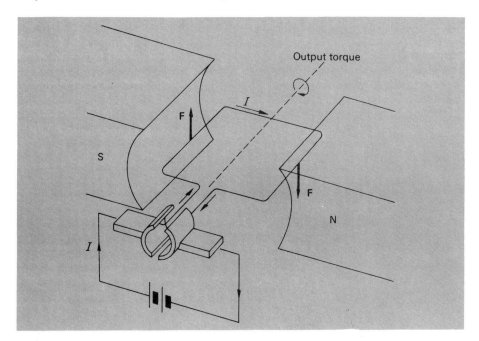

FIG. 18-18 The dc motor.

The current reversal is accomplished by using a *split-ring commutator,* as shown in Fig. 18-18. The commutator consists of two metal half-rings fused to each end of the conducting loop and insulated from each other. As the loop rotates, each brush touches first one half-ring and then the other. Thus the electric connections are reversed every half revolution at times when the loop is perpendicular to the magnetic field. In this manner, the torque acting on the loop is always in the same direction, and the loop will rotate continuously.

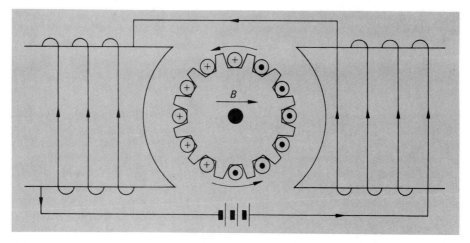

FIG. 18-19 Greater, more uniform torque is possible in commercial motors with many armature coils.

Although actual dc motors operate on the principle described for Fig. 18-18, there are a number of designs which increase the available torque and make it more uniform. One such design is shown in Fig. 18-19. A greater magnetic field is established by replacing the permanent magnets with electromagnets. Additionally, the torque can be increased and made more uniform by adding a number of different coils, each having a large number of turns around a slotted iron core called the *armature*. The commutator is an automatic switching arrangement which maintains the currents in the directions shown in the figure, regardless of the orientation of the armature. More will be said about the dc motor in the following chapter.

Summary

We have seen that a current flowing through a wire causes the formation of a magnetic field. If the wire is bent in a circle or wound around in a coil of several turns, the strength of the magnetic field is increased. The interaction of such fields with other magnetic fields forms the basis for many modern electric instruments. The galvanometer, voltmeter, ammeter, and dc motor were discussed in this chapter. The important concepts are summarized below:

- The law of magnetic forces states that *like magnetic poles repel and unlike poles attract.*
- Magnetic fields are represented by *flux lines* whose direction at any point is the same as the force on an imaginary north pole placed at that point.

- The modern theory of magnetism holds that magnetic properties arise from charges in motion and that the atoms of magnetic material are grouped into microscopic magnetic regions called *domains*.
- The direction of the magnetic field surrounding a current-carrying conductor is given by the *right-hand-thumb rule:* If the wire is grasped so that the thumb points in the current direction, the curled fingers will point in the direction of the magnetic field.
- For a *solenoid,* the magnetic poles can be determined by grasping the coil with the right hand and wrapping the fingers around the coil in the current direction. The thumb will point toward the north pole of the solenoid.
- The *torque* on a solenoid or current loop placed between two stationary magnets will be a maximum when the axis of the solenoid or loop is perpendicular to the field. It is proportional to the current and to the number of turns.
- The sensitivity of a galvanometer is the current in amperes required to deflect the pointer one scale division.
- The multiplier resistance R_m which must be placed in series with a voltmeter to give full-scale deflection for V_B is found from

$$R_m = \frac{V_B}{I_g} - R_g \qquad \text{\textit{Multiplier Resistance}}$$

where I_g is the galvanometer current and R_g is its resistance.
- The shunt resistance R_s which must be placed in parallel with an ammeter to give full-scale deflection for a current I is

$$R_s = \frac{I_g R_g}{I - I_g} \qquad \text{\textit{Shunt Resistance}}$$

- A simple dc motor consists of two stationary *field poles,* a rotating *armature,* a split-ring *commutator,* and two fixed *brushes* which serve to carry current to the armature in alternating directions.

Questions

18-1. Define the following terms:

a. magnetism
b. domains
c. magnetic poles
d. law of magnetic force
e. magnetic flux lines
f. permeability
g. magnetic induction
h. magnetic saturation
i. solenoid

j. right-hand-thumb rule
k. galvanometer
l. voltmeter
m. ammeter
n. multiplier resistance
o. shunt resistance
p. split-ring commutator
q. dc motor
r. armature

18-2. How can you positively determine whether a piece of metal is magnetized?

18-3. In general magnetic materials with high permeability have low retentivity. Why do you think this is true?

18-4. The earth acts as a huge magnet with one pole at the Arctic circle and the other in the Antarctic region. Can you justify the following statement? *The geographic North Pole is actually near a magnetic south pole?* Explain.

18-5. If an iron bar is placed parallel to a north-south direction and one end is hammered on, the bar becomes a temporary magnet. Explain, using the modern theory of magnetism.

18-6. When a bar magnet is broken into several pieces, each part becomes a magnet with a north pole and a south pole. Apparently, an isolated pole cannot exist. Explain this, using the domain theory.

18-7. Heating magnets or passing electric currents through them will cause a reduction in field strength. Explain.

18-8. A wire lying along a north-south direction supports a current from south to north. What happens to the needle of a compass if the compass is placed (*a*) above the wire, (*b*) below the wire, and (*c*) on the right side of the wire?

18-9. Use the right-hand-thumb rule to explain why two adjacent wires experience a force of attraction when the currents are in the same direction and a force of repulsion when the currents are in opposite directions. Explain with drawings.

18-10. A circular coil whose axis is in the plane of the paper supports an electric current. Determine the direction of the magnetic flux near the center of the coil when the current is counterclockwise.

18-11. To shield them from magnetic effects, many sensitive electric instruments are surrounded with material having a large permeability. Explain.

18-12. Why is it necessary to provide a *radial* magnetic field for the coil of a galvanometer?

18-13. A coil of wire is suspended by a thread with the plane of the loop coinciding with the plane of the paper. If the coil is placed in a magnetic field directed from left to right, and if a clockwise current is sent through the coil, describe its motion.

18-14. How are the actions of a galvanometer and a motor similar? How are they different?

18-15. How does the core of a galvanometer coil affect the sensitivity of the instrument?

18-16. Suppose the range of an ammeter is to be increased *n*-fold. Show that the shunt resistance which must be placed across the terminals of the ammeter is given by

$$ R_s = \frac{R_a}{n-1} $$

where R_a is the ammeter resistance.

18-17. Suppose the range of a voltmeter is to be increased *n*-fold. Show that the multiplier resistance which must be added in series with the voltmeter is given by

$$ R_m = (n-1)R_v $$

where R_v is the voltmeter resistance.

18-18. Show by diagrams how an ammeter and a voltmeter should be connected in a circuit. Compare the resistances of the two instruments.

18-19. (*a*) Discuss the error caused by the insertion of an ammeter into an electric circuit, and explain how the error can be minimized. (*b*) Discuss the error caused by the insertion of a voltmeter, and explain ways to minimize such errors.

18-20. A voltmeter is connected to a battery, and the reading is taken. An accurate resistance box is then placed in the circuit and adjusted until the voltmeter reading is one-half its previous value. Show that the voltmeter resistance must be approximately equal to the added resistance. (This is called the *half-deflection method* for determining voltmeter resistance.)

18-21. Explain what happens when a voltmeter is erroneously placed in series in a circuit. What happens if an ammeter is mistakenly placed in parallel?

18-22. Plot a graph of the torque as a function of time for a single-loop dc motor.

Problems

18-1. A laboratory galvanometer has a sensitivity of 5 μA per division. What current is required to deflect the pointer through 30 scale divisions to the right of the equilibrium position?

Answer 150 μA.

18-2. A current of 20 mA causes a galvanometer pointer to deflect 25 scale divisions. What is its sensitivity?

18-3. A certain galvanometer has an internal resistance of 20 Ω and gives a full-scale deflection for a current of 10 mA. Calculate the multiplier resistance required to convert this galvanometer to a voltmeter whose maximum range is 50 V. What is the total resistance of the resulting voltmeter?

Answer 4980 Ω; 5000 Ω.

18-4. A galvanometer having an internal resistance of 35 Ω is to be converted to a voltmeter which will read a maximum voltage of 30 V. What multiplier resistance is needed if full-scale deflection of the galvanometer requires 1 mA?

18-5. What shunt resistance is needed to convert the galvanometer of Prob. 18-3 to an ammeter which reads a maximum current of 50 mA?

Answer 50 Ω.

18-6. What shunt resistance is required to convert the galvanometer of Prob. 18-4 to an ammeter reading 10 mA at full scale?

18-7. A certain voltmeter draws 0.02 A for full-scale deflection of 50 V. (*a*) What is the resistance of the voltmeter? (*b*) What is the resistance per volt? (*c*) What multiplier resistance must be added to permit the measurement of 150 V?

Answer (*a*) 2.5×10^6 Ω; (*b*) 50,000 Ω/V; (*c*) 5×10^6 Ω.

18-8. A certain galvanometer has a sensitivity of 20 μA per division. What current is required to give full-scale deflection for a galvanometer with 25 scale divisions on either side of the equilibrium position?

18-9. The coil of an ammeter will burn out if a current of more than 40 mA is sent through it. If the coil resistance is 0.5 Ω, what shunt resistance should be added to permit a maximum current of 4 A before burnout?

Answer 0.00505 Ω.

18-10. An ammeter which has a resistance of 0.1 Ω is connected in a circuit and indicates a current of 10 A. A shunt resistance of 0.01 Ω is then connected across the meter terminals. What is the new reading on the meter?

18-11. A certain voltmeter reads 150 V full scale. The galvanometer coil has a resistance of 50 Ω and produces a full-scale deflection on 20 mV. Find the multiplier resistance in the voltmeter.

Answer 374,950 Ω.

18-12. A galvanometer has a coil resistance of 50 Ω and a current sensitivity of 1 mA (full scale). What shunt resistance is needed to convert this galvanometer to an ammeter reading 2 A full scale?

18-13. A laboratory ammeter has a resistance of 0.01 Ω and reads 5 A full scale. What shunt resistance is needed to increase the range of the ammeter 10-fold?

Answer 0.00111 Ω.

18-14. A commercial 3-V voltmeter requires a current of 0.02 mA to produce full-scale deflection. How can it be converted to an instrument with a range of 150 V?

18-15. A voltmeter of range 150 V and 15 kΩ is connected in series with another voltmeter of range 100 V and 20 kΩ. What will each meter read when they are connected across a 120-V battery?

Answer 51.45 V; 68.6 V.

18-16. A laboratory voltmeter has a resistance of 50 kΩ. What multiplier resistance must be added to increase the range of the voltmeter 10-fold?

19

INDUCED AND ALTERNATING CURRENTS

As a result of completing this chapter, you should be able to

1. Explain with drawings how an electric current is *induced* by a conductor moving through a magnetic field, discussing factors which affect its magnitude and direction.
2. Describe the function of each of the parts of an alternating-current (ac) or dc generator in the laboratory; if the generators are not available, explain with drawings.
3. Discuss and apply the relationship among applied voltage, back emf, and net voltage for a dc motor.
4. Explain the operation of a *transformer* and solve problems involving changes in voltage or power.
5. Define *inductance* and *capacitance* and discuss their effects on alternating currents.
6. Explain with diagrams the phase relationship for a circuit with pure resistance, pure capacitance, and pure inductance.
7. Write and apply the relationships for computation of *capacitive* and *inductive reactance*.
8. Calculate the *impedance, phase angle,* and *effective current* for a series ac circuit containing resistance, capacitance, and inductance.
9. Define *electrical resonance* and calculate the *resonant frequency* of a circuit when the capacitance and inductance are known.
10. Define *power factor,* and solve for it in ac circuits.

We have seen that an electric field can produce a magnetic field. In this chapter, you will learn that the reverse is also true: A magnetic field can give rise to an electric field. An electric current is *generated* by a conductor which is caused to move in a magnetic field. A rotating coil in a magnetic field *induces* an alternating emf, which produces an alternating current. About 99 percent of the energy generated in the United States is in ac form. There are good reasons for the predominant use of ac circuits. Such energy is conveniently transmitted and distributed by devices called *transformers* which minimize power losses even over long distances. For ac circuits, you will learn the effects of *capacitance* and *inductance*, which, in addition to the normal resistance, play important roles. A basic understanding of the concepts presented here is essential for any technical employee.

19-1 ■ INDUCED ELECTRIC CURRENTS

Michael Faraday discovered in 1831 that when magnetic flux lines are cut by a conductor, an emf is produced between the endpoints of the conductor. For example, an electric current is induced in the conductor of Fig. 19-1*a* as it is moved downward across the flux lines. (The lowercase symbol *i* will be used for induced currents and for varying currents.) The faster the movement, the more pronounced the galvanometer deflection. When the conductor is moved upward across the flux lines, a similar observation can be made, except that the current is reversed (see Fig. 19-1*b*). If no flux lines are cut, e.g., if the conductor is moved parallel to the field, no current is induced.

Suppose that a number of conductors are moved through a magnetic field, as illustrated by dropping a coil of *N* turns across the flux lines in Fig. 19-2. The magnitude of the induced current is directly proportional to the number of coils and to the rate of

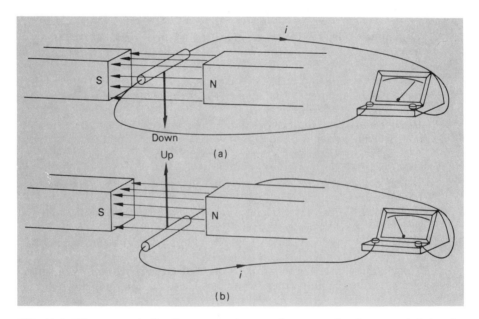

FIG. 19-1 When magnetic flux lines are cut by a conductor, an electric current is induced.

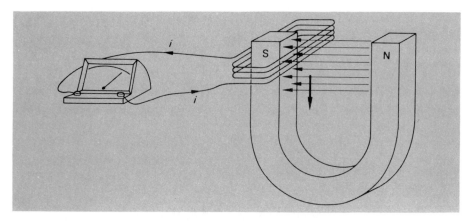

FIG. 19-2 Induced emf in a coil is proportional to the number of turns of wire passing through the field and the rate at which the magnetic lines are cut.

motion. The same effect is observed when the coil is held stationary and the magnet is moved upward. Evidently, *an emf is induced by the relative motion between the conductor and the magnetic field.*

Summarizing what we have learned from such experiments, we can state that
1. Relative motion between a conductor and a magnetic field induces an emf in the conductor.
2. The direction of the induced emf depends on the direction of motion of the conductor with respect to the field.
3. The magnitude of the emf is directly proportional to the number of turns of the conductor crossing the flux lines.
4. The magnitude of the emf is directly proportional to the rate at which magnetic lines are cut by the conductor.

As we pointed out earlier, it makes no difference whether the field is stationary and the conductor moves or whether the conductor is stationary and the field moves. It is the relative motion that induces the emf. Two examples of a changing flux density through a constant, stationary coil area are given in Fig. 19-3. In Fig. 19-3a the north pole of a magnet is moved through a circular coil. The changing flux density induces a current in the coil, as indicated by the galvanometer. In Fig. 19-3b no current is induced in coil *B* as long as the current in coil *A* is constant. However, by quickly varying the resistance in the left circuit, the magnetic flux density reaching coil *B* can be increased or decreased. While the flux density is changing, a current is induced in the coil on the right.

Note that when the north (N) pole of the magnet is moved into the coil in Fig. 19-3a, the current flows in a clockwise direction, as viewed toward the magnet. Therefore the end of the coil near the N pole of the magnet becomes an N pole (from the right-hand-thumb rule). The magnet and the coil will experience a force of repulsion, making it necessary to exert a force to bring them together. If the magnet is removed from the coil, a force of attraction will exist that makes it necessary to exert a force to separate them. This is an illustration of Lenz' law:

> **Lenz' law:** *An induced current will flow in such a direction that it will oppose by its magnetic field the motion of the magnetic field that is producing it.*

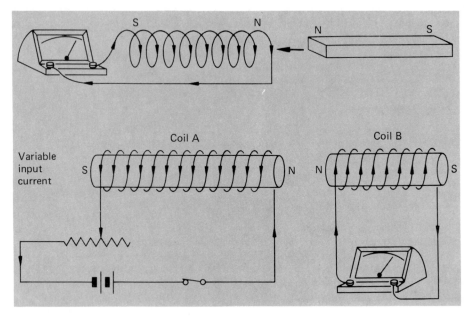

FIG. 19-3 (*a*) Inducing a current by moving a magnet into a coil. (*b*) A changing current in coil *A* induces a current in coil *B*.

The more work that is done in moving the magnet into the coil, the greater the induced current and hence the greater the resisting force. We might have expected this result from the law of conservation of energy. To produce a larger current, we must perform a greater amount of work.

The direction of the current induced in a straight conductor moving through a magnetic field can be determined from Lenz' law. However, there is an easier method, as illustrated in Fig. 19-4. It is known as *Fleming's rule,* or the *right-hand rule:*

> **Fleming's rule:** *If the thumb, forefinger, and middle finger of the right hand are held at right angles to each other, with the thumb pointing in the direction in which the wire is moving and the forefinger pointing in the field direction (north to south), the middle finger will point in the direction of induced conventional current.*

Fleming's rule is easy to apply and very useful for studying the currents induced by a simple generator. Students sometimes remember the rule by memorizing *motion—flux—current.* These are the directions given by the thumb, forefinger, and middle finger, respectively.

19-2 ■ AC GENERATOR

An electric generator converts mechanical energy to electric energy. We have seen that an emf is induced in a conductor when it experiences a change in flux linkage. When the conductor forms a complete circuit, an induced current can be detected. In a

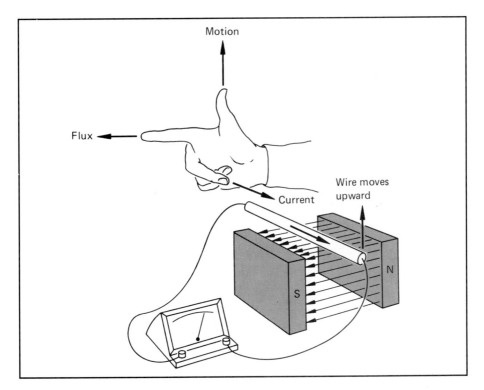

FIG. 19-4 Right-hand rule for determining the direction of induced current.

generator, a coil of wire is rotated in a magnetic field, and the induced current is transmitted by wires for long distances from its origin.

The construction of a simple generator is shown in Fig. 19-5. Essentially, there are three components: a *field magnet,* an *armature,* and *slip rings* with *brushes.* The field magnet may be a permanent magnet or an electromagnet. The armature for the generator in Fig. 19-5 consists of a single loop of wire suspended between the poles of the field magnet. A pair of slip rings is fused to each end of the loop; they rotate with the loop as it is turned in the magnetic field. Induced current is led away from the system by graphite brushes which ride on each slip ring. Mechanical energy is supplied to the generator by turning the armature in the magnetic field. Electric energy is generated in the form of an induced current.

The direction of the induced current must obey Fleming's rule of *motion—flux— current.* In Fig. 19-5, the downward motion of the left wire segment crosses a magnetic flux directed left to right. The induced current is, therefore, toward the slip rings. Similar reasoning shows that the current in the right loop, which is moving upward, will be away from the slip rings.

In order to understand the operation of an ac generator, let us follow the loop through a complete rotation, observing the current generated throughout the rotation. Figure 19-6 shows four positions of the rotating coil and the direction of the current delivered to the brushes in each case. Suppose that the loop is turned mechanically in a counterclockwise direction. In Fig. 19-6a the loop is horizontal, with side *M* facing the south pole of the magnet. At this point, a maximum current is delivered in the

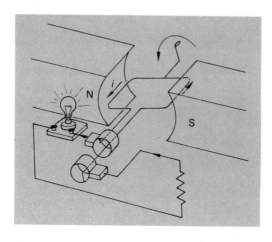

FIG. 19-5 Simple ac generator.

direction shown. In Fig. 19-6b the loop is vertical, with side M facing upward. At this point, no flux lines are being cut, and the induced current drops to zero. When the loop becomes horizontal again, as in Fig. 19-6c, side M is now facing the north pole of the magnet. Therefore, the current delivered to the slip ring M' has changed direction. An induced current flows through the external resistor in a direction opposite to that experienced earlier. In Fig. 19-6d the loop is vertical again, but now side M faces downward. No flux lines are cut, and the induced current again drops to zero. The loop next returns to horizontal as in Fig. 19-6a, and the cycle repeats itself. Thus the current delivered by such a generator alternates periodically, changing direction twice each rotation.

If we plot a graph of the change in emf as a function of time, a continuous wave is described as the emf builds to a maximum in one direction and then reverses to a maximum in the other direction. (Refer to Fig. 19-7.) This graph is called a *sine wave,* and the variation is said to be *sinusoidal.* This important fact is stated as follows:

> *If the armature is rotating with a constant angular velocity in a constant magnetic field, the magnitude of the induced emf or current varies sinusoidally with respect to time.*

The maximum current or the maximum emf will appear as peaks alternating in positive and negative directions.

One complete rotation constitutes a complete *cycle,* in which the current or emf rises to a maximum $+i$, $+\varepsilon$, then decreases to $-i$, $-\varepsilon$, and finally returns to zero. The number of complete cycles per second is called the *frequency* of the alternating current. The SI unit for frequency is the *hertz* (Hz), defined as

$$1 \text{ Hz} = 1 \text{ cycle/s}$$

Thus a 60-cycle/s alternating current has a frequency of 60 Hz.

19-3 ■ DC GENERATOR

A simple ac generator can easily be converted to a dc generator by substituting a split-ring commutator for the slip rings, as shown in Fig. 19-8. The operation is just the reverse of that discussed earlier for a dc motor (Chap. 18). In the motor, an electric

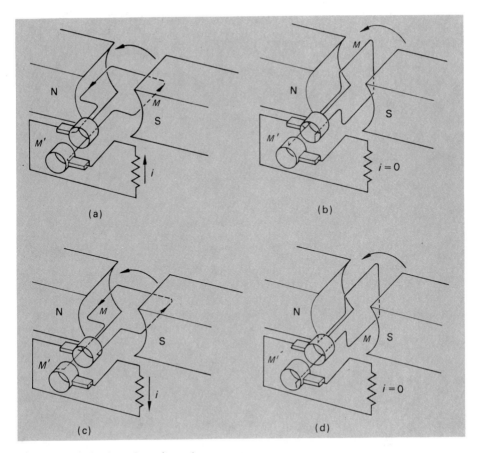

FIG. 19-6 Production of an alternating current.

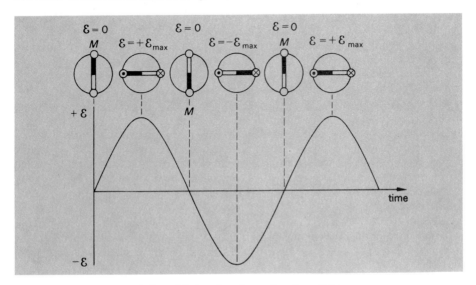

FIG. 19-7 Sinusoidal variation of induced emf as a function of time.

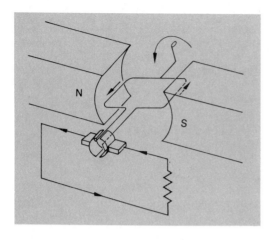

FIG. 19-8 Simple dc generator.

current gives rise to an external torque. In the dc generator, an external torque generates an electric current. The commutator reverses the connections to the brushes twice per revolution. As a result, the current pulsates but never reverses direction. The emf of such a generator varies with time, as shown in Fig. 19-9. Note that the emf is always in the positive direction, but that it rises to a maximum and falls to zero twice per complete rotation. Practical dc generators are designed with many coils in several planes so that the generated emf is large and nearly constant.

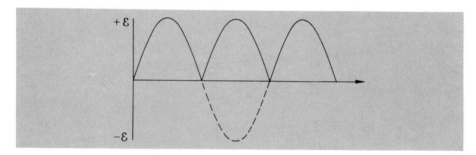

FIG. 19-9 Pulsating emf produced by a simple dc generator.

19-4 ■ BACK EMF IN A MOTOR

In an electric motor, a magnetic torque turns a current-carrying loop in a constant magnetic field. We have just seen that a coil rotating in a magnetic field will induce an emf that opposes the cause which gave rise to it (Lenz' law). This is true even if a current already exists in the loop. Thus *every motor is also a generator*. Since the induced emf *opposes* the generated emf, we call the induced emf the *back emf* or *counter emf*.

   The effect of a back emf is to reduce the net voltage delivered to the armature coils of the motor. Consider the circuit shown in Fig. 19-10. The net voltage delivered to the armature coils is equal to the applied voltage V less the induced voltage or back emf ε_b:

$$\text{Applied voltage} - \text{induced voltage} = \text{net voltage}$$

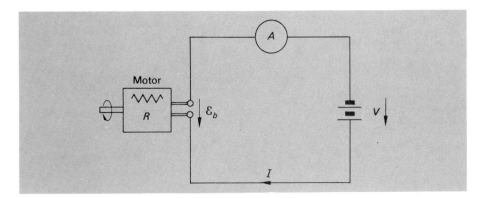

FIG. 19-10 Back emf in a dc motor.

According to Ohm's law, the net voltage across the armature coils is equal to the product of the coil resistance R and the current I. Symbolically, we write

$$V - \varepsilon_b = IR \qquad (19\text{-}1)$$

Equation (19-1) tells us that the current through a circuit containing a motor is determined by the magnitude of the back emf. The magnitude of this induced emf, of course, depends on the speed of rotation of the armature. We can show this experimentally by connecting a motor, an ammeter, and a battery in series, as shown in Fig. 19-11. When the armature is rotating freely, the current is low. The back emf reduces the effective voltage. If the armature is held stationary, stalling the motor, the back emf will drop to zero. The increased net voltage results in a large circuit current and can cause the motor to overheat and even burn out.

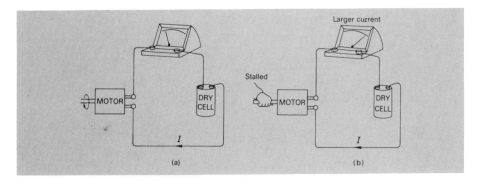

FIG. 19-11 Demonstrating the existence of a back emf in a dc motor.

EXAMPLE 19-1   A 120-V dc motor has an open-circuit resistance of 3.0 Ω as read by an ohmmeter. If a back emf of 110 V occurs while the motor is running at operating speed, (a) what is the starting current and (b) what is the current drawn by the motor when it reaches normal speed?

Solution (a)   To determine the operating current, we must recognize that the net voltage delivered is equal to 120 V less the back emf of the motor:

$$V - \varepsilon_b = IR$$
$$120 \text{ V} - 110 \text{ V} = IR$$

The net emf is only 10 V, and the operating current is

$$I = \frac{V}{R} = \frac{10 \text{ V}}{3 \text{ }\Omega} = 3.33 \text{ A}$$

Solution (b)   At the instant the motor is turned on, the armature is not yet turning, and $\varepsilon_b = 0$. In this case the current will be

$$I = \frac{V}{R} = \frac{120 \text{ V}}{3 \text{ }\Omega} = 40 \text{ A}$$

In the preceding example, it was assumed that the entire motor resistance was a part of a series circuit. Often motors are designed with armature and field windings which may be connected in parallel. In such cases the solution for the operating current is more complex, but it can be calculated by the methods shown in Chap. 17 for dc circuits.

19-5 ■ TYPES OF MOTORS

Dc motors are classified according to how the field coils and the armature are connected. When the armature coils and the field coils are connected in series, as in Fig. 19-12, the motor is said to be *series-wound*. In this type of motor, the current energizes both the field windings and the armature windings. When the armature turns slowly, the back emf is small and the current is large. Consequently, a large torque is developed at low speeds.

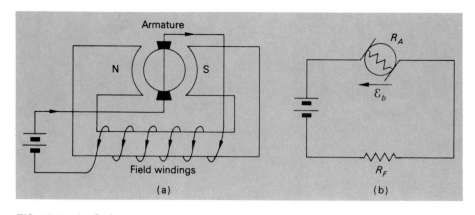

FIG. 19-12 (*a*) Series-wound dc motor; (*b*) schematic diagram.

In a *shunt-wound* motor, the field windings and the armature windings are connected in parallel, as shown in Fig. 19-13. The entire voltage is applied across both windings. The primary advantage of a shunt-wound motor is that it produces more constant torque over a range of speeds. However, the starting torque is usually lower than that of a similar series-wound motor.

In some applications, the field windings are in two parts, one connected in series with the armature and the other in parallel with it. Such a motor is called a *compound-wound* motor. The torque produced by a compound-wound motor varies between that of a series and that of a shunt motor.

In permanent-magnet motors, no field current is necessary. These motors have torque characteristics similar to those of shunt-wound motors.

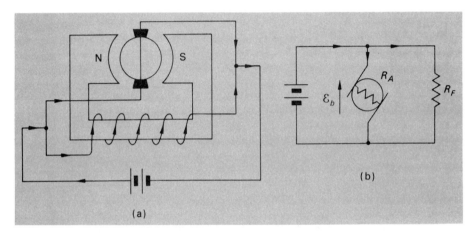

FIG. 19-13 (*a*) Shunt-wound dc motor; (*b*) schematic diagram.

19-6 ■ *TRANSFORMER*

It was noted earlier that a changing current in one wire loop will induce a current in a nearby loop. The induced current arises from the changing magnetic field associated with a changing current. Alternating current has a distinct advantage over direct current because of the inductive effect of a current which constantly varies in magnitude and direction. The most common application of this principle is offered by the *transformer,* a device which increases or decreases the voltage in an ac circuit.

A simple transformer is illustrated in Fig. 19-14. There are three essential parts: a primary coil connected to an ac source, a secondary coil, and a soft-iron core. As an alternating current is sent through the primary coil, magnetic flux lines move back and forth through the iron core, inducing an alternating current in the secondary coil.

The constantly changing magnetic flux is established throughout the core of the transformer and passes through both primary and secondary coils. Remember that the induced emf is proportional to the number of turns. Therefore, the ratio of the voltages in the two coils is the same as the ratio of the turns. Using the subscript p to refer to the primary coil and the subscript s to refer to the secondary coil, we may write

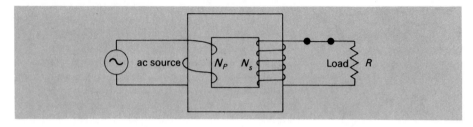

FIG. 19-14 Simple transformer.

$$\boxed{\frac{\mathcal{E}_p}{\mathcal{E}_s} = \frac{N_p}{N_s}} \qquad (19\text{-}2)$$

$$\frac{\text{Primary voltage}}{\text{Secondary voltage}} = \frac{\text{primary turns}}{\text{secondary turns}}$$

The induced voltage is in direct proportion to the number of turns. If the ratio of secondary turns N_s to primary turns N_p is varied, an input (primary) voltage can provide any desired output (secondary) voltage. For example, if there are 40 times as many turns in the secondary coil, an input voltage of 120 V will be increased to $40 \times 120 = 4800$ V in the secondary coil. A transformer which produces a larger output voltage is called a *step-up transformer*.

A *step-down transformer* can be constructed by making the number of primary turns greater than the number of secondary turns. A step-down transformer gives a lower output voltage.

The *efficiency* of a transformer is defined as the ratio of the power output to the power input. Recalling that electric power is equal to the product of voltage and current, we can write the efficiency E of a transformer as

$$E = \frac{\text{power output}}{\text{power input}} = \frac{\mathcal{E}_s i_s}{\mathcal{E}_p i_p} \qquad (19\text{-}3)$$

where i_p and i_s are the currents in the primary coil and the secondary coil, respectively. Most electric transformers are carefully designed for extremely high efficiencies, normally above 90 percent.

It is important to recognize that there is no power gain as a result of transformer action. When the voltage is stepped up, the current must be stepped down, so that the product $\mathcal{E}i$ does not increase. To see this more clearly, let us assume that a given transformer is 100 percent efficient. For this perfect transformer, Eq. (19-3) becomes

$$\mathcal{E}_s i_s = \mathcal{E}_p i_p \qquad \text{or} \qquad \frac{i_p}{i_s} = \frac{\mathcal{E}_s}{\mathcal{E}_p}$$

This equation clearly shows the inverse relationship between current and induced voltage.

EXAMPLE 19-2   An ac generator which delivers 20 A at 6000 V is connected to a step-up transformer. What is the output current at 120,000 V if the transformer efficiency is 100 percent?

Solution   From Eq. (19-3),

$$i_s = \frac{\varepsilon_p i_p}{\varepsilon_s}$$

$$= \frac{(6000 \text{ V})(20 \text{ A})}{120,000 \text{ V}} = 1 \text{ A}$$

Note in the preceding example that the current was reduced to 1 A, whereas the voltage was increased 20-fold. Since heat losses in transmission lines vary directly with the square of the current ($i^2 R$), this means that if this electric power is transmitted, more of it will reach its destination. Step-down transformers are used to provide the desired current at lower voltages.

19-7 ■ ALTERNATING CURRENT AND VOLTAGE

In the discussion of a simple ac generator, we have seen that an alternating current is not constant as the direct currents we studied in earlier chapters. The voltage and the current vary sinusoidally between maximum values in alternating directions. A convenient way to show the variation in emf or current for an ac circuit is with a rotating vector or by a sine wave. These representations are compared in Fig. 19-15. The vertical component of the rotating vector at any instant is the instantaneous value of the voltage or current. One complete revolution of the rotating vector or one complete sine wave on the curve represents one *cycle*. As mentioned earlier, the number of cycles per second is the frequency of the current or voltage in hertz.

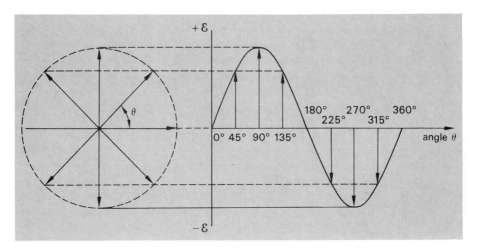

FIG. 19-15 A rotating vector and its corresponding sine wave can be used to represent alternating current or voltage.

Note that the average value of voltage and current for any cycle is zero, since the magnitude alternates between $\mathcal{E}_{max}$, i_{max} and $-\mathcal{E}_{max}$, $-i_{max}$. However, even though there is no *net* current, charge is in motion, and electric energy can be released in the form of heat or useful work. The best method of measuring the effective strength of alternating currents is to find the dc value which will produce the same *heating* effect or develop the same *power* as the alternating current. This current value, called the *effective current* i_{eff}, is found to be 0.707 times the maximum current. A similar relation holds for the effective voltage in an ac circuit. Thus

$$i_{eff} = 0.707\ i_{max}$$
$$\mathcal{E}_{eff} = 0.707\ \mathcal{E}_{max}$$

$$(19\text{-}4)$$

One **effective ampere** *is that alternating current which will develop the same power as one ampere of direct current.*

One **effective volt** *is that alternating voltage which will produce an effective current of one ampere through a resistance of one ohm.*

Ac meters are calibrated to show effective values. For example, if an ac meter measures household voltage to be 120 V at 10 A, Eq. (19-4) will show that the maximum values of current and voltage, respectively, are

$$i_{max} = \frac{10\ A}{0.707} = 14.14\ A$$

$$V_{max} = \frac{120\ V}{0.707} = 170\ V$$

Therefore, the house voltage actually varies between +170 and −170 V, while the current ranges from +14.14 to −14.14 A. The usual frequency of voltage variation is 60 Hz.

19-8 ■ INDUCTOR AND INDUCTIVE REACTANCE

The only important element in the dc circuit (besides the source of emf) was the resistor. Since alternating currents behave differently from direct currents, other quantities besides resistance impede current flow. A coil of wire, called an *inductor,* and a *capacitor* play important roles in ac circuits. We will discuss the inductor first.

An inductor is a continuous loop or coil of wire, such as that shown in Fig. 19-16 with its circuit diagram. When an alternating current flows through such a coil, the changing magnetic flux induces an opposing emf in the wires. This back emf is referred to as *self-inductance,* or simply *inductance.* The inductance of a coil can be increased by increasing the number of coils or by inserting a permeable material in the center of the coil. Inductance is measured in *henries,* defined as follows:

A given inductor has an inductance of one henry (1 H) if an emf of one volt is induced by a current changing at the rate of one ampere per second.

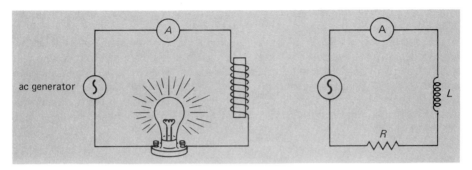

FIG. 19-16 Inductor with its circuit diagram.

The back emf resulting from inductance causes an opposition to the flow of alternating current. This opposition is called *inductive reactance*. The inductive reactance X_L of an inductor depends on the inductance L of the coil in henries and the frequency f of the applied voltage in hertz. The formula is

$$X_L = 2\pi f L \qquad \textit{Inductive Reactance} \quad (19\text{-}5)$$

Inductive reactance, as ordinary resistance, is measured in ohms.

We can see the effect of inductive reactance if we replace the ac generator in Fig. 19-16 with a dc source at the same voltage. If we set the ammeter to function on direct current, the current will show a higher reading than would be observed for the ac voltage. Restoring the ac generator would again cause the current to be reduced due to the inductance of the coil. Inserting a permeable core would cause even greater reduction in the current.

The effective current i in an inductor is determined from its inductive reactance X_L and the effective voltage V by an equation analogous to Ohm's law:

$$V = iX_L \qquad (19\text{-}6)$$

EXAMPLE 19-3 A coil having an inductance of 0.5 H is connected to a 120-V, 60-Hz power source. If the resistance of the coil is neglected, what is the effective current through the coil?

Solution The inductive reactance is

$$X_L = 2\pi f L = (2\pi)(60 \text{ Hz})(0.5 \text{ H})$$
$$= 188 \; \Omega$$

The current is found from $V = iX_L$:

$$i = \frac{V}{X_L} = \frac{120 \text{ V}}{188.4 \; \Omega} = 0.637 \text{ A}$$

19-9 ■ CAPACITOR AND CAPACITIVE REACTANCE

Direct current will not flow through a capacitor because of the plate separation. However, when a capacitor is inserted into an ac circuit, the alternating current will alternately charge and discharge the capacitor, permitting work to be accomplished in the circuit. The charge will pass from one plate to the other through the external circuit. A circuit containing a single capacitor is shown in Fig. 19-17. Just as the back emf in an inductor offered opposition to electric current, so will the back emf due to charge build up on a capacitor.

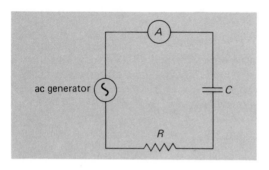

FIG. 19-17 Circuit containing a single capacitor.

Capacitive reactance is the nonresistive opposition to alternating current due to the back emf of a capacitor. The capacitive reactance X_C of a capacitor depends on its capacitance C and the frequency of the alternating current:

$$X_C = \frac{1}{2\pi fC} \qquad \text{Capacitive Reactance} \quad (19\text{-}7)$$

When the capacitance is measured in farads and the frequency is in hertz, the capacitive reactance will be in ohms.

The effective current in a capacitive circuit is also given by an equation similar to Ohm's law:

$$V = iX_C \qquad (19\text{-}8)$$

The quantity V is the alternating voltage applied to the capacitor, and i represents the alternating current.

19-10 ■ PHASE RELATION IN AC CIRCUITS

In most dc circuits, the voltage and the current reach maximum and zero values at the same time and are said to be *in phase*. The effects of inductance and capacitance in ac circuits prevent the voltage and current from reaching maxima and minima at the same time. In other words, the current and voltage in most ac circuits are *out of phase*.

To understand phase relations in an ac circuit, suppose we first consider a circuit containing a *pure resistor* in series with an ac generator, as in Fig. 19-18. This is an idealized circuit in which the inductive and capacitive effects are negligible. Many household devices, such as lights, heaters, and toasters, approximate a condition of pure resistance. In such devices, the instantaneous voltage *V* and current *i* are in phase. That is, variations in voltage will result in simultaneous variations in current. When the voltage is a maximum, the current is also a maximum. When the voltage is a minimum, the current is also a minimum. When the voltage is zero, the current is zero.

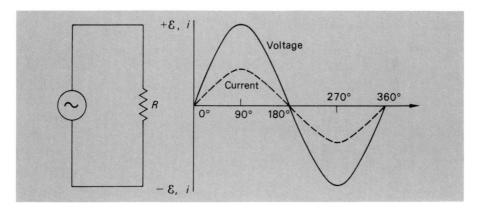

FIG. 19-18 In a circuit containing pure resistance, the voltage and the current are in phase.

Next, we consider the phase relation between current and voltage across a *pure inductor*. The circuit illustrated in Fig. 19-19 contains only an inductor in series with the ac generator. We have seen that the presence of inductance results in a back emf which *delays* the current in reaching its maximum. The voltage reaches a maximum while the current is still at zero. When the voltage reaches zero, the current is at a maximum. In a circuit containing only inductance, the voltage is said to *lead* (occur before) the current by one-fourth of a cycle (or 90°). Note the curve in Fig. 19-19.

In a circuit containing pure inductance, the voltage leads the current by 90°.

The effect of capacitance in an ac circuit is opposite to that of inductance. For the circuit shown in Fig. 19-20, the voltage must *lag* the current, since the flow of charge to the capacitor is necessary to build up an opposing emf. When the applied voltage is decreasing, charge flows from the capacitor. The rate of flow of this charge reaches a maximum when the applied voltage is zero.

In a circuit containing pure capacitance, the voltage lags the current by 90°.

This means that variations in voltage occur one-fourth of a cycle later than the corresponding variations in current.

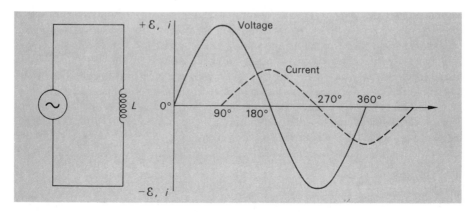

FIG. 19-19 In a circuit containing pure inductance, the voltage leads the current by 90°.

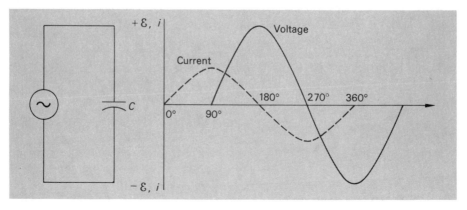

FIG. 19-20 In a circuit containing pure capacitance, the voltage lags the current by 90°.

19-11 ■ *SERIES AC CIRCUIT*

In general, an ac circuit contains resistance, capacitance, and inductance in varying amounts. A series combination of these elements is shown in Fig. 19-21. The total voltage drop in a dc circuit is the simple sum of the drops across each element in the circuit. However, in the ac circuit, the voltage and current are not in phase with each other. Recall that V_R is always in phase with the current, but V_L leads the current by 90° and V_C lags the current by 90°. Clearly, if we are to determine the effective voltage V of the entire circuit, we must develop a means of treating phase differences.

This can be accomplished by using a vector diagram, called the *phase diagram*. (See Fig. 19-22.) In this method, the effective values of V_R, V_L, and V_C are plotted as rotating vectors. The phase relationship is expressed in terms of the phase angle ϕ, which is a measure of how much the voltage leads the current in a particular circuit element. For example, in a pure resistor, the voltage and the current are in phase and $\phi = 0$. For an inductor $\phi = +90°$, and for a capacitor $\phi = -90°$. A negative phase angle occurs when the voltage lags the current. Following this scheme, V_R appears as a vector along the x axis, V_L is represented by a vector pointing vertically upward, and V_C is directed downward.

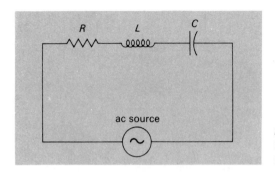

FIG. 19-21 Series ac circuit containing resistance, inductance, and capacitance.

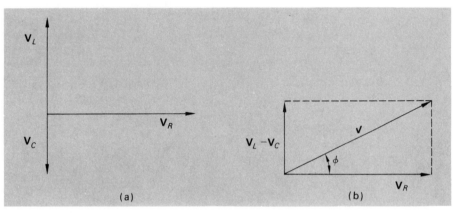

(a)

(b)

FIG. 19-22 Phase diagram.

The effective voltage V in an ac circuit can be defined as the vector sum of V_R, V_L, and V_C as they exist on the phase diagram. It can be seen from the figure that the magnitude of V is given by

$$V = \sqrt{V_R^2 + (V_L - V_C)^2} \qquad (19\text{-}9)$$

You should verify this equation by applying the pythagorean theorem to the vector diagram.

Note from the figure that a value of V_L which is greater than V_C results in a positive phase angle. In other words, the circuit is inductive, and the voltage leads the current. In a capacitive circuit, V_C is greater than V_L, and a negative phase angle will result, indicating that the voltage lags the current. If desired, the phase angle ϕ can be calculated by trigonometry:

$$\tan \phi = \frac{V_L - V_C}{V_R} \qquad (19\text{-}10)$$

A more useful form of the above equations can be found by recalling that

$$V_R = iR \qquad V_L = iX_L \qquad V_C = iX_C$$

Upon substitution, we find that Eq. (19-9) becomes

$$V = i\sqrt{R^2 + (X_L - X_C)^2} \qquad (19\text{-}11)$$

The quantity multiplied by the current in this relation is a measure of the combined opposition that the circuit offers to alternating current. It is called the *impedance,* denoted by the symbol Z, and is given by

$$Z = \sqrt{R^2 + (X_L - X_C)^2} \qquad \text{\it Impedance} \quad (19\text{-}12)$$

The higher the impedance in a circuit, the lower the current for a given voltage. Since R, X_L, and X_C are measured in ohms, the impedance is also expressed in ohms. Thus, the effective current i in an ac circuit is given by

$$i = \frac{V}{Z} \qquad \text{\it Effective Current} \quad (19\text{-}13)$$

where V is the applied ac voltage and Z is the impedance. Remember that Z depends on the frequency f of the alternating current as well as on the resistance, inductance, and capacitance.

Since the voltage across each element depends directly upon resistance or reactance, an alternative phase diagram can be constructed that treats R, X_L, and X_C as vector quantities. Such a diagram can be used to compute the impedance, as shown in Fig. 19-23. The phase angle ϕ in this diagram can be found from

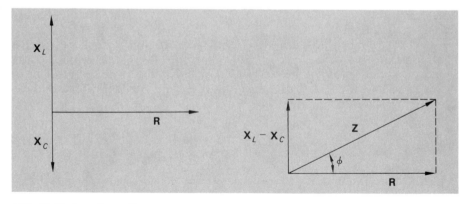

FIG. 19-23 Impedance diagram.

$$\tan \phi = \frac{X_L - X_C}{R} \qquad \textit{Phase Angle} \quad \text{(19-14)}$$

Of course, this is the same angle as that given by Eq. (19-10).

EXAMPLE 19-4 A 40-Ω resistor, a 0.4-H inductor, and a 10-μF capacitor are connected in series with an ac source which generates 120-V, 60-Hz alternating current. (*a*) Find the impedance of the circuit. (*b*) Determine the effective current.

Solution (a) We must first compute X_L and X_C as follows:

$$X_L = 2\pi f L = (2\pi)(60 \text{ Hz})(0.4 \text{ H}) = 151 \ \Omega$$

$$X_C = \frac{1}{2\pi f C} = \frac{1}{(2\pi)(60 \text{ Hz})(10 \times 10^{-6} \text{ F})} = 265 \ \Omega$$

The impedance of the circuit is found from Eq. (19-12):

$$Z = \sqrt{R^2 + (X_L - X_C)^2}$$
$$= \sqrt{(40 \ \Omega)^2 + (151 \ \Omega - 265 \ \Omega)^2} = 121 \ \Omega$$

Solution (b) Now we determine the effective current from the known impedance:

$$i = \frac{V}{Z} = \frac{120 \text{ V}}{121 \ \Omega} = 0.992 \text{ A}$$

If you know trigonometry, you should also show that the phase angle from Eq. (19-14) is $-71°$. This would indicate that the voltage will lag the current. The circuit is more capacitive than it is inductive.

19-12 ■ RESONANCE

Since inductance causes the current to lag the voltage and capacitance causes the current to lead the voltage, their combined effect is to cancel each other. The total reactance is given by $X_L - X_C$, and the impedance in a circuit is a minimum when $X_L = X_C$. When this is true, only the resistance R remains and the current will be a maximum. Setting $X_L = X_C$, we can write

$$2\pi f_r L = \frac{1}{2\pi f_r C}$$

from which

$$f_r = \frac{1}{2\pi \sqrt{LC}} \qquad \textit{Resonant Frequency} \quad \text{(19-15)}$$

When the applied voltage has this frequency, called the *resonant frequency*, the current in the circuit will be a maximum. Additionally, since the current is limited only by resistance, it will be in phase with the voltage. The antenna circuit in an AM radio contains a variable capacitor which acts as a tuner. The capacitance is varied until the resonant frequency is equal to a particular signal frequency. The current peaks when this happens, and the receiver responds to that incoming signal.

*19-13 ■ POWER FACTOR

In ac circuits, no power is consumed because of capacitance or inductance. Energy is merely stored at one instant and released at another, causing the current and voltage to be out of phase. Whenever the current and voltage are in phase, the average power P delivered is a maximum, given by

$$P = iV$$

where i is the effective current and V is the effective voltage. This condition is satisfied when the ac circuit contains only resistance R or when the circuit is in resonance $(X_L = X_C)$.

Normally, however, an ac circuit contains sufficient reactance to limit the effective power. In any case, the power delivered to the circuit is a function of only the component of voltage V which is in phase with the current. From Fig. 19-21, this component is V_R, and we can write

$$V_r = V \cos \phi$$

where ϕ is the phase angle. The quantity $\cos \phi$ is called the *power factor* of the circuit. The true power delivered in an ac circuit is found by multiplying this factor by the power if no reactance existed. Thus the true power is

$$P = iV \cos \phi \qquad (19\text{-}16)$$

The value of the power factor can also be calculated from

$$\cos \phi = \frac{R}{Z} = \frac{R}{\sqrt{R^2 + (X_L - X_C)^2}} \qquad (19\text{-}17)$$

EXAMPLE 19-5   (*a*) What is the power factor for the circuit described in Example 19-4? (*b*) What is the power absorbed in the circuit?

Solution (a)   The power factor is found from Eq. (19-17):

$$\cos \phi = \frac{R}{Z} = \frac{40 \ \Omega}{120 \ \Omega} = 0.33$$

Solution (b)   The power absorbed in the circuit is found from Eq. (19-16):

$$P = iV \cos \phi = (0.992 \text{ A})(120 \text{ V})(0.33)$$
$$= 39.3 \text{ W}$$

The power factor is sometimes expressed as a percentage instead of as a decimal. For example, the power factor of 0.33 in the above example could be expressed as 33 percent. Most commercial ac circuits have power factors from 80 to 90 percent because they usually contain more inductance than capacitance. Since this requires the electric power companies to furnish more current for a given power, the power companies extend a lower rate to users with power factors above 90 percent. Commercial users can improve their inductive power factors by adding capacitors, for instance.

Summary

Electromagnetic induction provides the basis for modern delivery and consumption of electric power. Large generators induce ac voltages that are stepped up by transformers and transmitted through high-tension wires to homes and to industry. The following concepts have been discussed in this unit:

- An emf is *induced* by the relative motion between a conductor and a magnetic field. The magnitude of the induced emf is directly proportional to the number of turns and to the rate at which magnetic lines are cut.
- The direction of induced current can be determined by using *Fleming's rule:* If the thumb, forefinger, and middle finger of the right hand are held at right angles to each other, with the thumb pointing in the direction in which the wire is moving and the forefinger pointing in the field direction (north to south), the middle finger will point in the direction of induced current.
- When the armature of a generator rotates with a constant angular velocity in a constant magnetic field, the magnitude of induced voltage or current varies sinusoidally with a frequency measured in cycles per second, or hertz.
- The *net voltage* delivered to the armature coils of a dc motor is equal to the *applied voltage* less the *induced voltage*. Symbolically,

$$\boxed{V - \varepsilon_b = IR}$$

- For a transformer with N_p primary turns and N_s secondary turns,

$$\frac{\text{Primary voltage}}{\text{Secondary voltage}} = \frac{\text{primary turns}}{\text{secondary turns}}$$

$$\boxed{\frac{\varepsilon_p}{\varepsilon_s} = \frac{N_p}{N_s}}$$

- The *efficiency* of a transformer is

$$E = \frac{\text{power output}}{\text{power input}} = \frac{\mathcal{E}_s i_s}{\mathcal{E}_p i_p}$$

- *Effective amperes* and *volts* for ac circuits are defined as those values which would produce the same heating effect as a direct current of that magnitude:

$$i_{\text{eff}} = 0.707 i_{\text{max}}$$
$$\mathcal{E}_{\text{eff}} = 0.707 \mathcal{E}_{\text{max}}$$

- Both capacitors and inductors offer resistance to the flow of alternating current (called *reactance*), calculated from

$$X_L = 2\pi f L$$ *Inductive Reactance X_L, Ω*

$$X_C = \frac{1}{2\pi f C}$$ *Capacitive Reactance X_C, Ω*

The symbol f refers to the frequency of the alternating current in hertz. One hertz is one cycle per second.

- The voltage, current, and resistance in a series ac circuit are studied with the use of phasor diagrams. Figure 19-23 illustrates such a diagram for X_C, X_L, and R. The resultant of these vectors is the effective resistance of the entire circuit, called the *impedance Z*. From the figure,

$$Z = \sqrt{R^2 + (X_L - X_C)^2}$$ *Impedance, Ω*

- The following equations are useful for problems involving ac circuits:

$$V = \sqrt{V_R^2 + (V_L - V_C)^2}$$

$$V_R = iR \qquad V_L = iX_L \qquad V_C = iX_C \qquad V = iZ$$

$$V = i\sqrt{R^2 + (X_L - X_C)^2}$$

- Because the voltage leads the current in an inductive circuit and lags the current in a capacitive circuit, the voltage and current maxima and minima usually do not coincide. The phase angle φ is given by

$$\tan \phi = \frac{V_L - V_C}{V_B}$$ or $$\tan \phi = \frac{X_L - X_C}{R}$$

- The resonant frequency occurs when the net reactance is zero ($X_L = X_C$):

$$f_r = \frac{1}{2\pi\sqrt{LC}}$$ *Resonant Frequency*

- No power is consumed because of capacitance or inductance. Since power is a function of the component of the impedance along the resistance axis, we can write

$$P = iV \cos \phi$$ power factor = cos φ

$$\cos \phi = \frac{R}{Z}$$ $$\cos \phi = \frac{R}{\sqrt{R^2 + (X_L - X_C)^2}}$$

Questions

19-1. Define the following terms:

a. induced emf
b. Lenz' law
c. Fleming's rule
d. generator (ac/dc)
e. dc motor
f. armature
g. slip rings
h. brushes
i. hertz
j. commutator
k. motor back emf
l. series-wound

m. shunt-wound
n. transformer
o. inductance
p. alternating current and voltage
q. reactance
r. effective current and voltage
s. inductor
t. henry
u. phase angle
v. impedance
w. power factor
x. resonant frequency

19-2. Discuss the various factors which influence the magnitude of an induced emf in a length of wire moving in a magnetic field.

19-3. A bar magnet is held in a vertical position with the north pole facing upward. If a closed-loop coil is dropped over the north end of the magnet, what is the direction of the induced current, viewed from the top of the magnet?

19-4. A circular loop is suspended with its plane perpendicular to a magnetic field directed from left to right. The loop is removed from the field by moving it upward quickly. What is the direction of the induced current viewed along the field direction? Is a force required to remove the loop from the field? Explain.

19-5. An induction coil is essentially a transformer which operates on direct current. As shown in Fig. 19-24, the induction coil consists of a few primary turns wound around an iron core with a large number of secondary turns surrounding the primary turns. A battery current magnetizes the core so that it attracts the armature of the vibrator and opens the circuit periodically. When the circuit is opened, the field collapses, and a large emf is induced in the secondary coil, producing a spark at the output terminals. What is the function of the capacitor C connected in parallel with the vibrator? Explain how an induction coil is used in the ignition system of an automobile.

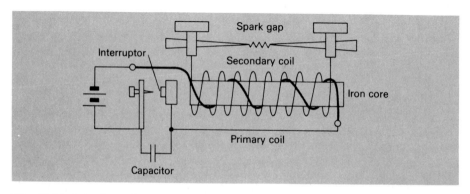

FIG. 19-24 Induction coil.

19-6. Explain clearly how an ac generator can be converted to a dc generator. How would you convert an ac generator to an ac motor?

19-7. When the electric motor in a plant is starting, a worker notices that the lights are momentarily dimmed. Why?

19-8. What type of dc motor should be purchased to operate a winch used to lift heavy objects? Why? What type of motor should be used to operate an electric fan where uniform torque is desired at relatively high speeds? Explain.

19-9. Explain how the existence of back emf in a motor helps to keep its speed constant. *Hint:* What happens to the back emf and current when the armature speed increases or decreases?

19-10. Explain, using diagrams, how transformers are used to transmit current economically from power installations to homes many miles away.

19-11. An ac generator produces 60-Hz alternating voltage. How many degrees of armature rotation will correspond to one-fourth of a cycle?

19-12. Why is it less expensive to provide alternating current to homes and industry rather than direct current?

19-13. The inductance in a circuit depends upon its geometry and upon the proximity of magnetic materials. Which of the following coils have higher inductances and why? (*a*) Closely or widely spaced turns; (*b*) long coil or short coil; (*c*) large cross section or small cross section; (*d*) iron core or air core.

19-14. Plot a curve of voltage vs. time for (*a*) charging a capacitor and (*b*) discharging a capacitor. Plot a curve of voltage as a function of time for an inductor.

19-15. A coil of wire is connected across the terminals of a 110-V dc battery. An ammeter connected in series with the coil indicates a current of 5 A. What happens to the current if an iron core is inserted into the coil? Now, disconnect the dc source, remove the iron core, and reconnect the system to a 110-V ac generator. Adjust the ammeter to read ac effective amperes. Has the current decreased, increased, or remained the same? What happens to the current if the iron core is inserted? Explain your observations.

19-16. Should someone interested in establishing the breakdown voltage of a capacitor in an ac circuit be concerned with the average voltage, maximum voltage, or effective voltage?

19-17. An incandescent lamp is connected in series with a 110-V ac generator and a variable capacitor. If the capacitance is increased, will the lamp glow more brightly or will it be dimmed? Explain. What will happen if the generator is replaced with a battery?

19-18. What happens to the resonant frequency of an ac circuit if the capacitance is increased?

19-19. As the frequency of an inductive circuit is increased, what happens to the alternating current in the circuit?

19-20. Inductive reactance depends *directly* on the frequency of the alternating current, whereas capacitive reactance varies *inversely* with frequency. Both oppose the flow of alternating current. Explain why their relationship to frequency differs.

19-21. When a circuit is tuned to its resonant frequency, what is the power factor? What steps can industry take to increase the power factor in its consumption of electric energy?

19-22. What is the power factor of a circuit containing (*a*) pure resistance, (*b*) pure inductance, (*c*) pure capacitance?

Problems

19-1. The resistance of a disconnected dc motor is measured to be 1.5 Ω. When connected to a 120-V source, the motor normally draws 6 A. What is the normal back emf of the motor?

Answer 111 V.

19-2. A 12-V dc motor is used to drive a small pulley. A back emf of 9-V occurs at full speed. What is the resistance of the motor if it draws 2 A while at operating speed?

19-3. A 120-V dc motor has an internal resistance of 2.5 Ω. When it is operated at normal speed, the back emf is 109 V. (*a*) What is the operating current? (*b*) What is the starting current?

Answer (*a*) 1.2 A; (*b*) 48 A.

19-4. A 220-V dc motor draws a current of 10 A in operation and has an armature resistance of 0.4 Ω. (*a*) What is the back emf when the motor is operating? (*b*) What is the starting current?

19-5. The armature coils of the starting motor in an automobile have a resistance of 0.05 Ω. The motor is driven by a 12-V battery, and when the armature is moving at its operating speed, a back emf of 6 V is generated. (*a*) What is the starting current? (*b*) What is the current at operating speed? Ignore the internal resistance and other circuit resistances.

Answer (*a*) 240 A; (*b*) 120 A.

19-6. A 120-V series-wound motor has a field resistance of 90 Ω and an armature resistance of 10 Ω. When it is operating at full speed, a back emf of 80 V is generated. (*a*) What is the total resistance of the motor? (*b*) What is the initial current drawn by the motor? (*c*) What is the operating current?

19-7. A step-up transformer has 400 secondary turns and only 100 primary turns. An alternating voltage of 120 V is connected to the primary coil. What is the output voltage?

Answer 480 V.

19-8. A step-down transformer is used to drop an alternating voltage from 10,000 to 500 V. What must be the ratio of secondary turns to primary turns? If the input current is 1 A and the efficiency of the transformer is 100 percent, what is the output current?

19-9. A step-up transformer has 80 primary turns and 720 secondary turns. The efficiency of the transformer is 95 percent. If the primary draws a current of 20 A at 120 V, what are the current and voltage for the secondary?

Answer 2.11 A; 1080 V.

19-10. A girl wishes to make a transformer that will lower the house voltage from 110 to 11 V in order to operate a small electric bell. She takes a piece of soft iron and some insulated copper wire. What must be the ratio of primary turns to secondary turns? If the current in the primary winding is 0.6 A, what current is delivered to the bell? Assume 100 percent efficiency.

19-11. The current supplied to a woodworking shop is rated at 10 A. What is the maximum value of the current supplied?

Answer 14.1 A.

19-12. A certain appliance is supplied with 220 V under an effective current of 20 A. What are the maximum values for the voltage and current?

19-13. A capacitor has a maximum voltage rating of 500 V. What is the highest effective ac voltage that can be supplied to it without danger of breakdown?

Answer 354 V.

19-14. A 0.05-H inductor of negligible resistance is connected to a 120-V, 60-Hz line. (*a*) What is the inductive reactance? (*b*) What is the current in the coil?

19-15. A 2-H inductor of negligible resistance is connected to a 50-V, 50-Hz line. (*a*) What is the reactance? (*b*) What is the current in the coil?

Answer (*a*) 628 Ω; (*b*) 79.6 mA.

19-16. An inductor has a reactance of 100 Ω at 200 Hz. What is the inductance?

19-17. A 3-μF capacitor connected to a 120-V ac line draws 0.005 A. What is the frequency of the source?

Answer 2.21 Hz.

19-18. Find the reactance of a 60-μF capacitor used in a 600-Hz ac circuit.

19-19. A 6-μF capacitor is connected to a 24-V, 50-Hz line. What is the alternating current?

Answer 45.2 mA.

19-20. When a 6-Ω resistor and an inductor are connected to a 110-V, 60-Hz line, the current is 10 A. What is the inductance of the coil?

19-21. A series ac circuit consists of a 100-Ω resistor, a 0.2-H inductor, and a 3-μF capacitor connected to a 110-V, 60-Hz line. (*a*) What is the inductive reactance?

(*b*) What is the capacitive reactance? (*c*) What is the current in the circuit? (*d*) What is the phase angle? (*e*) What is the power factor?

Answer (*a*) 75.4 Ω; (*b*) 884 Ω; (*c*) 0.135 A; (*d*) −83°; (*e*) 0.12.

19-22. Answer the questions of Prob. 19-21 for a circuit containing a 10-mH inductor, a 10-μF capacitor, and a 30-Ω resistor. The power source is 110 V at 400 Hz.

19-23. An inductor, resistor, and capacitor are connected in series with a 60-Hz power line. A voltmeter connected to each element in the circuit gives the following readings: $V_R = 60$ V, $V_L = 100$ V, and $V_C = 160$ V. (*a*) What is the total voltage drop in the circuit? (*b*) What is the phase difference between the voltage and the current?

Answer (*a*) 84.8 V; (*b*) −45°.

19-24. A resonant circuit has an inductance of 400 μH (1 μH = 1×10^{-6} H) and a capacitance of 100 pF (1 pF = 1×10^{-12} F). What is the resonant frequency?

19-25. A 50-μF capacitor and a 70-Ω resistor are connected in series across a 120-V, 60-Hz line. (*a*) What is the current in the circuit? (*b*) What power is dissipated in the circuit? (*c*) What is the power factor?

Answer (*a*) 1.36 A; (*b*) 131 W; (*c*) 0.796.

19-26. Refer to Prob. 19-25. (*a*) What is the phase angle? (*b*) What is the voltage across the resistor? (*c*) What is the voltage across the capacitor?

19-27. The antenna circuit in a radio receiver consists of a variable capacitor and a 9-mH coil. The resistance of the circuit is 40 Ω. A 980-kHz radio wave produces a potential difference of 0.2 mV across the circuit. (*a*) Determine the capacitance required for resonance. (*b*) What is the current in the circuit at resonance?

Answer (*a*) 2.93 pF; (*b*) 5 μA.

19-28. Find the resonant frequency of a circuit containing a 6-μF capacitor and a 2.6-μH inductor.

20

WAVE MOTION
AND SOUND

OBJECTIVES

As a result of completing this chapter, you should be able to

1. Define and describe *transverse* and *longitudinal* wave motion, indicating the measure of a wavelength for each type of wave.
2. Solve problems involving the *mass, length, tension,* and *wave speed* for transverse waves in a string.
3. Describe and apply the relationship among *wave speed, wavelength,* and *frequency* for periodic wave motion.
4. Define sound and discuss its nature and the factors which affect its propagation.
5. Solve problems involving the speed of sound as a function of air temperature.
6. Discuss the physical meaning of *loudness, pitch, quality, resonance,* and *beats* as the terms apply to sound waves.
7. Discuss the meaning of the *intensity level* for sound, and demonstrate your understanding of the *decibel* as a measure of the relative intensities of sounds.

One of the most important ways of transferring energy is through wave motion. Energy can be distributed from one point to other points without the physical transfer of material between the points. We are all familiar with water waves, which carry a disturbance through water, but the same principle can be applied to much more complicated events. Mechanical waves, which involve local disturbances in a mass medium, are the basis for the transmission of sound. Electromagnetic waves carry energy in the forms of light, heat, and radio waves through empty space. In all cases, you will need to understand the meaning of and the relationships among wave speed, wavelength, and wave frequency. The focus of this chapter is sound waves as a specific example of wave motion. Sound is used for burglar alarms, depth finders, machining operations, cleaning, delicate measurements, and many other specialized operations.

20-1 ■ MECHANICAL WAVES

When a stone is dropped into a pool of water, it creates a disturbance which spreads out in concentric circles, eventually reaching all parts of the pool. A small stick floating on the surface of the water bobs up and down as the disturbance passes. Energy has been transferred from the point of impact of the stone to the floating stick some distance away. This energy is passed along by the agitation of neighboring water particles. Only the disturbance moves through the water. The actual motion of any particular water particle is comparatively small. Energy propagation by means of a disturbance in a medium instead of the motion of the medium itself is called *mechanical wave motion*.

A water wave is a *mechanical wave* because its existence depends upon a mechanical source and a material medium.

A **mechanical wave** *is a physical disturbance in an elastic medium.*

It is important to recognize that not all disturbances are necessarily mechanical. For example, light waves, radio waves, and heat radiation carry energy by means of electric and magnetic disturbances. No physical medium is necessary for the transmission of electromagnetic waves. However, many of the basic ideas presented in this chapter for mechanical waves are also applicable to electromagnetic waves.

20-2 ■ TYPES OF WAVES

Waves are classified according to the motion of a local part of a medium with respect to the direction of the wave.

In a **transverse wave** *the vibration of the individual particles of the medium is perpendicular to the direction of wave propagation.*

For example, suppose we fasten one end of a rope to a post and grasp the other end, as shown in Fig. 20-1. By moving the free end up and down quickly, we send a single disturbance, called a *pulse*, down the length of rope. Three equally spaced knots at points *a, b,* and *c* demonstrate that the individual particles move up and down while the disturbance moves to the right with velocity *v*.

Another type of wave, which may occur in a coiled spring, is shown in Fig. 20-2. The coils near the left end are pinched closely together to form a *condensation*. When the distorting force is removed, a condensation pulse is sent down the length of string. No part of the spring moves very far from its equilibrium position, but the pulse continues to travel along the spring. Such a wave is called a *longitudinal wave* because the spring particles are displaced along the same direction in which the disturbance travels.

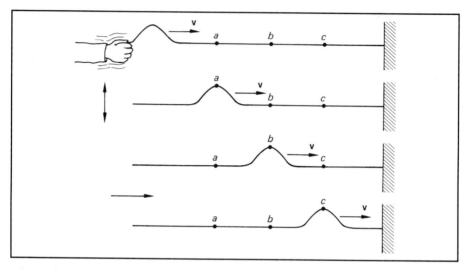

FIG. 20-1 In a transverse wave, the individual particles move perpendicular to the direction of propagation.

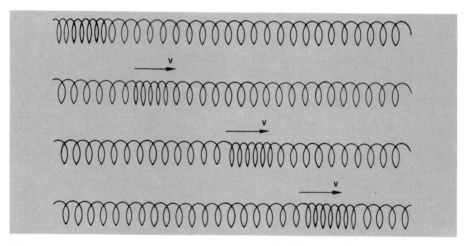

FIG. 20-2 In a longitudinal wave the motion of the individual particles is parallel to the direction of wave propagation. The illustration demonstrates the motion of a condensation pulse.

> In a **longitudinal wave** the vibration of the individual particles is parallel to the direction of the wave propagation.

If the coils of the spring in our example were forced apart at the left, a *rarefaction* would be formed, as shown in Fig. 20-3. Upon removal of the distorting force, a longitudinal rarefaction pulse would be sent along the spring. In general, a longitudinal wave consists of a series of condensations and rarefactions moving in a determined direction.

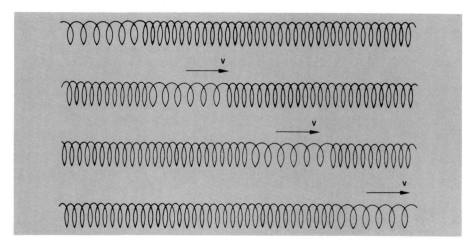

FIG. 20-3 Longitudinal motion of a rarefaction pulse in a coiled spring.

20-3 ■ *CALCULATING WAVE SPEED*

The speed with which a pulse moves through a medium depends upon the elasticity of the medium and the inertia of its particles. More elastic materials yield greater restoring forces when distorted. Less dense materials offer less resistance to motion. In either case, the ability of particles to pass on a disturbance to neighboring particles is improved, and a pulse will travel at a greater speed.

Let us consider the motion of a transverse pulse down a string of mass m and length l. (See Fig. 20-4.) The inertia of the individual particles is determined by the *mass per unit length* μ of the string, given by

$$\mu = \frac{m}{l}$$
Linear Density (20-1)

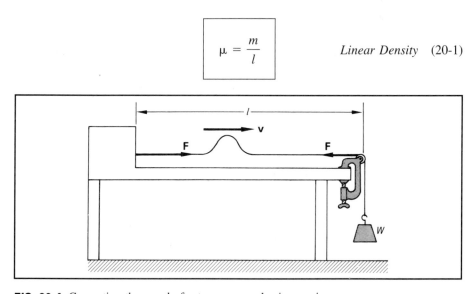

FIG. 20-4 Computing the speed of a transverse pulse in a spring.

The mass per unit length is usually called the *linear density* of the string. Its units are the kilogram per meter and the slug per foot (slug/ft).

Now let us suppose that the string is under a constant tension **F** due to the suspended weight. When the string is plucked near the left end, a transverse pulse will propagate along the string. The restoring force is measured by the tension **F** in the string, and the inertia of the rope particles is measured by the linear density μ. It can be shown that the speed of a transverse wave in a string is given by the square root of the ratio F/μ:

$$v = \sqrt{\frac{F}{\mu}} \qquad v = \sqrt{\frac{Fl}{m}} \qquad\qquad \textit{Wave Speed} \quad (20\text{-}2)$$

The second form of this equation results from substituting m/l for the linear density.

EXAMPLE 20-1 The length l of the string in Fig. 20-4 is 2 m, and the string has a mass of 3 g. Calculate the linear density and the wave speed if the string has a tension of 200 N.

Solution The linear density is found from Eq. (20-1):

$$\mu = \frac{m}{l} = \frac{0.003 \text{ kg}}{2 \text{ m}} = 0.0015 \text{ kg/m}$$

Then Eq. (20-2) gives the wave speed:

$$v = \sqrt{\frac{F}{\mu}} = \sqrt{\frac{200 \text{ N}}{0.0015 \text{ kg/m}}} = 365 \text{ m/s}$$

The same result could be obtained by substituting m, F, and l without first calculating the linear density. The important thing to remember in this equation is that the units must be consistent. Only *kilograms, newtons,* and *meters* or *slugs, pounds,* and *feet* may be used for m, F, and l, respectively.

20-4 ■ PERIODIC WAVE MOTION

So far we have been considering single, nonrepeated disturbances called *pulses*. What happens if similar disturbances are repeated periodically?

Suppose we attach the left end of a string to the endpoint of an electromagnetic vibrator, as shown in Fig. 20-5. The end of the metal vibrator is driven with harmonic motion by an oscillating magnetic field. Since the string is attached to the end of the vibrator, a series of periodic transverse pulses is sent down the string. The resulting waves consist of many crests and troughs, which move down the string at a constant speed. The distance between any two adjacent crests or troughs in such a wave train is called the *wavelength*, denoted λ.

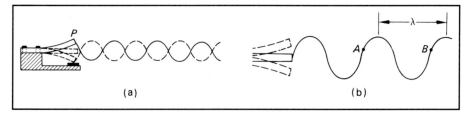

FIG. 20-5 (*a*) Production and propagation of a periodic transverse wave. (*b*) The wavelength λ is the distance between any two particles which are in phase, such as those at adjacent crests or at *A* and *B*.

As a wave travels along the string, each particle of the string vibrates about its equilibrium position with the same frequency and amplitude as the vibrating source. However, the particles of the string are not in corresponding positions at the same times. Two particles are said to be *in phase* if they are moving in the same direction and have the same displacement. Particles *A* and *B* in Fig. 20-5*b* are in phase. Since particles at the crests of a given wave train are also in phase, we can give a more general definition for the wavelength:

> The **wavelength** λ *of a periodic wave train is the distance between any two adjacent particles which are in phase.*

Each time the endpoint *P* of the vibrator makes a complete oscillation, the wave will move through a distance of one wavelength. The time required to cover this distance is therefore equal to the period *T* of the vibrating source. Hence the wave speed *v* can be related to the wavelength λ and period *T* by the equation

$$v = \frac{\lambda}{T} \tag{20-3}$$

The *frequency f* of a wave is the number of waves that pass a particular point in a unit of time. It is the same as the frequency of the vibrating source and is therefore equal to the reciprocal of the period ($f = 1/T$). The units of frequency may be expressed as waves per second, or cycles per second. The SI unit for frequency is the hertz, which is defined as a cycle per second:

$$1 \text{ Hz} = 1 \text{ cycle/s} = \frac{1}{\text{s}}$$

Thus, if 40 waves pass a point every second, the frequency is 40 Hz.

The speed of a wave is more often expressed in terms of its frequency than in terms of its period. Thus, Eq. (20-3) can be written

$$\boxed{v = f\lambda} \qquad\qquad \textit{Wave Speed} \quad (20\text{-}4)$$

This equation is an important physical relationship among the speed, frequency, and wavelength of *any* periodic wave. An illustration of each quantity is given in Fig. 20-6 for a periodic transverse wave.

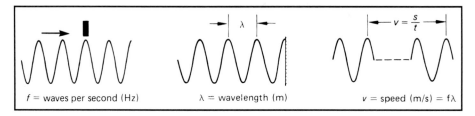

f = waves per second (Hz) λ = wavelength (m) v = speed (m/s) = $f\lambda$

FIG. 20-6 Relationship among the frequency, wavelength, and speed of a periodic transverse wave.

EXAMPLE 20-2   A man sits near the end of a fishing dock and counts the water waves as they strike a supporting post. In 1 min he counts 80 waves. If a particular crest travels 20 m in 8 s, what is the wavelength of the waves?

Solution   The frequency and velocity of the waves are calculated as follows:

$$f = \frac{80 \text{ waves}}{60 \text{ s}} = 1.33 \text{ Hz}$$

$$v = \frac{20 \text{ m}}{8 \text{ s}} = 2.5 \text{ m/s}$$

Now, since $v = f\lambda$, we solve for λ:

$$\lambda = \frac{v}{f} = \frac{2.5 \text{ m/s}}{1.33 \text{ Hz}} = 1.88 \text{ m}$$

A longitudinal wave can be generated by the apparatus shown in Fig. 20-7. The left end of a coiled spring is connected to a metal ball supported at the end of a clamped hacksaw blade. When the metal ball is displaced to the left and released, it vibrates with a consistent frequency and period. The resulting condensations and rarefactions are passed along the spring, producing a periodic longitudinal wave. Each particle of the coiled spring oscillates horizontally with the same frequency and amplitude as the metal ball. The distance between any two adjacent particles which are in phase is the wavelength. As indicated in Fig. 20-7, the distance between adjacent condensations or adjacent rarefactions is the wavelength. The relation $v = f\lambda$ also applies for a periodic longitudinal wave.

20-5 ■ SOUND

When a periodic disturbance takes place in air, longitudinal sound waves travel out from it. For example, if a tuning fork is struck with a hammer, the vibrating prongs send out longitudinal waves, as shown in Fig. 20-8. An ear, acting as a receiver for these periodic waves, interprets them as sound.

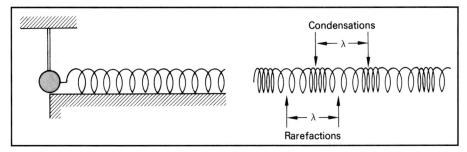

FIG. 20-7 Production of a periodic longitudinal wave.

FIG. 20-8 A tuning fork acts as a source of longitudinal sound waves in air.

Is the ear necessary for sound to exist? If the tuning fork were struck in the atmosphere of a distant planet, would there be sound even though no ear could interpret the disturbance? The answer depends on the definition of sound.

The term *sound* is used in two different ways. Physiologists define sound in terms of the auditory sensations produced by longitudinal disturbances in air. For them, sound does not exist on the distant planet. In physics, however, we refer to the disturbances themselves rather than to the sensations produced.

> **Sound** *is a longitudinal mechanical wave which travels through an elastic medium.*

In this case, sound does exist on the planet. It is this physical definition of sound that will apply in this chapter.

Two things must exist if a sound wave is to be produced. There must be a source of mechanical vibration, and there must be an elastic medium through which the disturbance can travel. The source may be a tuning fork, a vibrating string, or a vibrating air column in an organ pipe. *Sounds are transmitted by vibrating matter.* The need for an elastic medium can be demonstrated by placing an electric bell inside an evacuable flask, as shown in Fig. 20-9. The bell is connected to a battery so that it rings continuously, and then the flask is slowly evacuated. As more and more of the air is pumped from the flask, the sound of the bell becomes fainter and fainter until finally it cannot be heard at all. When air is allowed to reenter the flask, the sound returns. Thus, air is necessary to transmit sound.

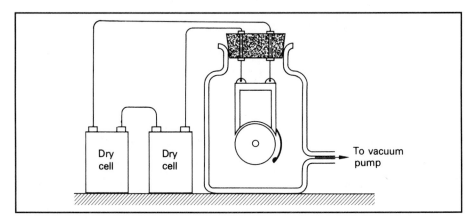

FIG. 20-9 A bell ringing in a vacuum cannot be heard. A material medium, such as air, is required for the production of a sound wave.

Let us now examine more closely the longitudinal sound waves in air as they proceed from a vibrating source. A thin strip of metal clamped tight at its base is pulled to one side and released. As the free end oscillates to and fro, a series of periodic, longitudinal sound waves spread through the air away from the source. The air molecules in the vicinity of the metal strip are alternately compressed and expanded, sending out a wave like that shown in Fig. 20-10a. The dense regions where many molecules are packed tightly together are called *compressions*. They are exactly analogous to the condensations in a coiled spring. The regions with relatively few molecules are referred to as *rarefactions*. Compressions and rarefactions alternate throughout the medium as the individual air particles oscillate to and fro in the direction of wave propagation. Since a compression corresponds to a high-pressure region and a rarefaction corresponds to a low-pressure region, a sound wave can also be represented by plotting the change in pressure P as a function of the distance x. (See Fig. 20-10b.) The distance between two successive compressions or rarefactions is the wavelength.

20-6 ■ SPEED OF SOUND

Anyone who has seen a weapon being fired at a distance has observed the smoke from the weapon before hearing the report. Similarly, we observe the flash of lightning before we hear the thunder. Even though both light and sound travel at finite speeds, the speed of light is so much greater in comparison that the transmission of light can be considered instantaneous. We can measure the speed of sound directly by observing the time required for the waves to move through a known distance. In air at 0°C, sound travels at a speed of 331 m/s, or 1087 ft/s.

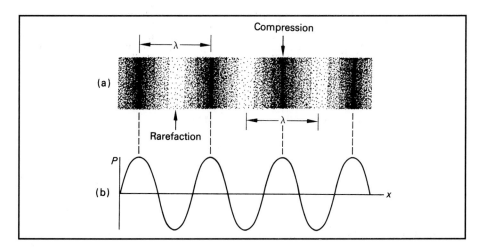

FIG. 20-10 (*a*) Compressions and rarefactions in a sound wave in air at a particular instant. (*b*) Sinusoidal variation in pressure of a function of displacement.

The speed of sound is significantly greater at 27 than at 0°C. For each Celsius degree rise in temperature (above 0°C), the speed of sound in air increases by approximately 0.6 m/s, or 2 ft/s. Hence the speed v of sound in air may be approximated by

$$v = 331 \text{ m/s} + \left(0.6 \, \frac{\text{m/s}}{\text{C}^\circ} \right)(t) \tag{20-5}$$

or

$$v = 1087 \text{ ft/s} + \left(2 \, \frac{\text{ft/s}}{\text{C}^\circ} \right)(t) \tag{20-6}$$

where t is the Celsius temperature of the air.

EXAMPLE 20-3   What is the approximate speed of sound in air at room temperature (20°C)?

Solution   From Eq. (20-5),

$$v = 331 \text{ m/s} + \left(0.6 \, \frac{\text{m/s}}{\text{C}^\circ} \right)(20°\text{C}) = 343 \text{ m/s}$$

Substitution into Eq. (20-6) would give 1127 ft/s.

Since the speed of sound increases with the elasticity of a medium, generally sound travels faster in most substances than it does in air. For example, the speed of sound in water at room temperature is approximately 1461 m/s, about 4 times its speed in air. In a steel rail, sound may travel at as much as 15 times its speed in air. The speed of sound in solids and liquids is not greatly changed by differences in temperature.

20-7 ■ FORCED VIBRATION AND RESONANCE

When a vibrating body is placed in contact with another body, the second body is forced to vibrate with the same frequency as the original vibrator. For example, if a tuning fork is struck with a hammer and then placed with its base against a wooden tabletop, the intensity of the sound will suddenly increase. When the tuning fork is removed from the table, the intensity decreases to its original level. The vibrations of the particles in the tabletop in contact with the tuning fork are called *forced vibrations*.

Elastic bodies have certain natural frequencies of vibration which are characteristic of the material and its conditions of support. The amplitude of a swing, for example, can be increased by pushing at definite intervals determined by the length of the rope. A taut wire of a particular length can produce sounds of characteristic frequencies. An open or a closed pipe also has natural frequencies of vibration. Whenever a body is acted on by a series of periodic impulses with a frequency nearly equal to one of the natural frequencies of the body, the body is set into vibration with a relatively large amplitude. This phenomenon is referred to as *resonance,* or *sympathetic vibration*.

An example of resonance is offered by a child sitting in a swing. Experience tells us that the swing can be set into motion with large amplitude by a series of small pushes at just the right intervals. Such resonance occurs only when the pushes are in phase with the natural frequency of vibration of the swing. A slight variation of the input pushes would result in little vibration.

The reinforcement of sound by resonance has many useful applications as well as many unpleasant consequences. The resonance of an air column in an organ pipe amplifies the weak sound of a vibrating air jet. Many musical instruments are designed with resonant cavities to produce varying sounds. Electrical resonance in radio receivers enables the listener to hear weak signals clearly. When the receiver is tuned to the frequency of a desired station, the signal is detected by electrical resonance. In poorly designed auditoriums or long hallways, music and voices may have a hollow sound which is unpleasant to the ear. Bridges have been known to collapse because of sympathetic vibrations set up by gusts of wind.

20-8 ■ AUDIBLE SOUND WAVES

We have defined sound as a *longitudinal mechanical wave traveling through an elastic medium*. This is a very broad definition that makes no restriction whatsoever on the frequencies of sound. The physiologist is concerned primarily with sound waves which are capable of affecting the sense of hearing. Thus it is useful to divide the spectrum of sound according to the following definitions:

> **Audible sound** *refers to sound waves in the frequency range from 20 to 20,000 Hz.*

Sound waves with frequencies below the audible range are termed **infra-sonic.**

Sound waves with frequencies above the audible range are called **ultrasonic.**

In the study of audible sound, the physiologist uses the terms *loudness, pitch,* and *quality* to describe the sensations produced. Unfortunately, these terms represent sensory magnitudes and are therefore subjective. What is loud to one person is moderate to another. What one person perceives as quality, another considers inferior. As always, the physicist must deal with explicit, measurable definitions. The physicist therefore attempts to correlate the sensory effects with the physical properties of waves. These correlations can be summarized as follows:

Sensory effect		Physical property
Loudness	$\longleftrightarrow$	Intensity
Pitch	$\longleftrightarrow$	Frequency
Quality	$\longleftrightarrow$	Waveform

The meanings of the terms on the left may vary considerably among individuals. The terms on the right are measurable and objective.

Sound waves constitute a flow of energy through matter. The intensity of a given sound wave is a measure of the rate at which energy is propagated through a given volume of space. A convenient method of specifying sound intensity is in terms of the rate at which energy is transferred across a unit of area perpendicular to the direction of wave motion. (See Fig. 20-11.) Since the rate at which energy flows is the *power* of a wave, the intensity can be related to the power per unit area passing a given point:

Sound **intensity** *is the power transferred by a sound wave through a unit of area perpendicular to the wave:*

$$I = \frac{P}{A} \qquad (20\text{-}7)$$

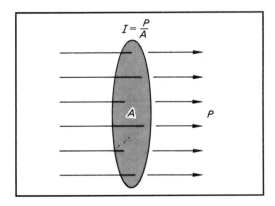

FIG. 20-11 The intensity of sound is a measure of the power transmitted per unit area perpendicular to the direction of wave motion.

The intensity I_0 of the faintest audible sound is of the order of magnitude of 1×10^{-12} W/m². This intensity is referred to as the *hearing threshold*. It has been adopted by acoustic experts as the zero of intensity.

The range of intensities to which the human ear is sensitive is enormous. It extends from the hearing threshold to an intensity which is 10^{12} times as great. The upper extreme represents the point at which the intensity is intolerable to the human ear. The sensation becomes one of feeling or pain instead of simply hearing. This upper limit, which is 1 W/m², is called the *pain threshold*.

An increase in sound intensity for a given frequency causes the sound to seem louder to an observer. However, the relationship between intensity and loudness is not a direct proportion. A sound that is 10 times as intense seems twice as loud, a sound 100 times as intense seems 3 times as loud, and a sound must be 1000 times as intense in order to seem 4 times as loud. (These are approximations, and they vary for individuals.)

Because of the way that loudness varies with intensity, and because of the wide range of intensities to which the ear is sensitive, a more convenient scale has been developed according to the following rule:

> When the intensity I_1 of one sound is 10 times as great as the intensity I_2 of another, the ratio of intensity is said to be 1 bel (B).

Thus, when we compare the intensities of two sounds, we refer to a difference in intensity levels.

In practice, the bel is too large for most comparisons. To obtain a more useful unit, we define a *decibel* (dB) as one-tenth of a bel. By using the standard zero of intensity I_0 as a standard for comparing all intensities, we can develop a general scale that can be used for rating any sound. The intensity level in decibels of any sound of intensity I can be found from the general relation

$$\beta = 10 \log \frac{I}{I_0} \qquad \text{decibels (dB)} \qquad (20\text{-}8)$$

where I_0 is the hearing threshold (10^{-12} W/m²).

Through the logarithmic decibel notation, the wide range of intensities has been reduced to *intensity levels* from 0 to 120 dB. However, we must remember that the scale is not linear. A 40-dB sound is much more than twice as intense as a 20-dB sound. A sound which is 100 times as intense as another is only 20 dB larger. Several examples of the intensity levels of common sounds are given in Table 20-1.

20-9 ■ PITCH AND QUALITY

The effect of intensity on the human ear is perceived as *loudness*. In general, sound waves which are more intense are also louder, but the ear is not equally sensitive to all frequencies. Therefore, a high-frequency sound may not seem as loud as one of lower frequency which has the same intensity.

TABLE 20-1 INTENSITY LEVELS FOR COMMON SOUNDS

Sound	Intensity level, dB
Hearing threshold	0
Rustling leaves	10
Whisper	20
Quiet radio	40
Normal conversation	65
Busy street corner	80
Subway car	100
Pain threshold	120
Jet engine	140–160

The frequency of a sound determines what the ear judges as the *pitch* of the sound. Musicians designate pitch by letters corresponding to key notes on the piano. For example, the C note, the D note, and the F note each refer to a specific pitch or frequency. A siren disk, shown in Fig. 20-12, can be used to demonstrate how the frequency of a sound determines the pitch. A stream of air is directed against a row of evenly spaced holes. By varying the rate of rotation of the disk, the pitch of the resulting sound can be increased or decreased.

Two sounds of the same pitch can easily be distinguished. For example, suppose we sound a C note (256 Hz) successively on a piano, a flute, a trumpet, and a violin. Even though each sound has the same pitch, there is a marked difference in the tones. This distinction is said to result from a difference in the *quality* of sound.

Regardless of the source of vibration in a musical instrument, several modes of oscillation are usually excited simultaneously. Therefore, the sound produced consists not only of the fundamental but also of many overtones. *The quality of a sound is determined by the number and relative intensities of the overtones present.* The difference in quality between two sounds can be observed objectively by analyzing the complex waveforms resulting from each sound. In general, all complex waves are made up of a number of waves of different frequencies and amplitudes.

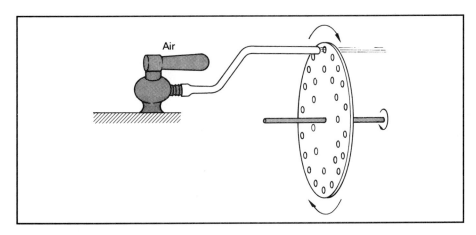

FIG. 20-12 Demonstrating the relationship between pitch and frequency.

20-10 ■ INTERFERENCE AND BEATS

We will now look at the combined sound of two different frequencies that are sounded together. The two longitudinal sound waves will interfere with each other as they leave their sources, producing a combined wave. To see how the waves combine, we will apply the following rule:

> *When two or more waves exist simultaneously in the same medium, each wave travels through the medium as though the other were not present.*

The resultant wave is a superposition of the component waves. In other words, the resultant displacement of a particular particle is the algebraic sum of the displacements each wave would produce independently of the other. This is the superposition principle:

> **Superposition principle:** *When two or more waves exist simultaneously in the same medium, the resultant displacement at any point and time is the algebraic sum of the displacements of each wave.*

The application of this principle is shown graphically in Fig. 20-13. A longitudinal sound wave is shown as a variation in pressure to give it a transverse appearance. Two waves, indicated by the solid and dotted lines, superpose to form the resultant wave, indicated by the heavy line. In Fig. 20-13*a* the superposition results in a wave of larger amplitude. These waves are said to *interfere constructively. Destructive interference* occurs when the resulting amplitude is smaller, as in Fig. 20-13*b*.

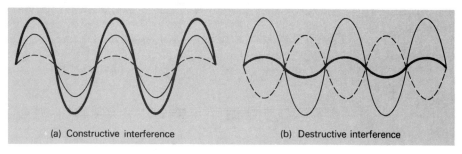

(a) Constructive interference   (b) Destructive interference

FIG. 20-13 Superposition principles. The heavy solid line gives the sum of the other two waves.

A common example of interference in sound waves occurs when two tuning forks (or other single-frequency sound sources) whose frequencies differ only slightly are struck simultaneously. The sound produced fluctuates in intensity, alternating between loud tones and virtual silence. These regular pulsations are referred to as *beats*. The *vibrato* effect obtained on some organs is an application of this principle. Every vibrato note is produced by two pipes tuned to slightly different frequencies.

To understand the origin of beats, let us examine the interference set up between sound waves proceeding from two tuning forks of slightly different frequencies, as shown in Fig. 20-14. The superposition of waves *A* and *B* illustrates the origin of beats.

The loud tones occur when the waves interfere constructively, and the quiet tones when the waves interfere destructively. Observation and calculation show that the two waves interfere constructively $f - f'$ times per second. Thus we can write

$$\text{Number of beats per second} = |f - f'| \qquad (20\text{-}9)$$

For example, if tuning forks of frequencies 256 and 259 Hz are struck at the same time, the resulting sound will pulsate 3 times every second.

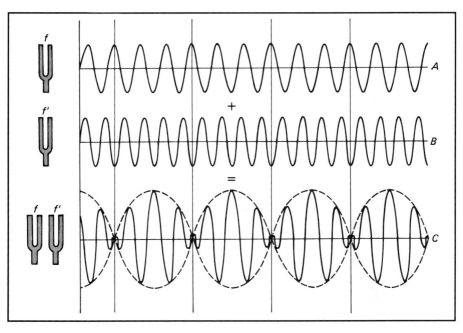

FIG. 20-14 Diagram illustrating the origin of beats. Wave C is a superposition of waves A and B.

Summary

A mechanical wave is a disturbance in an elastic medium. Such waves may be longitudinal, as in the case of sound waves, or they may be transverse, as in a vibrating string. Periodic mechanical waves have many applications in industry, and this chapter presented many of the basic ideas necessary for understanding these applications. The basic concepts which must be remembered are as follows:

■ The velocity v of a transverse wave in a string depends on the linear density $\mu = m/l$ and the string tension F. The following equations apply:

$$v = \sqrt{\frac{F}{\mu}} \qquad v = \sqrt{\frac{Fl}{m}} \qquad \qquad \textit{Wave Speed}$$

Units	Force F	Mass m	Length l	Speed v
SI	N	kg	m	m/s
USCS	lb	slug	ft	ft/s

■ For any wave of period T or frequency f, the speed v can be expressed in terms of the wavelength λ as follows:

$$v = \frac{\lambda}{T} \qquad v = f\lambda$$

Frequency in Hertz

■ Sound is a longitudinal wave traveling through an elastic medium. Its speed in air at 0°C is 331 m/s, or 1087 ft/s. At other temperatures the speed of sound is approximated by

$$v = 331 \text{ m/s} + \left(0.6 \, \frac{\text{m/s}}{\text{C}^\circ} \right)(t)$$

$$= 1087 \text{ ft/s} + \left(2 \, \frac{\text{ft/s}}{\text{C}^\circ} \right)(t)$$

■ *Audible* sound waves are defined as those in the frequency range from 20 to 20,000 Hz. *Infrasonic* refers to sound waves below these frequencies, and *ultrasonic* classifies sound waves which are above the audible range.
■ The *loudness* of a sound is determined by its intensity. A logarithmic scale of *intensity levels* from 0 to 120 dB defines the range of sound from the *hearing threshold* to the *pain threshold*.
■ The *superposition principle* is used to find the resultant wave when two or more waves exist simultaneously in the same medium. The resultant displacement at any point is the algebraic sum of the displacements of each wave.
■ *Beats* occur when two sources of slightly different frequencies are sounded together. They occur as a result of alternating *constructive* and *destructive* interference. The number of beats per second is equal to the difference in the two frequencies $|f - f'|$.

Questions

20-1. Define the following terms:

a. mechanical wave
b. transverse wave
c. longitudinal wave
d. wavelength
e. wave speed
f. frequency
g. linear density
h. sound
i. resonance
j. pitch
k. loudness
l. quality

m.	beats	**q.**	infrasonic
n.	intensity level	**r.**	ultrasonic
o.	decibel	**s.**	hearing threshold
p.	audible sound	**t.**	pain threshold

20-2. Explain how a water wave is both transverse and longitudinal in nature.

20-3. Describe an experiment to demonstrate that energy is associated with wave motion.

20-4. In a *torsional wave* the individual particles of the medium vibrate with angular harmonic motion about the axis of propagation. Give a mechanical example of such a wave.

20-5. Discuss the interference of waves. Is there a loss of energy when two waves interfere? Explain.

20-6. A transverse wave is sent down a string of mass m and length l under a tension F. How will the speed of the pulse be affected if (*a*) the mass of the string is quadrupled, (*b*) the length of the string is quadrupled, and (*c*) the tension is reduced to one-fourth its original value?

20-7. Draw graphs of a periodic transverse wave and a periodic longitudinal wave. Indicate on the figures the wavelength and amplitude of each wave.

20-8. Show graphically the superposition of two waves traveling in the same direction. One wave has three-fourths the amplitude and one-half the wavelength of the other wave.

20-9. What is the physiological definition of *sound?* What is the meaning of *sound* in physics? Give examples.

20-10. Why must astronauts on the surface of the moon communicate by radio? Can they hear another spacecraft as it lands nearby? Can they hear by touching helmets?

20-11. Vocal sounds originate with the vibration of the vocal cords. The mouth and nasal openings act as a resonant cavity to amplify and distinguish sounds. Suppose you hum at a constant pitch equal to the C note on the piano. If you open and close your mouth, what physiological property of the sound are you affecting?

20-12. The distance to a thunderstorm can be estimated by counting the number of seconds elapsing between the flash of lightning and the arrival of a clap of thunder and dividing the result by 5. Explain why this is a reasonable approximation for the distance in miles.

20-13. A store window is broken by an explosion several miles away. A glass of thin crystal shatters when a high note is reached on a violin. Are the causes of damage similar? What physical property of sound was principally responsible in each case?

20-14. Compare the speeds of sound in solids, liquids, and gases. Explain the reasons for the differences in observed speeds.

20-15. If the average ear cannot hear sounds of frequencies much in excess of 15,000 Hz, what is the advantage of building stereo music systems which have frequency responses much higher than 15,000 Hz?

Problems

20-1. A string 1.2 m long has a linear density of 0.0068 kg/m. What is the speed of a transverse wave in this string when it has a tension of 60 N?

Answer 93.9 m/s.

20-2. A wire 1 m in length has a mass of 20 g. One end is fixed, and the other passes over a light pulley and has an unknown mass suspended from it. What is the suspended mass if the speed of a transverse wave is 120 m/s?

20-3. A metal wire of mass 0.5 kg and length 50 cm is under a constant tension of 80 N. What is the speed of a transverse wave in the wire? If the length is reduced by one-half, what will the new mass of the wire be? Show that the speed of a transverse wave in the wire is unchanged.

Answer 8.94 m/s; 0.25 kg.

20-4. A 1.2-kg rope is stretched over 5.2 m and placed under a tension of 120 N. What is the speed of a transverse wave in the rope? What is its linear density?

20-5. A 20-ft length of cable has a weight of 8 lb and is under a tension of 600 lb. What is the mass of the cable? What is its linear density? What is the speed of a transverse wave in the cable?

Answer 0.25 slug; 0.0125 slugs/ft; 219 ft/s.

20-6. A 12-ft length of cable weighs 4 lb. What tension is required to produce a wave speed of 200 ft/s?

20-7. A 30-m cord under a tension of 200 N sustains a wave whose speed is 72 m/s. What is the mass of this cord?

Answer 1.16 kg.

20-8. A string 2 m long has a mass of 0.5 kg. What string tension is required for a wave speed of 1.2 m/s?

20-9. A longitudinal wave has a frequency of 200 Hz and a wavelength of 4.2 m. (*a*) What is the speed of the wave? (*b*) If the frequency is unchanged and the wavelength is reduced by one-half, what is the new wave speed?

Answer (a) 840 m/s; (b) 420 m/s.

20-10. A wooden float at the end of a fishing line makes 8 complete oscillations in 10 s. If it takes 3.6 s for a single wave to travel 11 m, what is the wavelength of the water waves?

20-11. Light consists of electromagnetic transverse waves. If the frequency of yellow light is 5×10^{14} Hz and the velocity of light is 3×10^8 m/s, what is the wavelength?

Answer 600 nm.

20-12. The wavelength of a wave traveling at 200 ft/s is measured to be 0.5 ft. What is its frequency?

20-13. A sound wave is sent from a ship to the ocean floor, where it is reflected and returned. If the round trip takes 0.6 s, how deep is the ocean floor? Consider that speed of sound in seawater is approximately 1430 m/s.

Answer 429 m.

20-14. On a day when the speed of sound is 343 m/s, you notice that 2 s passes between the firing of a gun and the hearing of the report. How far away is the gun?

20-15. What is the speed of sound on a day when the air temperature is 27°C? If the temperature drops to 10°C, what is the new speed of sound?

Answer 347 m/s; 337 m/s.

20-16. Steam is seen coming from a factory whistle on a day when the air temperature is 28°C. If the sound is heard 6 s later, how far away is the factory?

20-17. At what air temperature will the velocity of sound be 1100 ft/s? At what temperature will the velocity be 340 m/s?

Answer 6.5°C; 15°C.

20-18. The audible range of sound frequencies is from 20 to 20,000 Hz. On a day when the temperature is 20°C, what are the wavelengths of the lower and upper limits of the audible range?

20-19. The rangefinder of a camera works by measuring the time required to send ultrasonic waves to the subject being photographed and reflect them. (*a*) What will this time be for a subject 6 ft from the camera at room temperature? (*b*) By how much will this change when the temperature drops by 20 C°?

Answer (a) 1.06×10^{-2} s; (b) 3.92×10^{-4} s.

21

LIGHT

As a result of completing this chapter, you should be able to

1. Discuss the nature of light and work problems involving its *energy, frequency, velocity,* and *wavelength.*
2. Illustrate with drawings your understanding of the formation of shadows, labeling the *umbra* and *penumbra.*
3. Demonstrate your understanding of the concepts of *luminous flux, luminous intensity,* and *illumination,* and solve problems similar to those in the text.
4. Demonstrate by example and definition your understanding of the concepts of *reflection* and *refraction.*

Not all forms of wave motion require the existence of a medium, as sound waves do. In fact, many useful transfers of energy occur in the absence of any carrier material. *Heat* and *light* are transferred through practically empty space to the earth from the sun. *Radio* waves also travel through empty space. These are examples of self-sustaining waves called *electromagnetic waves.*

In this chapter, you will study light as an electromagnetic wave capable of affecting the sense of sight. Light which enters our eyes is the principal means by which we receive information. Light intensity and the effective illumination of surfaces are important concerns in industrial workplaces. Applications of the principles of reflection, refraction, diffraction, and polarization are commonly encountered. Some of the most frequent uses of these concepts are found in the many modern optical instruments, such as microscopes, telescopes, and spectrometers.

21-1 ■ *NATURE OF LIGHT*

An iron bar resting on a table is in thermal equilibrium with its surroundings. From its outward appearance one would never guess that it was very active internally. All objects continually radiate energy which is related to their temperature. The bar is in thermal equilibrium only because it is absorbing energy at the same rate that it is radiating energy. If we upset the balance by placing one end of the bar in a hot flame, the bar becomes more active internally and emits heat energy at a greater rate. As the heating continues to around 600°C, some of the radiation emitted becomes *visible*. That is, it affects our sense of sight. The color of the bar becomes a dull red, turning brighter as more heat is supplied. The radiation emitted before visibility occurs is called *heat*. The radiation which produces the visible color is called *light*. The two forms of energy are fundamentally the same, but they differ in energy content. The characteristic which distinguishes light from other forms of electromagnetic radiation is its energy.

> **Light** *is electromagnetic radiation which is capable of affecting the sense of sight.*

The exact nature of light has been the subject of scientific debate for many years. When light travels through a material, such as air or glass, it behaves very much as a wave. Reflection occurs at the surface of a medium, light bends as it passes from one medium to another, and light also bends around obstacles placed in its path. However, light also behaves as a particle in some instances. When it interacts with matter, such as when it is absorbed or emitted, it behaves very much as a particle with zero rest mass. This "particle" is called a *photon*. The energy of photons of visible light varies from about 2.8×10^{-19} to around 5×10^{-19} J.

The propagation of light is best explained in terms of a wave theory of light. However, since light travels in a vacuum without a medium to "vibrate," the wave must be fundamentally different from a mechanical wave, such as a sound wave. Light must be a self-sustaining wave. James Clerk Maxwell demonstrated how this is possible when he showed in 1865 that *an accelerated charge can radiate electromagnetic waves in space*. Maxwell explained that the energy in an electromagnetic wave is equally divided between electric and magnetic fields which are mutually perpendicular. Both fields oscillate perpendicular to the direction of wave propagation, as shown in Fig. 21-1. A collapsing magnetic field generates an electric field, and a collapsing electric field generates a magnetic field. Thus a light wave does not have to depend on the vibration of matter. It can be propagated by changing transverse fields. Such a wave can "break off" from the region around an accelerating charge and fly off into space at a constant velocity. Maxwell's equations predicted that heat and electric action, as well as light, were propagated at the same speed.

Maxwell's equations have been experimentally confirmed, and it is known today that all electromagnetic radiation travels at a constant speed of 3×10^8 m/s (186,000 mi/s):

$$c = 3 \times 10^8 \text{ m/s} \qquad \textit{Speed of Light} \quad (21\text{-}1)$$

The symbol c will be used to represent the speed of any electromagnetic wave in a vacuum. Such waves will travel at slower speeds in a material medium.

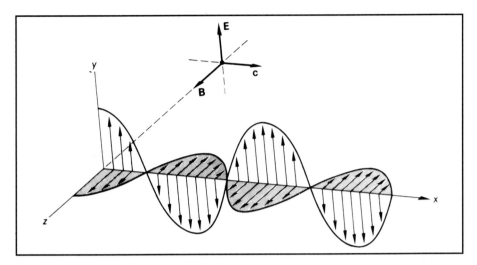

FIG. 21-1 Electromagnetic theory holds that light is propagated as an oscillating transverse field. The energy is equally divided between electric **E** and magnetic **B** fields, which are mutually perpendicular.

21-2 ■ ELECTROMAGNETIC SPECTRUM

Today, the electromagnetic spectrum is known to spread over a tremendous range of frequencies. A chart of the electromagnetic spectrum is given in Fig. 21-2. The wavelength λ of electromagnetic radiation is related to its frequency f by the general equation

$$c = f\lambda \qquad (21-2)$$

where c is the velocity (3×10^8 m/s). In terms of wavelengths, the tiny portion of the electromagnetic spectrum referred to as the *visible region* lies between 0.00004 and 0.00007 cm.

Because of the small wavelengths of light radiation, it is more convenient to define smaller units of measure. The SI unit is the *nanometer* (nm):

$$1 \text{ nm} = 1 \times 10^{-9} \text{ m} = 1 \times 10^{-7} \text{ cm}$$

The visible region, in this unit, extends from 400 nm for violet light to 700 nm for red light. Other units still in use today are the *millimicron* (mμ), which is the same size as the *nanometer,* and the *angstrom* (Å), which is 0.1 nm.

EXAMPLE 21-1 The wavelength of yellow light from a sodium flame is 589 nm. Compute its frequency.

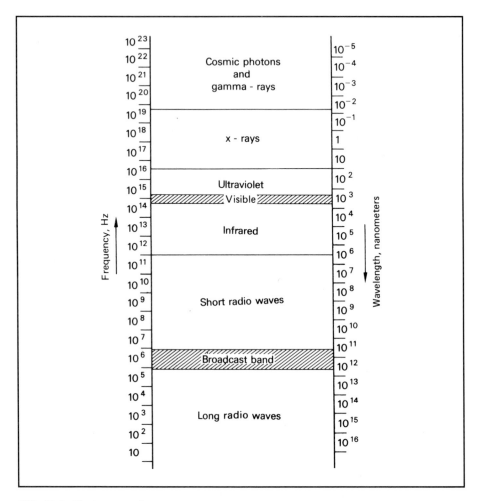

FIG. 21-2 Electromagnetic spectrum.

Solution   The frequency is found from $c = f\lambda$:

$$f = \frac{c}{\lambda} = \frac{3 \times 10^8 \text{ m/s}}{589 \times 10^{-9} \text{ m}} = 5.09 \times 10^{14} \text{ Hz}$$

Newton was the first to make detailed studies of the visible region by dispersing "white light" through a prism. In order of increasing wavelength, the spectral colors are violet (450 nm), blue (480 nm), green (520 nm), yellow (580 nm), orange (600 nm), and red (640 nm). Anyone who has seen a rainbow has seen the effects that different wavelengths of light have on the human eye.

The electromagnetic spectrum is continuous; there are no gaps between one form of radiation and another. The boundaries given in Fig. 21-2 are somewhat arbitrary. In theory, there is no upper or lower limit to the spectrum. For convenience we have divided the spectrum into eight major regions: (1) long radio waves, (2) short radio waves, (3) the infrared region, (4) the visible region, (5) the ultraviolet region, (6) x-rays, (7) gamma rays, and (8) cosmic photons.

21-3 ■ QUANTUM THEORY OF LIGHT

The wave theory of light is enormously successful in describing the propagation of light. However, when light interacts with matter, the wave description is lacking. As early as 1887, H. R. Hertz noticed that an electric spark would jump between two spheres more readily when their surfaces were illuminated by the light from another spark. The phenomenon, known as the *photoelectric effect,* is demonstrated by the apparatus shown in Fig. 21-3. A beam of light strikes a metal surface A in an evacuated tube. Electrons ejected by the light are drawn to the collector B by external batteries. The flow of electrons is indicated by the ammeter. The photoelectric effect defied explanation in terms of a wave theory. In fact, the ejection of electrons could be accounted for more easily in terms of a particle theory. Still, there could be no doubt of the wave properties either. Science faced a remarkable paradox.

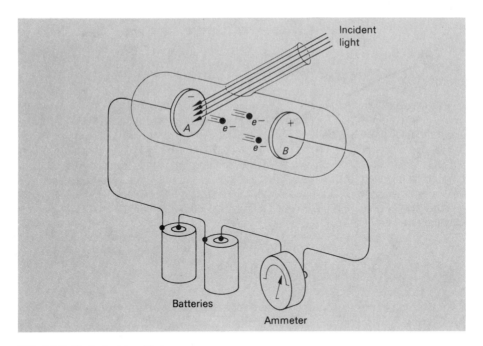

FIG. 21-3 Photoelectric effect.

In an attempt to bring observation into agreement with theory, Max Planck, a German physicist, published his *quantum hypothesis.* It postulated that electromagnetic energy is absorbed or emitted in discrete packets, or *quanta.* The energy content of these quanta, or *photons,* as they were called, is proportional to the frequency of the radiation. Planck's equation can be written

$$E = hf \qquad h = 6.626 \times 10^{-34} \text{ J/Hz} \qquad\qquad (21\text{-}3)$$

where E is the energy of the photon in joules, f is the frequency of the photon in hertz, and h is Planck's constant of proportionality. Einstein used Planck's equation to explain the photoelectric effect in terms of light photons.

Thus it appears that light is *dualistic* in nature. The wave theory is retained, for the photon is considered to have a frequency and an energy proportional to the frequency. The modern practice is to use the wave theory in studying the propagation of light. The particle theory, however, is necessary to describe the interaction of light with matter. We may think of light as radiant energy transported in photons carried along by a wave field.

EXAMPLE 21-2 Calculate the energy of a photon of blue light whose wavelength is 480 nm.

Solution Since $c = f\lambda$, we know that $f = c/\lambda$. Substitution for f in Eq. (20-3) gives an alternative form for Planck's equation:

$$E = \frac{hc}{\lambda} = \frac{(6.626 \times 10^{-34} \text{ J/Hz})(3 \times 10^8 \text{ m/s})}{480 \times 10^{-9} \text{ m}}$$

$$= 4.14 \times 10^{-19} \text{ J}$$

The origin of light photons was not understood until Niels Bohr, in 1913, devised a model of the atom based on quantum ideas. Bohr postulated that electrons can move about the nucleus of an atom only in certain orbits or *discrete energy levels,* as shown in Fig. 21-4. Atoms were said to be *quantized.* If the atoms are somehow energized, as by heat, the orbital electrons may be caused to jump into a higher orbit. At some later time, these excited electrons will fall back to their original level, releasing as photons the energy which they had originally absorbed. Although Bohr's model was not strictly correct, it provided a basis for understanding the emission and absorption of electromagnetic radiation in quantum units.

21-4 ■ LIGHT RAYS AND SHADOWS

One of the most important properties of light is the fact that it travels in a straight line. Instinctively, we rely quite heavily on this property for estimating distances, directions, and shapes. As an aid to the study of light, we will use the concept of a ray to denote the general direction in which light travels. Consider the point source of light in Fig. 21-5. Light travels radially outward in waves, represented by concentric circles. Note that these spherical waves become essentially plane wave fronts in any specific direction at a long distance away from the source. An imaginary straight line drawn perpendicular to the wave fronts in the direction of the moving wave is called a *ray*. There are, of course, an infinite number of rays starting from the point source of light.

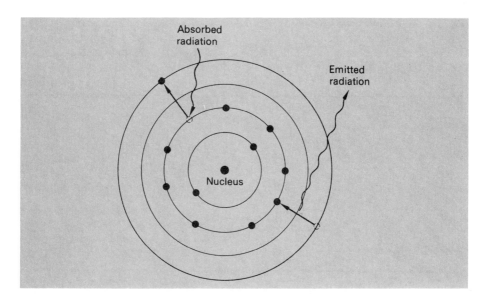

FIG. 21-4 Bohr's theory of the atom.

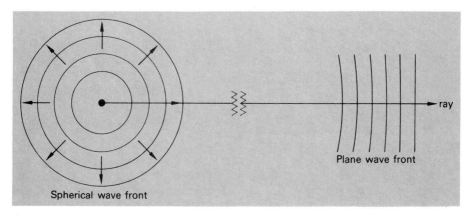

FIG. 21-5 A ray is an imaginary line drawn perpendicular to advancing wavefronts which indicates the direction of light propagation.

Any dark-colored object absorbs light, but a black one absorbs nearly all the light it receives. Light that is not absorbed upon striking a surface is either reflected or transmitted. If all the light incident upon an object is reflected or absorbed, the object is said to be *opaque*. Since light cannot pass through an opaque body, a shadow will be produced in the space behind the object. The shadow formed by a point source of light is shown in Fig. 21-6. Since light is propagated in straight lines, rays drawn from the source past the edges of the opaque object form a sharp shadow proportional to the shape of the object. Such a region in which no light has entered is called an *umbra*.

If the source of light is an extended one rather than a point, the shadow will consist of two portions, as shown in Fig. 21-7. The inner portion receives no light from the source and is therefore the umbra. The outer portion is called the *penumbra*. An observer within the penumbra would see a portion of the source but not all of it. An observer located outside both regions would see all the source. Solar and lunar eclipses can be studied by similar construction of shadows.

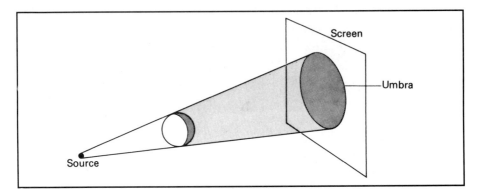

FIG. 21-6 Shadow formed by a point source of light.

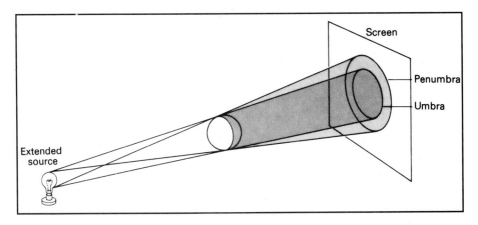

FIG. 21-7 Shadow formed by an extended light source.

21-5 ■ LUMINOUS FLUX AND LUMINOUS INTENSITY

Most sources of light emit electromagnetic energy distributed over many wavelengths. Electric power is supplied to a lamp, and radiation is emitted. The radiant energy emitted per unit of time by the lamp is called the *radiant power*, or the *radiant flux*. Only a small portion of this radiant power is in the visible region, i.e., between 400 and 700 nm. The sensation of sight depends on only the visible, or *luminous*, energy radiated per unit of time.

>The **luminous flux** F *is that part of the total radiant power emitted from a light source which affects the sense of sight.*

The unit of luminous flux is the *lumen*, which is equivalent to about ⅟₆₈₀ W of radiant power in the middle of the visible spectrum (555 nm). In a common incandescent bulb, only about 10 percent of the energy radiated is luminous flux. The bulk of radiant power is nonluminous. Other light sources are more efficient in producing visible radiant power. The luminous flux emitted from a 40-W light bulb is around 500 lm, whereas the flux from a 40-W fluorescent tube is about 2300 lm.

Light travels outward in straight lines from a source which is small in comparison with its surroundings. For such a source of light, the luminous flux included in a solid angle Ω remains the same at all distances from the source, as shown by Fig. 21-8. It is therefore convenient to speak of the *flux per unit solid angle* rather than to talk of the total flux. The quantity which expresses this relationship is called the *luminous intensity*.

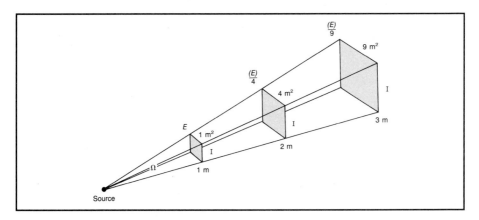

FIG. 21-8 The luminous intensity I is constant for a given solid angle Ω. However, the illumination E (flux per unit area) decreases with the square of the distance from the source.

The **luminous intensity** I *of a source of light is the luminous flux* F *emitted per unit solid angle* Ω:

$$I = \frac{F}{\Omega} \qquad\qquad (21\text{-}4)$$

The unit for intensity is the *lumen per steradian* (lm/sr), called the *candela* (cd). The candela (formerly called the *candle*) is the SI unit for measuring light intensity.

When luminous flux strikes a surface, that surface is said to be *illuminated*. The radiant flux falling on a unit of area is called the *illumination* of that surface:

The **illumination** E *of a surface* A *is defined as the luminous flux* F *per unit area:*

$$E = \frac{F}{A} \qquad\qquad \text{illumination (lx) \quad (21\text{-}5)}$$

When the flux F is measured in lumens and the area A in square meters, the illumination E has the units of *lumens per square meter*, or *lux*.

Direct application of Eq. (21-5) requires a knowledge of the luminous flux falling on a surface. Unfortunately, the flux of common light sources is difficult to determine. For this reason most applications involving the illumination of a surface make use of the inverse-square relationship between the illumination and the distance to the source. Refer to Fig. 21-8. An object located twice as far away from a source will have one-fourth the illumination; an object 3 times as far away will have one-ninth the illumination. For surfaces located at distances R_1 and R_2, the ratio of surface illuminations is given by

$$\frac{E_1}{E_2} = \frac{R_2^2}{R_1^2} \tag{21-6}$$

EXAMPLE 21-3 The incandescent lamp of Fig. 21-9 produces an illumination of 60 lx when a tabletop is 3 m below the bulb. What is the illumination if the light is lowered to a position 1 m above the table?

Solution Solving for E_2 in Eq. (21-6), we have

$$E_2 = \frac{E_1 R_1^2}{R_2^2} = \frac{(60 \text{ lx})(3 \text{ m})^2}{(1 \text{ m})^2}$$

$$= \frac{(60 \text{ lx})(9 \text{ m}^2)}{1 \text{ m}^2} = 540 \text{ lx}$$

Notice that the illumination is 9 times as large due to decreasing the distance by one-third.

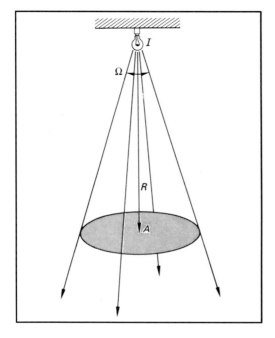

FIG. 21-9 Computing the illumination of a surface perpendicular to the incident flux.

21-6 ■ REFLECTION OF LIGHT

When a ray of light strikes a surface, as shown in Fig. 21-10a, part of the light will be reflected into the incident medium. The ray that strikes the surface is called the *incident ray,* and the ray that bounces back is called the *reflected ray.* The reflection of light follows the same general law of mechanics that governs other bouncing phenomena; i.e., the angle of incidence equals the angle of reflection. Consider the pool table in Fig. 21-10b. In order to hit the black ball on the right, we must aim at a point on the rail such that the incident angle θ_i is equal to the reflected angle θ_r. Similiarly, light reflected from a smooth surface will have an incident angle θ_i equal to the reflected angle θ_r. Two basic laws of reflection can be stated:

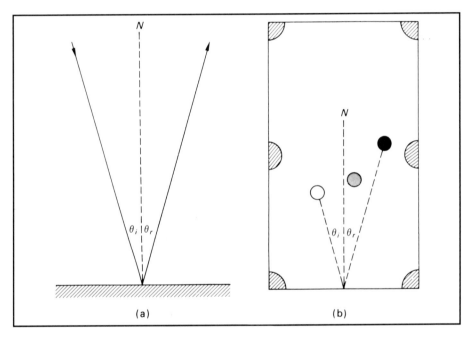

(a)   (b)

FIG. 21-10 (*a*) For light reflection, the angle of incidence is equal to the angle of reflection. (*b*) Comparison with a bouncing pool ball.

> *The angle of incidence is equal to the angle of reflection.*
> *The incident ray, the reflected ray, and the perpendicular to the surface all lie in the same plane.*

Light reflection from a smooth surface, in Fig. 21-11a, is called *regular* or *specular reflection.* Light striking the surface of a mirror or glass is specularly reflected. If all the incident light which strikes a surface were reflected in this manner, we could not see the surface. We would see only the images of other objects. It is *diffuse reflection* (Fig. 21-11b) that enables us to see a surface. A rough or irregular surface will spread out and scatter the incident light, resulting in illumination of the surface. Reflection of light from brick, concrete, or newsprint provides examples of diffuse reflection.

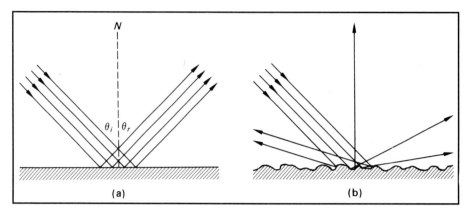

FIG. 21-11 (*a*) Specular reflection. (*b*) Diffuse reflection.

21-7 ■ *REFRACTION OF LIGHT*

Light travels in straight lines at a constant speed in a uniform medium. If the medium changes, the speed will also change, and the light will travel in a straight line along a new path. The bending of a light ray as it passes obliquely from one medium to another is known as *refraction*. To see how light travels in different media, consider a ray of light which passes from air to water to glass and back to air again. (See Fig. 21-12.) The angles shown are all measured with respect to an imaginary line perpendicular to the boundaries. Notice that the angle of refraction into the water θ_w is less than the angle of incidence from the air θ_a. This bending toward the perpendicular is due to the slower speed of light in water. When the light enters the glass from the water, it bends still farther toward the normal, indicating that light travels still more slowly in glass. As the light finally returns from the glass to the air, the speed increases to its original value. The emerging light bends away from the perpendicular and continues parallel with the entering ray. However, the light has been displaced laterally. We may summarize what we have learned as follows:

> When a ray of light enters a denser optical medium at an angle, it bends toward the normal; when it enters a less dense medium at an angle, it bends away from the normal. The incident ray, the refracted ray, and the perpendicular to the surface all lie in the same plane.

We can control the direction of light by taking advantage of the principle of refraction. In Fig. 21-13a and b, you can see how the refraction of light by a prism and by a lens may be used to control the direction of light. Such devices play important roles for many optical instruments, such as microscopes, telescopes, and spectrometers.

Summary

An understanding of the nature of light and how light behaves is important for many industrial applications. We have learned that light is electromagnetic radiation capable of affecting the sense of sight. The major concepts are summarized below:

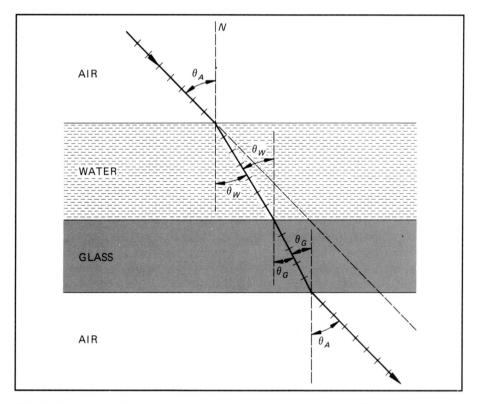

FIG. 21-12 A ray of light passes from air to water and back into air. The light is refracted toward the normal when it enters a denser optical medium and away from the normal when it enters a less dense optical medium.

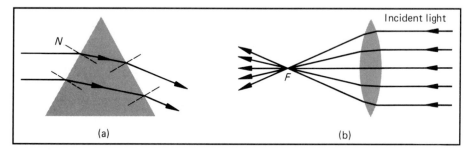

FIG. 21-13 (a) A prism can be used to change the direction of light rays. (b) A lens refracts parallel light rays to a point, called the *focal point* of the lens.

■ The wavelength λ of electromagnetic radiation is related to its frequency f by the general equation

$$c = f\lambda \qquad c = 3 \times 10^8 \text{ m/s}$$

- The range of wavelengths for visible light goes from 400 nm for violet to 700 nm for red:

$$1 \text{ nm} = 10^{-9} \text{ m}$$

The nanometer is used for wavelengths.

- The energy of light photons is proportional to the frequency:

$$E = hf \qquad E = \frac{h\lambda}{c} \qquad h = 6.626 \times 10^{-34} \text{ J/Hz}$$

The constant h is *Planck's constant.*
- The *luminous intensity I* of a light source is the *luminous flux F* per unit solid angle Ω. Luminous flux is the radiant power in the visible region. It is measured in *lumens*.

$$I = \frac{F}{\Omega}$$

Luminous Flux, 1 cd = 1 lm/sr

- For an *isotropic* source, one emitting light in all directions, the luminous flux F is

$$F = 4\pi I$$

Isotropic Source, $\Omega = 4\pi$ sr

- The *illumination E* of a surface A is defined as the luminous flux F per unit area:

$$E = \frac{F}{A}$$

Illumination, lx

- Whenever light strikes the boundary between two media, part of the light is *reflected*, part is *absorbed*, and some may be *refracted*. The meanings of these terms are evident from Fig. 21-14. For reflection, the angle of incidence θ_i is equal to the angle of reflection θ_r. For refraction, the angle of refraction θ_r is less than θ_i when the light enters a denser optical medium, and the angle θ_r is greater than θ_i when light enters a medium of smaller optical density.

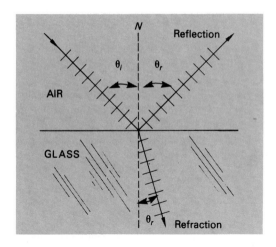

FIG. 21-14 When light strikes the boundary between two media, it may be reflected, refracted, or absorbed.

Questions

21-1. Define the following terms:
  a. light
  b. electromagnetic waves
  c. visible region
  d. nanometer
  e. quantum theory
  f. photons
  g. light ray
  h. umbra
  i. penumbra
  j. luminous flux
  k. luminous intensity
  l. steradian
  m. illumination
  n. isotropic source
  o. reflection
  p. refraction

21-2. What is meant by the dualistic nature of light? In what ways does light behave as particles? In what ways does light behave as a wave?

21-3. Explain how the energy of an electromagnetic wave depends on its frequency and how it depends on the wavelength.

21-4. When light enters glass from air, its energy in glass is the same as its energy in air. Is its frequency the same? What about its wavelength? Explain.

21-5. Microwave ovens, television, and radar utilize electromagnetic waves between infrared and radio waves. Compare the energy, frequency, and wavelengths of these waves with the energy, frequency, and wavelengths of visible radiation.

21-6. Review the definition of a radian, and discuss how the steradian for solid angles is similar to the radian for plane angles. How many radians are in a complete circle? How many steradians are in a complete sphere?

21-7. Draw a diagram illustrating a solar eclipse, labeling the umbra and penumbra regions. If you view a partial eclipse of the sun, are you standing in the umbra or the penumbra region?

21-8. Can you justify the following definition for a lumen? *A lumen is equal to the luminous flux falling on a surface of one square meter, all points of which are one meter from a uniform point source of one candela?*

21-9. An older unit for the illumination was the *footcandle,* that illumination E received by a surface 1 ft^2 located at a distance of 1 ft from a 1-cd source of light. Explain how this definition is the same as that given in this text.

21-10. Describe the distribution of luminous flux from an incandescent lamp. Why is the lamp not an isotropic source?

21-11. Discuss the factors which will affect the illumination of a table in a machine shop.

21-12. Illumination is sometimes referred to as *flux density*. Explain why such a term might be appropriate.

21-13. Photometry is the science of measuring light. The intensity of a light source can be determined with the photometer, shown in Fig. 21-15. The luminous intensity I_x of an unknown source is found by visually comparing the unknown with a standard source of known intensity I_s. If the distances from each source are adjusted so that the grease spot is equally illuminated by each source, the unknown intensity I_x can be calculated from the inverse-square law. Derive the photometer equation

$$\frac{I_x}{r_x^2} = \frac{I_s}{r_s^2}$$

Photometry Equation (21-7)

where r_s is the distance to the standard and r_x is the distance to the unknown source.

21-14. If two 40-W light bulbs are compared by using the photometer, will they necessarily be located at equal distances from the grease spot?

21-15. Match the two columns below by placing in the blank the letter which is best related to each numbered item.

   a. Luminous flux   _____ 1. The surface

   b. Luminous intensity   _____ 2. The source

   c. Illumination   _____ 3. The space between the source and the surface

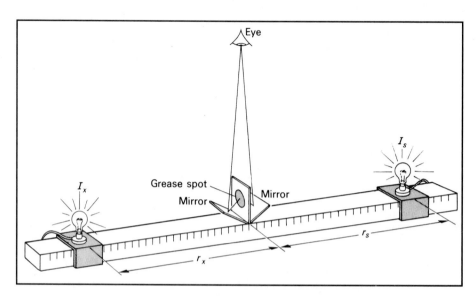

FIG. 21-15 Grease-spot photometer used to measure the intensity of an unknown source of light.

21-16. Discuss this statement: *One cannot "see" the surface of a perfect mirror.*

21-17. Explain how the day is lengthened by atmospheric refraction of the sun's rays.

21-18. A coin is placed on the bottom of a bucket so that it is just out of sight when it is viewed at an angle from the top. Show by the use of diagrams why the coin becomes visible if the bucket is filled with water.

Problems

21-1. A television signal is an electromagnetic wave. How many kilometers will such a signal travel in 0.2 s? How many miles?

Answer 60,000 km; 37,200 mi.

21-2. The sun is approximately 93,000,000 mi from the earth. How many minutes are required for light to reach us from the sun?

21-3. An astronaut communicates with an engineer on earth. If the astronaut is on the surface of the moon and it takes 1.3 s for the signal to reach the earth, how far away is the moon from the earth?

Answer 3.9×10^9 m.

21-4. An infrared spectrophotometer scans the wavelengths from 1 to 16 μm (1 μm = 1×10^{-6} m). Express this range in terms of the frequencies of the infrared rays.

21-5. A microwave radiator used in measuring automobile speeds emits radiation of frequency 1.2×10^9 Hz. What is the wavelength?

Answer 0.25 m.

21-6. What is the range of frequencies for visible light? What is the range of energies for visible light? What is the frequency of green light whose wavelength is 520 nm?

21-7. A ray of yellow light has a wavelength of 600 nm. What is the energy of this light? What is the frequency of light whose photon energy is 3.5×10^{-19} J?

Answer 3.31×10^{-19} J; 5.28×10^{14} Hz.

21-8. A woman is 1.4 m tall, and she stands in front of a large plane mirror. What is the shortest mirror length required to enable her to see her entire image?

21-9. A ray of light makes an angle of 20° with the normal to the surface. The mirror is then turned through 5°, causing a new angle of incidence equal to 25°. Through what angle is the reflected ray rotated?

Answer 10°.

21-10. When light of wavelength 550 nm passes from air into a thin glass plate and out again into the air, the frequency never changes. If the speed of light inside the glass is reduced to 2×10^8 m/s, what is the wavelength inside the glass?

21-11. A tabletop 1 m wide and 2 m long is located 4 m from a lamp. If 40 lm of flux fall on this surface, what is the illumination E of the surface? What should be the location of the lamp in order to produce twice the illumination?

Answer 20 lx; 2.83 m.

21-12. The illumination of a given surface is 80 lx when it is 3 m away from the light source. At what distance will the illumination be 20 lx?

21-13. For a given surface area, the intensity also decreases proportionally with the square of the distance. Refer to Question 21-13. The grease-spot photometer of Fig. 21-15 is used to compare a lamp of unknown intensity I_x with a standard 30-cd source of light. The two sources of light are 3 m apart, and the grease spot is moved toward the standard light. When the grease spot is 80 cm from the standard light source, the illumination is judged to be equal on each side. Compute the intensity of the unknown light source.

Answer 422 cd.

21-14. Two light sources with intensities of 200 and 50 cd are positioned 1 m apart and compared by using the grease-spot photometer (Fig. 21-15). At what distance from the 200-cd source will the illumination appear equal?

22

NUCLEAR PHYSICS

OBJECTIVES

As a result of completing this chapter, you should be able to

1. Describe the structure of an atom and its nucleus, stating what is currently known about the mass, charge, and size of the fundamental nuclear particles.
2. Define the mass number A and the atomic number Z and write and apply a relationship between them.
3. Demonstrate your understanding of the equivalence of mass and energy by interchanging units in kilograms, atomic mass units, joules, and electron volts.
4. Define and calculate the mass defect and the binding energy for a given atomic nucleus.
5. Write a brief description of alpha particles, beta particles, and gamma rays, listing their properties.
6. Demonstrate your understanding of radioactive decay and nuclear reactions by writing nuclear equations for these events.
7. Calculate the activity and the quantity of a radioactive material remaining after a period of time if the half-life and the initial quantity are known.
8. Draw a rough diagram of a nuclear reactor, describing the various components and their function in the production of nuclear power.

Nuclear technology has grown enormously since its beginning in the early 1940s. The study of the atomic nucleus once was a subject reserved mainly for physicists, but today there are few people whose lives are not touched by some aspect of nuclear science. As patients, we see the doctor use radioactive materials to diagnose or treat a condition. As citizens, we are concerned with the promises and dangers of large-scale nuclear power production. More than ever before technicians and engineers need a better understanding of the atomic nucleus and its potential. In this unit you will study the fundamental particles contained in the nucleus of atoms, you will learn about radioactivity, you will study nuclear reactions, and you will become familiar with the production of nuclear energy.

22-1 ■ ATOMIC NUCLEUS

All matter is composed of different combinations of at least three fundamental particles: protons, neutrons, and electrons. For example, a beryllium atom (Fig. 22-1) has a nucleus which contains four protons and five neutrons. Since beryllium is electrically neutral, it must have four electrons surrounding the nucleus. The two inner electrons are at a different energy level ($n = 1$) than the outer two electrons ($n = 2$). The symbol n is used to enumerate the various energy levels.

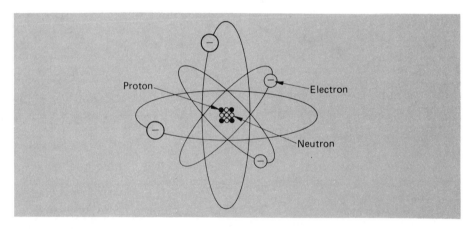

FIG. 22-1 Model of a beryllium atom. The nucleus consists of four protons and five neutrons surrounded by four electrons. The positive charge of the protons is exactly balanced by the negatively charged electrons in the neutral atom.

It has been determined that the nucleus contains most of the mass of an atom and that the nucleus is only about one ten-thousandth of the diameter of the atom. Thus, a typical atom with a diameter of 1×10^{-10} m (100 pm) would have a nucleus about 1×10^{-14} m (10 fm) in diameter. [The prefixes *pico* (10^{-12}) and *femto* (10^{-15}) are useful for atomic and nuclear dimensions.] Since the diameter of the atom is 10,000 times that of its nucleus, the atom, and therefore matter, consists largely of space that is almost empty.

Let us review what is known about the fundamental particles. The electron has a mass of 9.1×10^{-31} kg and a charge e of -1.6×10^{-19} C. The proton is the nucleus of a hydrogen atom. It has a mass of 1.673×10^{-27} kg and a positive charge equal in magnitude to the charge of an electron $(+e)$. Since the mass of an electron is extremely small, the mass of a proton is approximately the same as the mass of a hydrogen atom, which consists of one proton and one electron. The proton has a diameter of approximately 3 fm. An electron's size is not precisely known, but it is believed to be "point-like" and to have a diameter near zero.

The other nuclear particle, the neutron, is present in the nuclei of all elements except hydrogen. It has a mass of 1.675×10^{-27} kg, slightly greater than that of the proton, but it has no charge. Thus, while neutrons contribute to the mass of a nucleus, they do not affect the net positive charge of the nucleus, which is due only to protons. The neutron also has a diameter of approximately 3 fm. Table 22-1 summarizes the data we have been discussing for the three fundamental particles.

TABLE 22-1 FUNDAMENTAL PARTICLES

Particle	Symbol	Mass, kg	Charge, C	Diameter, fm
Electron	e	9.1×10^{-31}	-1.6×10^{-19}	≈ 0
Proton	p	1.673×10^{-27}	$+1.6 \times 10^{-19}$	3
Neutron	n	1.675×10^{-27}	0	3

From what we know about the fundamental particles, it is clear that diagrams like Fig. 22-1 cannot be taken too seriously. Distances are not normally presented to scale in such diagrams. Moreover, classical laws of physics often do not apply for the microworld of the nucleus.

A true understanding of atomic and nuclear events will require a new way of thinking. For example, one might ask what holds the nucleus together. Clearly if Coulomb's law of electrostatic repulsion applies in the nucleus, it must be overcome by a much larger force. Both this much larger force and the electrostatic force are immense when compared with the gravitational force. This third force is called the *nuclear force*.

The nuclear force is a very strong, short-range force. If two nucleons (which are protons *or* neutrons) are separated by approximately 1 fm, a strong attractive force occurs which quickly drops to zero as their separation becomes larger. The force appears to be the same, or nearly the same, between two protons, two neutrons, or a proton and a neutron. It is an attractive force until the nucleons get too close, when it becomes strongly repulsive, so that the nucleons cannot occupy the same space at the same time. If one nucleon is completely surrounded by other nucleons, its nuclear force field will be saturated, and it cannot exert any force on nucleons outside those surrounding it.

22-2 ■ ELEMENTS

For many centuries, scientists have been studying the various elements found on the earth. A number of attempts have been made to organize the different elements according to their chemical and/or physical properties. The modern grouping of elements is the *periodic table*. One form of the periodic table is printed in Table 22-2.

Each element is assigned a number that distinguishes it from any other element. For example, the number for hydrogen is 1, the number for helium is 4, and the number for oxygen is 8. These numbers equal the number of protons in the nucleus of the element. The number is given the symbol Z and is called the *atomic number*.

The **atomic number** Z *of an element is equal to the number of protons in the nucleus of an atom of that element.*

TABLE 22-2 THE PERIODIC TABLE *(Adapted from* General Chemistry *by J.B. Russell. Copyright 1980 by McGraw-Hill, Inc. Used with permission of McGraw-Hill Book Company.)*

Atomic weight values listed in parentheses are approximate.

The atomic number indirectly determines the chemical properties of an element because Z determines the number of electrons needed to balance the positive charge of the nucleus. The chemical nature of an atom depends on the number of electrons, in particular the outermost, or *valence,* electrons.

As the number of protons in a nucleus increases, so does the number of neutrons. In lighter elements, the increase is approximately one to one, but heavier elements may have more than 1½ times more neutrons than protons. For example, oxygen has 8 protons and 8 neutrons, whereas uranium has 92 protons and 146 neutrons. The total number of nucleons in a nucleus is called the *mass number A.*

The **mass number** A *of an element is equal to the total number of protons and neutrons in its nucleus.*

If we represent the number of neutrons by N, we can write the mass number in terms of the atomic number Z and the number of neutrons:

$$A = Z + N$$

Mass Number (22-1)

Thus, the mass number for uranium is 92 + 146, or 238. The mass number for oxygen is 16.

A general way of describing the nucleus of a particular atom is to write the symbol for the element with its mass number and atomic number shown as follows:

$$\text{mass number} \atop \text{atomic number} [\text{Symbol}] = {}_Z^A X \qquad (22\text{-}2)$$

For example, the uranium atom has the symbol ${}_{92}^{238}U$.

The structures and the symbols for the first four elements are shown in Fig. 22-2. An alphabetical listing of all the elements is given in Table 22-3.

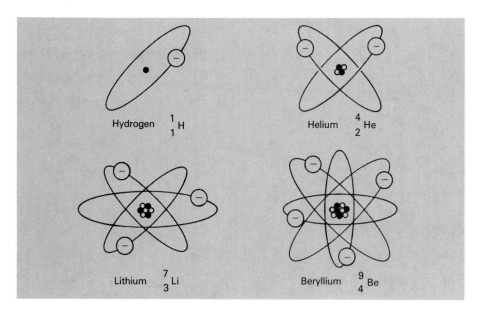

Hydrogen ${}_1^1 H$ Helium ${}_2^4 He$

Lithium ${}_3^7 Li$ Beryllium ${}_4^9 Be$

FIG. 22-2 Structure and symbols for the first four elements.

EXAMPLE 22-1 How many neutrons are in the nucleus of an atom of mercury ${}_{80}^{201}Hg$?

Solution The symbol shows that the atomic number is 80 and the mass number is 201. Thus, Eq. (22-1) gives

$$N = A - Z = 201 - 80 = 121$$

It is possible for two atoms of the same element to have nuclei that contain different numbers of neutrons. Such atoms are called *isotopes:*

Isotopes *are atoms which have the same atomic number Z but different mass numbers A.*

For example, naturally occurring carbon is a mixture of two isotopes. The most abundant form, ${}_6^{12}C$, has 6 protons and 6 neutrons in its nucleus. Another form, ${}_6^{13}C$, has an extra neutron. Some elements have as many as 10 different isotopic forms.

TABLE 22-3 INTERNATIONAL ATOMIC WEIGHTS (BASED ON CARBON 12)*

Element	Symbol	Atomic number	Atomic weight
Actinium	Ac	89	(227)
Aluminum	Al	13	26.9815
Americium	Am	95	(243)
Antimony	Sb	51	121.75
Argon	Ar	18	39.948
Arsenic	As	33	74.9216
Astatine	At	85	(210)
Barium	Ba	56	137.34
Berkelium	Bk	97	(247)
Beryllium	Be	4	9.0122
Bismuth	Bi	83	208.980
Boron	B	5	10.811
Bromine	Br	35	79.904
Cadmium	Cd	48	112.40
Calcium	Ca	20	40.08
Californium	Cf	98	(251)
Carbon	C	6	12.01115
Cerium	Ce	58	140.12
Cesium	Cs	55	132.905
Chlorine	Cl	17	35.453
Chromium	Cr	24	51.996
Cobalt	Co	27	58.9332
Columbium	(see Niobium)		
Copper	Cu	29	63.546
Curium	Cm	96	(247)
Dysprosium	Dy	66	162.50
Einsteinium	Es	99	(254)
Erbium	Er	68	167.26
Europium	Eu	63	151.96
Fermium	Fm	100	(257)
Fluorine	F	9	18.9984
Francium	Fr	87	(223)
Gadolinium	Gd	64	157.25
Gallium	Ga	31	69.72
Germanium	Ge	32	72.59
Gold	Au	79	196.967
Hafnium	Hf	72	178.49
Hahnium	Ha	105	(260)
Helium	He	2	4.0026
Holmium	Ho	67	164.930
Hydrogen	H	1	1.00797
Indium	In	49	114.82
Iodine	I	53	126.9044
Iridium	Ir	77	192.2
Iron	Fe	26	55.847
Krypton	Kr	36	83.80
Lanthanum	La	57	138.91
Lawrencium	Lw	103	(257)
Lead	Pb	82	207.19
Lithium	Li	3	6.939
Lutetium	Lu	71	174.97
Magnesium	Mg	12	24.312
Manganese	Mn	25	54.9380
Mendelevium	Md	101	(256)
Mercury	Hg	80	200.59
Molybdenum	Mo	42	95.94
Neodymium	Nd	60	144.24
Neon	Ne	10	20.183
Neptunium	Np	93	(237)
Nickel	Ni	28	58.71
Niobium (Columbium)	Nb	41	92.906
Nitrogen	N	7	14.0067
Nobelium	No	102	(254)
Osmium	Os	76	190.2
Oxygen	O	8	15.9994
Palladium	Pd	46	106.4
Phosphorus	P	15	30.9738
Platinum	Pt	78	195.09
Plutonium	Pu	94	(244)
Polonium	Po	84	(209)
Potassium	K	19	39.102
Praseodymium	Pr	59	140.907
Promethium	Pm	61	(145)
Protactinium	Pa	91	(231)
Radium	Ra	88	(226)
Radon	Rn	86	(222)
Rhenium	Re	75	186.22
Rhodium	Rh	45	102.91
Rubidium	Rb	37	85.47
Ruthenium	Ru	44	101.07
Rutherfordium	Rf	104	(260)
Samarium	Sm	62	150.35
Scandium	Sc	21	44.956
Selenium	Se	34	78.96
Silicon	Si	14	28.086
Silver	Ag	47	107.868
Sodium	Na	11	22.9898
Strontium	Sr	38	87.62
Sulfur	S	16	32.064
Tantalum	Ta	73	180.948
Technetium	Tc	43	(97)
Tellurium	Te	52	127.60
Terbium	Tb	65	158.924
Thallium	Tl	81	204.37
Thorium	Th	90	232.038
Thulium	Tm	69	168.934
Tin	Sn	50	118.69
Titanium	Ti	22	47.90
Tungsten (Wolfram)	W	74	183.85
Uranium	U	92	238.03
Vanadium	V	23	50.942
Wolfram (Tungsten)	W	74	183.85
Xenon	Xe	54	131.30
Ytterbium	Yb	70	173.04
Yttrium	Y	39	88.905
Zinc	Zn	30	65.37
Zirconium	Zr	40	91.22

*Values in parentheses are mass numbers of longest-lived or best-known isotopes.

22-3 ■ *ATOMIC MASS AND ENERGY*

The very small masses of nuclear particles call for an extremely small unit of mass. Scientists normally express the atomic and nuclear masses in *atomic mass units:*

> One **atomic mass unit** (*u*) *is exactly equal to one-twelfth of the mass of the most abundant form of the carbon atom.*

In terms of the kilogram, the atomic mass unit is

$$1 \text{ u} = 1.6606 \times 10^{-27} \text{ kg} \qquad (22\text{-}3)$$

The mass m_p of the proton and the mass m_n of the neutron can now be given in atomic mass units:

$$m_p = 1.007276 \qquad m_n = 1.008665 \text{ u} \qquad (22\text{-}4)$$

EXAMPLE 22-2   The periodic table shows the average atomic mass of barium to be 137.34 u. What is the average mass of the barium nucleus?

Solution   The mass of the nucleus is the atomic mass less the mass of the surrounding cloud of electrons. Thus, we must first determine the mass of an electron in atomic mass units:

$$m_e = \left(9.1 \times 10^{-31} \text{ kg} \right) \left(\frac{1 \text{ u}}{1.6606 \times 10^{-27} \text{ kg}} \right)$$
$$= 5.5 \times 10^{-4} \text{ u}$$

Since the atomic number Z of barium is 56, there must be the same number of electrons. The total mass of the electrons is

$$m_T = 56(5.5 \times 10^{-4} \text{ u}) = 3.08 \times 10^{-2} \text{ u}$$

The average atomic mass was given as 137.34 u. Thus, the nuclear mass is

$$137.34 \text{ u} - 0.0308 \text{ u} = 137.31 \text{ u}$$

The masses given for atoms in the periodic table include the electron masses.

Nuclear reactions are frequently accompanied by the liberation of large amounts of energy. In 1905, Albert Einstein proposed his now famous equation showing the equivalence of mass and energy. The conversion factor is the square of the velocity of light ($c = 3 \times 10^8$ m/s):

$$E = mc^2 \qquad\qquad\qquad (22\text{-}5)$$

Since the speed of light is so large, its square c^2 is much larger. It is easy to see that large amounts of energy can be released by the conversion of small amounts of matter to energy. For example, a 1-kg mass is equivalent to 9×10^{16} J of energy. This represents around 25 billion kilowatthours, enough to operate a typical home for a million years.

EXAMPLE 22-3   Convert a mass of 1 u to energy expressed in joules.

Solution   Recalling that 1 u = 1.66×10^{-27} kg, we obtain

$$E = mc^2 = (1.66 \times 10^{-27} \text{ kg})(3 \times 10^8 \text{ m/s})^2 = 1.49 \times 10^{-10} \text{ J}$$

Clearly, we need a smaller unit for measuring energy equivalents for such small masses. A new unit called an *electron volt* (eV) can be defined as the energy required to move an electron between two points having a potential difference of 1 V. The relationship between an electron volt and a joule is

$$1 \text{ eV} = 1.6 \times 10^{-19} \text{ J}$$

One million electron volts would be written as 1 MeV by using the metric prefix M. Thus, 1 MeV = 1.6×10^{-13} J.

In Example 22-3, we determined that a mass of 1 u is equivalent to an energy of 1.49×10^{-10} J. To see how much energy 1 u represents in millions of electron volts, we convert as follows:

$$(1.49 \times 10^{-10} \text{ J})\left(\frac{1 \text{ MeV}}{1.6 \times 10^{-13} \text{ J}}\right) = 931 \text{ MeV}$$

We are now in a position to write a conversion factor which will change atomic mass units to millions of electron volts. From Einstein's equation, we have seen that

$$(1 \text{ u})(c^2) = 931 \text{ MeV}$$

Solving for c^2, we can write

$$c^2 = 931 \text{ MeV/u} \qquad\qquad\qquad (22\text{-}6)$$

As an exercise you should verify that the electron has a rest mass energy of 0.511 MeV and the proton has a rest mass energy of 938 MeV.

22-4 ■ MASS DEFECT AND BINDING ENERGY

One startling result of the study of the nucleus is that the total mass of an atom is not exactly equal to the sum of its parts. Consider, for example, the helium atom, $^4_2$He, which has two electrons about a nucleus containing 2 protons and 2 neutrons. The atomic mass of the $^4_2$He atom is 4.0026 u.

Now let us compare this value with the total mass of all the individual particles which make up the atom:

$$
\begin{aligned}
2\ p &= 2(1.0007276\ \text{u}) = 2.014552\ \text{u} \\
2\ n &= 2(1.008665\ \text{u}) = 2.017330\ \text{u} \\
2\ e &= 2(0.00055\ \text{u}) = \underline{0.001100\ \text{u}} \\
\text{Total mass} &= 4.032982\ \text{u}
\end{aligned}
$$

The mass of the parts (4.0330 u) is apparently greater than the mass of the atom (4.0026 u):

$$
\begin{aligned}
m_{\text{parts}} - m_{\text{atom}} &= 4.0330\ \text{u} - 4.0026\ \text{u} \\
&= 0.0304\ \text{u}
\end{aligned}
$$

When protons and neutrons join to form a helium nucleus, the mass is decreased in the process. This difference is called the *mass defect M_D*. A mass defect can be shown to exist for all elements.

> The **mass defect** M_D *is defined as the difference between the rest mass of a nucleus and the sum of the rest masses of its constituent parts.*

We have seen from $E = mc^2$ that mass and energy are equivalent. We might suppose, then, that the decrease in mass when the nucleons join together will result in an energy decrease. Since energy is conserved, a decrease in the energy of the system means that energy must be released in joining the system together. In the case of helium, this energy would form a mass defect of 0.0304 u and would be equivalent to

$$
E = mc^2 = (0.0304\ \text{u})(931\ \text{MeV/u}) = 28.3\ \text{MeV}
$$

The total energy which would be released if we could build a nucleus from protons and neutrons is called the *binding energy E_B* of the nucleus. As we have just seen, the binding energy of $^4_2$He is 28.3 MeV, as shown by Fig. 22-3a.

We can also reverse the above process and state that the binding energy is the energy required to break a nucleus apart into its constituent particles:

> The **binding energy** E_B *of a nucleus is defined as the energy required to separate a nucleus into its constituent nucleons.*

In our example, an energy of 28.3 MeV must be supplied to $^4_2$He in order to separate the nucleus into 2 protons and 2 neutrons (Fig. 22-3b).

An isotope of atomic number Z and mass number A consists of Z protons, Z electrons, and $N = A - Z$ neutrons. If we neglected the binding energy of the electrons, a neutral isotope would have the same mass as Z neutral hydrogen atoms plus the mass of the neutrons. The masses of $^1_1$H and m_n are, respectively,

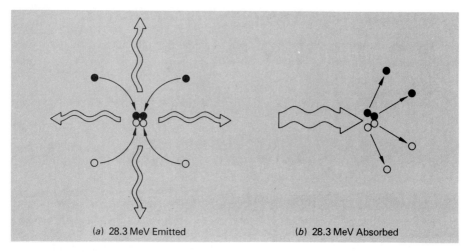

(a) 28.3 MeV Emitted   (b) 28.3 MeV Absorbed

FIG. 22-3 (a) When two protons and two neutrons are fused to form a helium nucleus, energy is released. (b) The same amount of energy is required to break the nucleus apart into its constituent nucleons.

$$m_H = 1.007825 \text{ u} \qquad m_n = 1.008665 \text{ u} \qquad (22\text{-}7)$$

If we represent the atomic mass by M, the binding energy E_B can be approximated by

$$E_B = [(Zm_H + Nm_n) - M]c^2 \qquad \textit{Binding Energy} \quad (22\text{-}8)$$

In applying this equation, we should remember that $N = A - Z$ and that $c^2 = 931$ MeV/u.

EXAMPLE 22-4 Determine the total binding energy and the binding energy per nucleon for the $^{14}_{7}N$ nucleus.

Solution For nitrogen, $Z = 7$, $N = 7$, and $M = 14.003074$ u. Thus

$$
\begin{aligned}
E_B &= (Zm_H + Nm_n - M)c^2 \\
&= [7(1.007825 \text{ u}) + 7(1.008665 \text{ u}) - 14.003074 \text{ u}](931 \text{ MeV/u}) \\
&= (0.112356 \text{ u})(931 \text{ MeV/u}) = 104.6 \text{ MeV}
\end{aligned}
$$

Since $^{14}_{7}N$ contains 14 nucleons, the binding energy per nucleon is

$$\frac{E_B}{A} = \frac{104.6 \text{ MeV}}{14 \text{ nucleons}} = 7.47 \text{ MeV per nucleon}$$

The atomic mass M used in Eq. (22-8) must be the mass for the particular isotope of the element, not the mass taken from Table 22-2 or Table 22-3. These tables give the atomic masses of the naturally occurring mixture of isotopes for each element. The atomic mass of $^{12}_{7}C$, for example, is exactly 12.00 u by definition. The periodic table gives a value of 12.01115 u because naturally occurring carbon contains very small amounts of $^{13}_{7}C$ in addition to $^{12}_{7}C$ atoms.

The binding energy per nucleon, as computed in Example 22-4, is an important way of comparing the nuclei of various elements. A plot of the binding energy per nucleon as a function of increasing mass number is shown in Fig. 22-4 for many stable nuclei. Note that the mass numbers toward the center (50 to 80) yield the highest binding energy per nucleon. Elements ranging from $A = 50$ to $A = 80$ are the most stable.

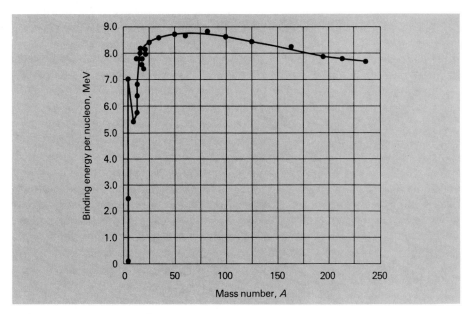

FIG. 22-4 Average binding energy per nucleon for the most stable nucleus at each mass number.

22-5 ■ RADIOACTIVITY

The strong nuclear force holds the nucleons tight in the nucleus, overcoming the coulomb repulsion of protons. However, the balance of forces is not always maintained, and sometimes particles or photons are emitted from the nuclei of atoms. Such unstable nuclei are said to be *radioactive* and have the property of *radioactivity*.

All naturally occurring elements with atomic numbers greater than 83 are radioactive. They are slowly decaying and disappearing from the earth. Uranium and radium are two of the better-known examples of naturally radioactive elements. A few other naturally occurring, lighter, and less active elements have also been discovered.

Unstable nuclei are also produced artificially as by-products of nuclear reactors, for study in laboratories, or for other purposes. Additionally, some elements are made radioactive naturally by bombardment with high-energy photons.

There are three major forms of radioactive emission from atomic nuclei:

1. *Alpha particles* (α)

An alpha particle is the nucleus of a helium atom and consists of two protons and two neutrons. It has a charge of $+2e$ and a mass of 4.001506 u. Because of their positive charges and relatively low speeds ($\approx 0.1c$), alpha particles do not have great penetrating power.

2. *Beta particles* (β)

There are two kinds of beta particles, a beta plus minus particle (β^-) and a beta plus particle (β^+). The beta minus particle is simply an electron with charge $-e$ and mass equal to 0.00055 u. A beta plus particle, also called a *positron,* has the same mass as an electron but the opposite charge ($+e$). These particles are generally emitted at speeds near the velocity of light. The beta minus particles are much more penetrating than alpha particles, but beta particles combine easily with electrons; then rapid annihilation of both the positrons and the electrons occurs, with the emission of gamma rays.

3. *Gamma rays* (γ)

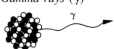

A gamma ray is a high-energy electromagnetic wave similar to heat and light but of much higher frequency. These rays have no charge or rest mass and are the most penetrating radiation emitted by radioactive elements.

To understand why these forms of radiation are emitted, it is helpful to look at nuclei which are relatively stable. Plotting the number of neutrons N vs. the number of protons Z for these stable nuclei gives a rough graph like that shown in Fig. 22-5. Note that the light elements are stable when the ratio of Z to N is close to 1. More neutrons are required for stability in the heavier elements. The additional nuclear forces of the extra neutrons are needed to balance the higher electric forces which result as more protons are collected. Whenever a nucleus occurs which deviates very much from the line, it is unstable and will emit some form of radiation, thereby achieving stability.

22-6 ■ RADIOACTIVE DECAY

Let us look at radioactive decay by alpha, beta, and gamma radiation and see what occurs during each process. The emission of an alpha particle $^4_2\alpha$ reduces the number of protons in the parent nucleus by 2 and the number of nucleons by 4. Symbolically, we write

$$^A_Z X \rightarrow \, ^{A-4}_{Z-2}Y + \, ^4_2\alpha + \text{energy} \tag{22-9}$$

The energy term results from the fact that the rest energy of the products is less than that of the parent atom. The difference in energy is carried away primarily by the kinetic energy imparted to the alpha particle. The recoil kinetic energy of the much more massive daughter atom is small by comparison.

EXAMPLE 22-5 Write the reaction which occurs when $^{226}_{88}$Ra decays by alpha emission.

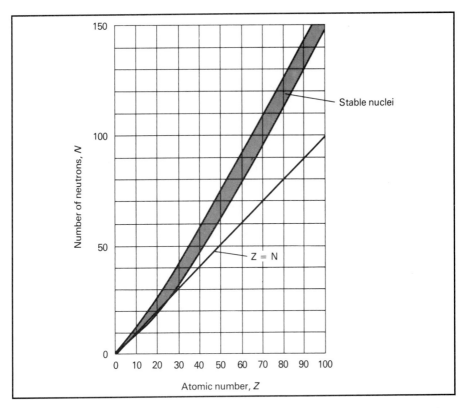

FIG. 22-5 Comparison of the number of neutrons as a function of the atomic number. Notice that nuclei with higher Z have a greater proportion of neutrons.

Solution Applying Eq. (22-9), we write

$$^{226}_{88}\text{Ra} \rightarrow \, ^{222}_{86}\text{Rn} + \, ^{4}_{2}\alpha + \text{energy}$$

Notice that the unstable element radium has been transformed to a new element, radon, which is closer to the stability line.

Next consider the emission of beta minus particles from the nucleus. If beta minus particles are electrons, how can an electron come from a nucleus containing only protons and neutrons? This can be answered, at least in part, by analogy to the Bohr atom. We have seen that photons, which do not exist in the atom, are emitted by atoms when they change from one state to another. Similarly, electrons, which do not exist in nuclei, can be emitted as a form of radiation when the nucleus changes from one state to another. When such a change does occur, the total charge must be conserved. This requires the conversion of a neutron to a proton and an electron:

$$^{1}_{0}n \rightarrow \, ^{1}_{1}p + \, ^{0}_{-1}e$$

Thus, in beta minus emission a neutron is replaced by a proton. The atomic number Z increases by 1, and the mass number is unchanged. Symbolically,

$$\frac{A}{Z}X \rightarrow \ _{Z+1}^{A}Y + \ _{-1}^{0}\beta + \text{energy} \qquad (22\text{-}10)$$

An example of beta emission is the decay of an isotope of neon into sodium:

$$_{10}^{23}\text{Ne} \rightarrow \ _{11}^{23}\text{Na} + \ _{-1}^{0}\beta + \text{energy}$$

The increase in Z is necessary to conserve charge.

Similarly, in positron (beta plus) emission, a proton in the nucleus decays to a neutron and a positron:

$$_{1}^{1}p \rightarrow \ _{0}^{1}n + \ _{+1}^{0}e$$

The atomic number Z decreases by 1, and the mass number A is unchanged. Symbolically,

$$\frac{A}{Z}X \rightarrow \ _{Z-1}^{A}Y + \ _{+1}^{0}\beta + \text{energy} \qquad (22\text{-}11)$$

An example of positron emission is the decay of an isotope of nitrogen into an isotope of carbon:

$$_{7}^{13}\text{N} \rightarrow \ _{6}^{13}\text{C} + \ _{+1}^{0}\beta + \text{energy}$$

In both types of beta emission, the kinetic energy is shared mostly by the beta particle and another particle called a *neutrino*. The neutrino has no rest mass and no electric charge, but it can have both energy and momentum.

In gamma emission, the parent nucleus maintains the same atomic number Z and the same mass number A. The gamma photon simply carries away energy from an unstable nucleus. Frequently, a succession of alpha and beta decays is accompanied by gamma decays, which carry off excess energy.

The radioactive disintegration of $_{92}^{238}\text{U}$ is shown in Fig. 22-6 as a series of decays through a number of elements until it becomes a stable $_{82}^{206}\text{Pb}$ nucleus.

22-7 ■ HALF-LIFE

A radioactive material continues to emit radiation until all the unstable atoms have decayed. The number of unstable nuclei decaying or disintegrating every second for a given isotope can be predicted on the basis of probability. This number is referred to as the *activity R*, given by

$$R = \frac{-\Delta N}{\Delta t}$$

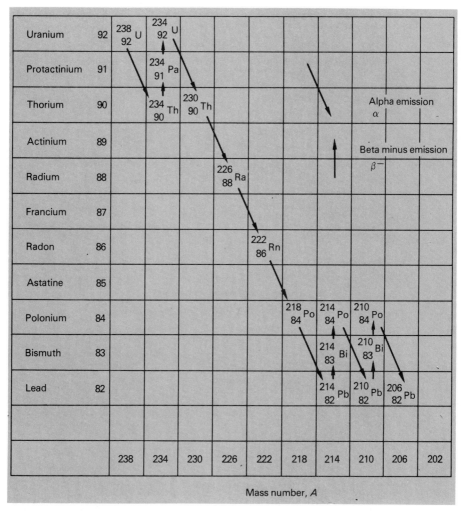

FIG. 22-6 Uranium series of disintegration. Uranium decay, through a series of alpha and beta minus emissions, from $^{238}_{92}U$ to $^{206}_{82}Pb$.

where N is the number of undecayed nuclei. The negative sign is included because N is decreasing with time. The units for R are inverse seconds (s^{-1}).

In practice, the activity in disintegrations per second is so large that a more convenient unit, the *curie* (Ci), is defined as follows:

> One **curie** (Ci) is the activity of a radioactive material which decays at the rate of 3.7×10^{10} disintegrations per second:

$$1 \text{ Ci} = 3.7 \times 10^{10} \text{ s}^{-1} \tag{22-12}$$

The activity of 1 g of radium is slightly less than 1 Ci.

The random nature of nuclear decay means that the activity R at any time is directly proportional to the number of nuclei remaining; i.e., as the number of nuclei remaining decreases with time, the activity must also decrease with time. Therefore, if we plot the number of remaining nuclei as a function of time, as illustrated in Fig. 22-7, we see that radioactive decay is not linear. The time it takes for this curve to drop to one-half its original value is different for each radioactive isotope; this is called the *half-life:*

*The **half-life** $T_{1/2}$ of a radioactive isotope is the length of time in which one-half of its unstable nuclei will decay.*

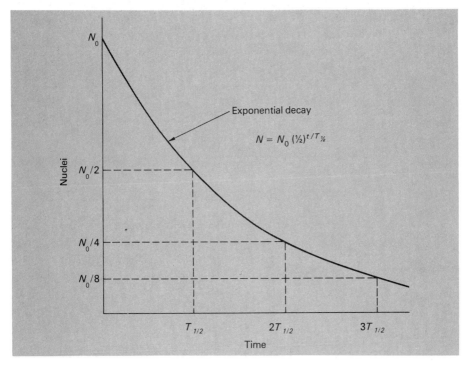

FIG. 22-7 Radioactive decay curve illustrating the half-life at time $T_{1/2}$ required for one-half of the unstable nuclei present at time $t = 0$ to decay.

For example, the half-life of radium 226 is 1620 yr; 1 g of this isotope will decay to 0.5 g in 1620 yr, to 0.25 g in 2(1620 yr), to 0.125 g in 3(1620 yr), and so on.

We can use this definition of the half-life to determine how many nuclei are present at a time t. If we start out at time $t = 0$ with a number N_0 of unstable nuclei, then after n half-lives have passed, there will be left a number of nuclei N given by

$$N = N_0 \left(\frac{1}{2}\right)^n \tag{22-13}$$

The number n of half-lives in time t is, of course, $t/T_{1/2}$. Thus, a more applicable form of the above relation is

$$N = N_0\left(\frac{1}{2}\right)^{t/T_{1/2}}$$

Since the amount of radioactive material is determined by the number of nuclei present, an equation similar to Eq. (22-13) can be used to compute the mass of radioactive material remaining after a number of half-lives.

The same idea applied to the activity R of a radioactive sample yields the relation

$$R = R_0\left(\frac{1}{2}\right)^{t/T_{1/2}} \tag{22-14}$$

EXAMPLE 22-6   The worst by-product of ordinary nuclear reactors is the radioactive isotope plutonium 239, which has a half-life of 24,400 yr. Suppose the initial activity of a sample containing 1.64×10^{20} $^{239}_{94}$Pu nuclei is 4 mCi. (*a*) How many of these nuclei remain after 73,200 yr? (*b*) What will be the activity at that time?

Solution (a)   Substitution into Eq. (22-12) yields

$$N = N_0\left(\frac{1}{2}\right)^{t/T_{1/2}} = (1.64 \times 10^{20})\left(\frac{1}{2}\right)^{73,200 \text{ yr}/24,400 \text{ yr}}$$

$$= (1.64 \times 10^{20})\left(\frac{1}{2}\right)^3 = (1.64 \times 10^{20})\left(\frac{1}{8}\right)$$

$$= 2.05 \times 10^{19} \text{ nuclei}$$

Solution (b)   We obtain the remaining activity from Eq. (22-14):

$$R = R_0\left(\frac{1}{2}\right)^{t/T_{1/2}} = (4 \text{ mCi})\left(\frac{1}{8}\right) = 0.5 \text{ mCi}$$

Both these calculations assume that no new $^{239}_{94}$Pu nuclei are being created by other processes. It is easy to see from this example why disposal of some radioactive materials is such a difficult problem.

22-8 ■ NUCLEAR REACTIONS

In a chemical reaction the atoms of two molecules react to form different molecules. In a *nuclear reaction,* nuclei, radiation, and/or nucleons collide to form different nuclei, radiation, and nucleons. If the colliding objects are charged, at least one of the colliding masses must be accelerated to a relatively high velocity. Normally the bombarding particle is very light, e.g., a proton $\frac{1}{1}p$ or an alpha particle $\frac{4}{2}\alpha$. These nuclear projectiles are accelerated with many different devices, e.g., Van de Graaff generators, cyclotrons, and linear accelerators.

In the nuclear reactions we shall study, several conservation laws must be observed, primarily *conservation of charge, conservation of nucleons,* and *conservation of mass-energy.*

> **Conservation of charge** *requires that the total charge of a system neither increase nor decrease in a nuclear reaction.*

> **Conservation of nucleons** *requires that the total number of nucleons in the interaction remain unchanged.*

> **Conservation of mass-energy** *requires that the total mass-energy of a system remain unchanged in a nuclear reaction.*

Now let us observe what happens when an alpha particle $\frac{4}{2}\alpha$ strikes a nucleus in a sample of nitrogen gas $^{14}_{7}N$. (Refer to Fig. 22-8.) The first step is the entry of the alpha particle, which adds 2 protons and 2 neutrons to the nucleus. The atomic number Z is increased by 2, and the mass number A is increased by 4. The resulting nucleus is an *unstable* compound nucleus of fluorine $^{18}_{9}F$. This unstable nucleus quickly disintegrates into the final products, oxygen $^{17}_{8}O$ and hydrogen $^{1}_{1}H$. The overall reaction can be written

$$\frac{4}{2}\alpha + {}^{14}_{7}N \rightarrow {}^{1}_{1}H + {}^{17}_{8}O \tag{22-15}$$

Note how charge and nucleons are conserved in these reactions. There was a net charge of $+9e$ before the reaction and a net charge of $+9e$ after the reaction, and there are 18 nucleons before and after the reaction.

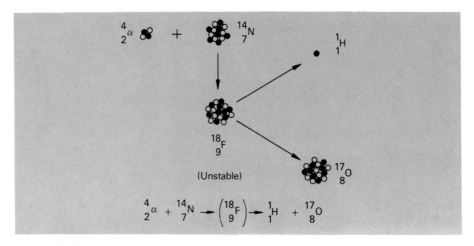

FIG. 22-8 Striking a nitrogen 14 nucleus with an alpha particle.

22-9 ■ NUCLEAR FISSION

Before the discovery of the neutron in 1932, alpha particles and protons were the primary particles used to bombard atomic nuclei, but as charged particles they have the disadvantage of being repelled electrostatically by the nucleus. Consequently, very large energies are required before nuclear reactions occur.

Since neutrons have zero electric charge, they can easily penetrate the nucleus of an atom with no coulomb repulsion. Very fast neutrons may pass completely through a nucleus or may cause it to disintegrate. Slow neutrons may be captured by a nucleus, creating an unstable isotope which may disintegrate.

Whenever the absorption of an incoming neutron causes a nucleus to split into two smaller nuclei, the reaction is called *fission* and the product nuclei are called *fission fragments*.

> **Nuclear fission** *is the process by which heavy nuclei are split into two or more nuclei of intermediate mass numbers.*

Whenever a slow neutron is captured by a uranium nucleus $^{235}_{92}U$, an unstable nucleus ($^{236}_{92}U$) is produced which may decay in several ways into smaller product nuclei (Fig. 22-9). Such fission reactions may produce fast neutrons, beta particles, and gamma rays in addition to the product nuclei. For this reason the products of a fission process, including fallout from a nuclear explosion, are highly radioactive.

The fission fragments have a smaller mass number and therefore about 1 MeV more binding energy per nucleon (see Fig. 22-4). As a result, fission releases a large amount of energy. In the above example approximately 200 MeV per fission is produced.

Because each nuclear fission releases more neutrons, which may lead to additional fission, a *chain reaction* is possible. As seen in Fig. 22-10, the three neutrons released from the fission of $^{235}_{92}U$ produce three additional fissions. Thus, starting with one neutron, we have liberated nine after only two steps. If such a chain reaction is not controlled, it can lead to an explosion of enormous magnitude.

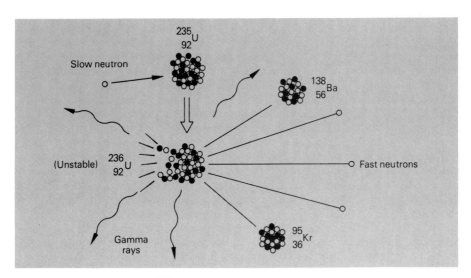

FIG. 22-9 Nuclear fission of $^{235}_{92}U$ by capture of a slow neutron.

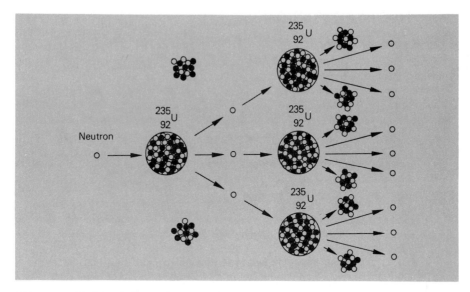

FIG. 22-10 Nuclear chain reaction.

22-10 ■ *NUCLEAR REACTORS*

A *nuclear reactor* is a device which controls the nuclear fission of radioactive material, producing new radioactive substances and large amounts of energy. These devices are used to furnish heat for electric power generation, propulsion, and industrial processes; they are used to produce new elements or radioactive materials for a multitude of applications; and they are used as a supply of neutrons for scientific experimentation.

A schematic diagram of a typical reactor is given in Fig. 22-11. The basic components are (1) a *core* of nuclear fuel, (2) a *moderator* for slowing down fast neutrons, (3) *control rods* or other means for regulating the fission process, (4) a heat exchanger for removing heat generated in the core, and (5) radiation shielding. Steam produced by the reactor is used to drive a turbine, which generates electricity. The spent steam is changed to water in the condenser and pumped back to the heat exchanger for another cycle.

The essential ingredient in the reactor is the fissionable material, or nuclear fuel. About the only naturally occurring fissionable material is $^{235}_{92}U$, which constitutes about 0.7 percent of naturally available uranium. The remaining 99.3 percent is $^{238}_{92}U$. Fortunately, $^{238}_{92}U$ is a *fertile* material, i.e., it changes to a fissionable material when struck by neutrons. Plutonium $^{239}_{94}Pu$ produced in this manner can provide new fuel for the reactor.

The production of additional fuel as a part of the reactor's operation has led to the design of *breeder reactors,* in which there is a net increase in fissionable material. In other words, the reactor produces more fuel than it consumes. This does not violate the law of conservation of energy. It only provides for the production of fissionable material from fertile materials.

The fissionable fuel in most reactors depends on the availability of slow neutrons, which are more likely to produce fission. Fast neutrons liberated by the fission of nuclei must therefore be slowed down. For this reason reactor fuel is embedded in a suitable substance called a *moderator*. The function of this substance is to slow neutrons without capturing them.

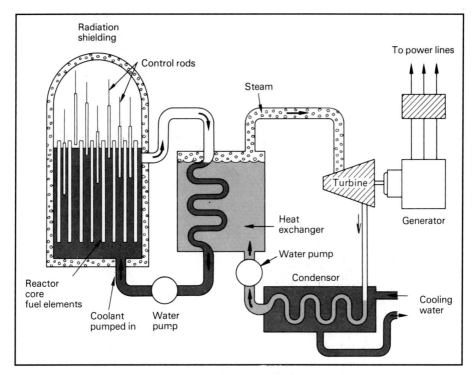

FIG. 22-11 Schematic diagram of a nuclear reactor. Water heated under pressure in the reactor core is pumped into a heat exchanger, where it produces steam to operate a turbine.

Neutrons have a mass about the same as that of a hydrogen atom. It might be expected, then, that substances containing hydrogen atoms would be effective as moderators of neutrons. The neutron is analogous to a marble, which can be stopped by a collision with another marble but will merely bounce off a cannonball because of the great mass difference. Water (H_2O) and heavy water, containing ${}_1^2H$ instead of ${}_1^1H$, are often used as moderators. Other suitable materials are graphite and beryllium.

In order to control the nuclear furnace, it is necessary to regulate the number of neutrons which initiate the fission process. Substances like boron and cadmium capture neutrons efficiently and are excellent control materials. A typical reactor has control rods which can be inserted into the reactor at variable distances. By adjusting the position of these rods, the activity of the nuclear furnace is controlled. A supplementary set of rods is available to allow the reactor to be shut down completely in an emergency.

22-11 ■ NUCLEAR FUSION

In our earlier discussions on mass defect, we calculated that 28.3 MeV of energy is released in the formation of ${}_2^4He$ from its component nucleons. This joining of light nuclei into a single heavier nucleus is called *nuclear fusion*. This process provides the fuel for stars like our own sun, and it is also the principle behind the hydrogen bomb. Many consider the fusion of hydrogen into helium as the ultimate fuel.

The use of nuclear fusion as a controlled source of energy is not without problems. For one thing, the fusing nuclei must be given millions of electron volts of kinetic energy to overcome their coulomb repulsion. In the hydrogen bomb, this enormous amount of energy is supplied by an atomic explosion, which then triggers the fusion process. However, for the peaceful production of energy, it poses a serious problem of containment. The nuclear fuel would have to be so hot that it would instantly disintegrate any known substance. Present research methods involve containment by magnetic fields or rapid heating by powerful lasers.

If and when the problems of heat generation and containment are solved, fusion could provide a solution to our formidable problem of dwindling resources. The deuterium commonly found in seawater could provide us with a virtually inexhaustible supply of fuel. It would represent more than a billion times the energy available in all our coal and oil reserves. In addition, it appears that fusion reactors would have much less of a problem with radioactive residue than fission reactors.

Summary

The production and use of nuclear energy become more important every day. In this chapter, you have reviewed the basic concepts of nuclear structure, binding energy, radioactivity, nuclear reactions, and nuclear power production. The major points which must be remembered are as follows:

- The fundamental nuclear particles discussed in this chapter are summarized in the following table. The masses are given in atomic mass units, and the charge is in terms of the electronic charge $+e$ or $-e$, which is 1.6×10^{-19} C.

FUNDAMENTAL PARTICLES

Particle	Symbol	Mass, u	Charge
Electron	$_{-1}^{0}e$, $_{-1}^{0}\beta$	0.00055	$-e$
Proton	$_{1}^{1}p$	1.007276	$+e$
Neutron	$_{0}^{1}n$	1.008665	0
Positron	$_{+1}^{0}e$, $_{+1}^{0}\beta$	0.00055	$+e$
Alpha particle	$_{2}^{4}\alpha$, $_{2}^{4}\mathrm{He}$	4.001506	$+2e$

The atomic masses of the various elements are given in the text.
- The atomic number Z of an element is the number of protons in its nucleus. The mass number A is the sum of the atomic number and the number of neutrons N. These numbers are used to write the nuclear symbol:

$$A = Z + N \qquad \text{symbol: } _{Z}^{A}\mathrm{X}$$

- One *atomic mass unit* (u) is equal to one-twelfth the mass of the most abundant carbon atom. Its value in kilograms is given below. Also, since $E = mc^2$, we can write the conversion factor from mass to energy as c^2:

$$1 \text{ u} = 1.6606 \times 10^{-27} \text{ kg} \qquad c^2 = 931 \text{ MeV/u}$$
$$1 \text{ MeV} = 1 \times 10^6 \text{ eV} = 1.6 \times 10^{-13} \text{ J}$$

■ The *mass defect* is the difference between the rest mass of a nucleus and the sum of the rest masses of its nucleons. The *binding energy* is obtained by multiplying the mass defect by c^2:

$$E_B = (Zm_H + Nm_n - M)c^2$$

Binding Energy

where $m_H = 1.007825$ u
$m_n = 1.008665$ u
$c^2 = 931$ MeV/u
$M = $ atomic mass
$N = A - Z$
$Z = $ atomic number

■ Several general equations for radioactive decay are

$$^A_Z X \rightarrow \, ^{A-4}_{Z-2}Y + \, ^4_2\alpha + \text{energy}$$

Alpha Decay

$$^A_Z X \rightarrow \, ^A_{Z+1}Y + \, ^0_{-1}\beta + \text{energy}$$

Beta Minus Decay

$$^A_Z X \rightarrow \, ^A_{Z-1}Y + \, ^0_{+1}\beta + \text{energy}$$

Beta Plus Decay

■ The *activity* R of a sample is the rate at which the radioactive nuclei decay. It is generally expressed in curies:

One curie (1 Ci) $= 3.7 \times 10^{10}$ disintegrations per second (s^{-1})

■ The *half-life* of a sample is the time $T_{1/2}$ in which one-half the unstable nuclei will decay.
■ The number of unstable nuclei remaining after a time t depends on the number n of half-lives that have passed. If N_0 nuclei exist at time $t = 0$, then a number N exists at time t. We have

$$N = N_0\left(\frac{1}{2}\right)^n$$

where $n = \dfrac{t}{T_{1/2}}$

■ The activity R and mass m of the radioactive portion of a sample are found from similar relations:

$$R = R_0\left(\frac{1}{2}\right)^n$$

$$m = m_i\left(\frac{1}{2}\right)^n$$

■ In any nuclear equation, the number of nucleons on the left side must equal the number of nucleons on the right side. Similarly, the net charge must be the same on each side.

Questions

22-1. Define the following terms:
a. nuclear force
b. nucleon
c. atomic number
d. mass number
e. atomic mass unit
f. isotopes
g. mass spectrometer
h. mass defect
i. binding energy
j. radioactivity
k. alpha particles

l. beta particles
m. gamma rays
n. half-life
o. activity
p. curie
q. nuclear fission
r. chain reaction
s. nuclear reactor
t. nuclear fusion
u. moderator

22-2. Write the symbol $_Z^A X$ for the most abundant isotopes of (a) cadmium, (b) silver, (c) gold, (d) polonium, (e) magnesium, and (f) radon.

22-3. From the curve describing the binding energy per nucleon (Fig. 22-4), would you expect the mass defect to be greater for chromium $_{24}^{52}Cr$ or uranium $_{92}^{238}U$? Why?

22-4. The binding energy is greater for the mass numbers in the central part of the periodic table. Discuss the significance of this in relation to nuclear fission and nuclear fusion. How do you explain the release of energy in both fusion and fission in view of the fact that one process brings nuclei together and the other tears them apart?

22-5. How is the stability of an isotope affected by the ratio of the mass number A to the atomic number Z? Does the element whose ratio is closest to 1 always appear to be the more stable?

22-6. Define and compare alpha particles, beta particles, and gamma rays. Which are likely to do the most damage to human tissue?

22-7. Given a source which emits alpha, beta, and gamma radiation, draw a diagram showing how you could demonstrate the charge and penetrating power of each type of radiation. Assume that you have at your disposal a source of a magnetic field and several thin sheets of aluminum.

22-8. Describe and explain, step by step, the decay of $_{92}^{238}U$ to the stable isotope of lead $_{82}^{206}Pb$. (Refer to Fig. 22-6.)

22-9. Write in the missing symbol, in the form $_Z^A X$, for the following nuclear disintegrations:

a. $_{90}^{234}Th \rightarrow \, _{91}^{234}Pa + \underline{\hspace{1cm}}$

b. $_{94}^{239}Pu \rightarrow \, _{90}^{234}Th + \underline{\hspace{1cm}}$

c. $_{15}^{32}P \rightarrow \underline{\hspace{1cm}} + \, _{+1}^{0}e$

d. $_{92}^{238}U \rightarrow \underline{\hspace{1cm}} + \, _{2}^{4}\alpha$

22-10. Write the missing symbol for the following nuclear reactions:

    a. $^2_1H + ^3_1H \rightarrow ^4_2He +$ _____

    b. $^9_4Be + ^4_2\alpha \rightarrow ^{12}_6C +$ _____

    c. $^{25}_{12}Mg +$ _____ $\rightarrow ^{28}_{13}Al + ^1_1H$

    d. $^1_1H +$ _____ $\rightarrow ^{12}_6C + ^4_2He$

22-11. Explain the function of the following components of a nuclear reactor: (*a*) uranium, (*b*) radiation shielding, (*c*) moderator, (*d*) control rods, (*e*) heat exchanger, and (*f*) condenser.

22-12. Give examples to show how beta decay and alpha decay tend to bring unstable nuclei closer to the stability curve of Fig. 22-5.

22-13. Radon has a half-life of 3.8 days. Consider a sample of radon having a mass *m* and an activity *R*. What mass of radioactive radon remains after 3.8 days? Does this mean that the activity is reduced to one-half in the time of one half-life?

22-14. Radioactive carbon $^{14}_6C$ has a half-life of 5570 yr. In a living organism, the relative concentration of this isotope is the same as it is in the atmosphere because of the interchange of materials between the organism and the air. When an organism dies, this interchange stops and radioactive decay begins without replacement from the living organism. Explain how this principle can be used to determine the age of fossil remains.

Problems

22-1. How many neutrons are in the nucleus of $^{208}_{82}Pb$? How many protons? What is the ratio *N/Z*?

    Answer 126; 82; 1.54.

22-2. From a stability curve it is determined that the ratio of neutrons to protons for a cesium nucleus is 1.49. What is the mass number for this isotope of cesium (*Z* = 55)?

22-3. Calculate the binding energy of tritium ^3_1H. What are the total binding energy and the binding energy per nucleon? How much energy in joules is required to tear the nucleus apart into its constituent nucleons? (The atomic mass of tritium is 3.016049 u.)

    Answer 8.48 MeV; 2.83 MeV; 1.36×10^{-18} J.

22-4. Calculate the mass defect of ^7_3Li. What is the binding energy per nucleon? (The atomic mass of ^7_3Li is 7.0160 u.)

22-5. Determine the binding energy per nucleon for (*a*) $^{12}_6C$ and (*b*) tin, $^{120}_{50}Sn$ (atomic mass = 119.90220 u).

    Answer (*a*) 7.68 MeV/nucleon; (*b*) 8.5 MeV/nucleon.

22-6. Calculate the energy required to separate the nucleons in $^{204}_{80}Hg$ if the atomic mass is 203.9735 u.

22-7. The $^{60}_{27}Co$ nucleus emits gamma rays of approximately 1.2 MeV of energy. How much mass is lost by the nucleus when it emits a gamma ray of this energy?

    Answer 0.00129 u.

22-8. The half-life of the radioactive isotope indium 109 is 4.3 h. If the activity of a sample is 1 mCi at the start, how much activity remains after 4.3, 8.6, and 12.9 h?

22-9. The initial activity of a sample containing 7.7×10^{11} bismuth 212 nuclei is 4.0 mCi. The half-life of this isotope is 6.0 min. (*a*) How many bismuth 212 nuclei remain after ½ h? (*b*) What is the activity then?

    Answer (*a*) 2.41×10^{10}; (*b*) 0.125 mCi.

22-10. Strontium 90 is produced in appreciable quantities in the atmosphere during the test of an atom bomb. If this isotope has a half-life of 28 yr, how long will it take for the initial radioactivity to drop to one-fourth its original activity?

22-11. Consider a sample of 24 mg of radioactive bismuth $^{210}_{83}$Bi. If its half-life is 5 days, how much of the sample will remain after 15 days?

Answer 3 mg.

22-12. How long will it take for 40 mg of the unstable isotope $^{206}_{81}$Tl to decay to only 10 mg if the half-life is 9.0 min?

22-13. Determine the minimum energy released in the nuclear reaction $^{19}_{9}$F + $^{1}_{1}$H → $^{16}_{8}$O + $^{4}_{2}\alpha$. (Assume that $^{19}_{9}$F = 18.99840 u and $^{16}_{8}$O = 15.99492 u.)

Answer 9.12 MeV.

22-14. Determine the approximate kinetic energy imparted to the alpha particle when $^{226}_{88}$Ra decays to form $^{222}_{86}$Rn. Neglect the energy imparted to the radon nucleus. The nuclidic masses of $^{266}_{88}$Ra and $^{222}_{86}$Rn are 226.02536 and 222.01753 u, respectively.

22-15. Find the energy evolved in the production of two alpha particles in the reaction

$$^{7}_{3}\text{Li} + ^{1}_{1}\text{H} \rightarrow ^{4}_{2}\text{He} + ^{4}_{2}\text{He}$$

Assume the atomic mass of $^{7}_{3}$Li to be 7.01600.

Answer 0.735 MeV.

22-16. Compute the kinetic energy released in the beta minus decay of $^{233}_{90}$Th. The nuclidic masses of $^{233}_{90}$Th and $^{233}_{91}$Pa are 233.04147 and 233.04013 u, respectively.

22-17. A nuclear reactor operates at a power level of 2 MW. Assuming that approximately 200 MeV of energy is released for a single fission of $^{235}_{92}$U, how many fission processes are occurring each second in the reactor?

Answer 6.25 × 10^6 fission/s.

APPENDIX

Answers to Exercises (Chapters 1–3)

CHAPTER 1.
TECHNICAL
MATHEMATICS

Exercises 1–1
(Pages 5–6)

1. $+7$
2. $+4$
3. $+2$
4. -2
5. -10
6. -33
7. -5
8. -17
9. $+6$
10. -32
11. -36
12. $+24$
13. -48
14. $+144$
15. $+2$
16. -2
17. -4
18. -3
19. $+2$
20. -4
21. -3
22. $+12$
23. $+8$
24. -4
25. 0
26. $+220$
27. 32 ft
28. -32 J
29. (a) $-6°C$
    (b) $-17°C$
    (c) $36°C$
30. -50 mm
    decrease
    in length

Exercises 1–2
(Pages 9–10)

1. $x = -3$
2. $x = +7$

3. $x = -7$
4. $x = -12$
5. $x = -\frac{1}{2}$
6. $x = +\frac{1}{2}$
7. $x = 5$
8. $x = -\frac{3}{4}$
9. $x = +\frac{4}{3}$
10. $x = 17$
11. $x = \sqrt{17}$
12. $x = -96$
13. $x = -\frac{5}{4}$
14. $x = +8$
15. $x = +1$
16. $x = -32$
17. $m = 6$
18. $p = +4$
19. $m = -4$
20. $m = 8$
21. $x = 36$
22. $p = 1$
23. $x = 2$
24. $b = 14$
25. $R = 5$
26. $P = 3$
27. $R = \dfrac{V}{I}$
28. $\dfrac{PV}{nR} = T$
29. $a = F/m$
30. $d = s - vt$
31. $R = \dfrac{mv^2}{F}$
32. $a = \dfrac{2s}{t^2}$
33. $a = \dfrac{V_f^2 - V_o^2}{2s}$
34. $V = \dfrac{Q^2}{2C}$
35. $R = \dfrac{R_1 R_2}{R_2 + R_1}$
36. $t = \dfrac{MV}{F}$
37. $V_2 = \dfrac{Ft + mV_1}{m}$

38. $T_2 = \dfrac{T_1 P_2 V_2}{P_1 V_1}$
39. $a = \dfrac{v - v_o}{t}$
40. $b = \sqrt{c^2 - a^2}$

Exercises 1–3
(Page 13)

1. 2^{12}
2. $3^5 2^3$
3. x^{10}
4. x^5
5. $\dfrac{1}{a}$
6. $\dfrac{a}{b^2}$
7. $\dfrac{1}{2^2}$
8. $\dfrac{a^2}{b^2}$
9. $2x^5$
10. $\dfrac{1}{a^2 b^2}$
11. m^6
12. $\dfrac{c^4}{n^6}$
13. 64×10^6
14. $\frac{1}{36} \times 10^4$
15. 4
16. 3
17. x^3
18. $a^2 b^3$
19. 2×10^2
20. 2×10^{-9}
21. $2a^2$
22. $x + 2$

Exercises 1–4
(Pages 15–16)

1. 4×10^4
2. 6.7×10^1
3. 4.8×10^2
4. 4.97×10^5
5. 2.1×10^{-3}

6. 7.89×10^{-1}
7. 8.7×10^{-2}
8. 9.67×10^{-4}
9. $4{,}000{,}000$
10. 4670
11. 37
12. $140{,}000$
13. 0.0367
14. 0.4
15. 0.006
16. 0.0000417
17. 8×10^6
18. 7.4×10^4
19. 8×10^2
20. 1.8×10^{-8}
21. 2.68×10^9
22. 7.4×10^{-3}
23. 1.6×10^{-5}
24. 2.7×10^{19}
25. 1.8×10^{-3}
26. 2.4×10^1
27. 2.0×10^6
28. 2×10^{-3}
29. 2.0×10^{-9}
30. 5.71×10^{-1}
31. 2.3×10^5
32. 6.4×10^2
33. 2.4×10^3
34. 5.6×10^{-5}
35. -6.9×10^{-2}
36. -3.3×10^{-3}
37. 6×10^{-4}
38. 6.4×10^6
39. -8×10^6
40. -4×10^{-6}

Exercises 1–5
(Pages 18–19)

1. $V = 144$ ft/s
   $t = 3.1$ s
2. 88 turns
3. 857 m
   375 m
4. 10.4 W
   64.8 W

Exercises 1–6
(Pages 23–24)

3. (*a*) $B = 35°$
$A = 17°$
(*b*) $B = 40°$
$A = 50°$
4. (*a*) $A = 50°$
$B = 130°$
(*b*) $B = 70°$
$A = 42°$

**CHAPTER 2.
TECHNICAL
MEASUREMENT**

Exercises 2–1
(Page 35)

1. (*a*) 1 mV =
0.001 volt
(*b*) 1 μA =
0.000001
ampere
(*c*) 1 kW =
1000 watts
(*d*) 1 Mm =
1,000,000
meters
(*e*) 1 nm =
0.000000001
meter
(*f*) 1 Gm =
1,000,000,000
meters
2. (*a*) 1 cV =
centivolt =
0.01 volt
(*b*) 1 Mg =
megagram =
1,000,000 grams
(*c*) 1 mA =
milliampere =
0.001 ampere
(*d*) 1 kW =
kilowatt =
1000 watts
(*e*) 1 μJ =
microjoule =
0.000001 joule

(*f*) 1 nN =
nanonewton =
0.000000001
newton
3. (*a*) 40 mm
(*b*) 40 km
(*c*) 6 mm
(*d*) 300 μm
(*e*) 700 nm
(*f*) 20 m

Exercises 2–2
(Page 40)

(Ans. rounded to 3
significant digits)
1. (*a*) 0.4 m
(*b*) 50.8 cm
(*c*) 143 mm
(*d*) 28 m
(*e*) 141 ft
(*f*) 0.000333 yd
(*g*) 18,500 in.
(*h*) 86,000 km
2. 6340 ft
3. 49.3 mm
4. $8.58
5. 40 gal
6. 366 m/min
7. 2100 cal/min

Exercises 2–3
(Pages 44–45)

1. (*a*) 8 in.^{2}
(*b*) 159 m^2
(*c*) 800 cm^2
2. 25.9 cm^2
3. $87.75
4. (*a*) 867 yd^2
(*b*) 38.1 mm^2
(*c*) 236 in.^{2}
5. 28.2 in.^{2}
6. 6912 in.^{2}
7. 9,730,000 mm^2
8. 9.82 Btu/h

Exercises 2–4
(Page 48)

1. 22.4 gal
2. 0.154 L

3. 33.5 m^3
4. (*a*) 0.0795 m^3
(*b*) 79.5 L
5. 3.7 yd^3
6. 75,400 L

**CHAPTER 3.
FORCE
AND VECTORS**

Exercises 3–1
(Pages 64–66)

1. 539 km,
68.2°S of W
2. 25.6 mi/h,
51.3°N of E
3. 473 N; 369 N
4. 226 ft, W;
197 ft, N
5. 173 lb
6. 51.4 N
7. 662 km,
4.3°N of E
8. 124 N,
46.3 N of W
9. $R = 380$ lb

Exercises 3–2
(Pages 72–73)

1. 0.921
2. 0.669
3. 1.66
4. 0.559
5. 0.875
6. 0.268
7. 19.3
8. 143
9. 267
10. 32.4
11. 235
12. 2425
13. 684
14. 346
15. 803
16. 266
17. 2191
18. 1620
19. 54.2°

20. 6.37°
21. 50.2°
22. 27.1°
23. 76.8°
24. 35.9°
25. 36.9°
26. 76.0°
27. 31.2°
28. $\theta = 35.8°$
$R = 30.8$ ft
29. $\phi = 56.3°$
$R = 721$ m
30. 233 m,
607 m
31. $\phi = 23.6°$
$x = 458$ km
32. $\theta = 31.7°$
340 m
33. 164 in.
202 in.
34. $\theta = 26.6°$
$R = 89.4$ lb
35. 260 m,
150 m, left
36. $\theta = 26.6°$
$R = 447$ km
37. $R = 78.9$ mm
38. 179 ft
$\theta = 26.7°$
39. $\theta = 43.3°$
40. $R = 14.2$ m
41. 473 lb,
369 lb
42. $\theta = 33°$,
$R = 23.9$ N

Exercises 3–3
(Pages 80–81)

1. 41°N of E
2. 65°N of W
3. 15°N of W
4. 47°N of W
5. 30°S of E
6. 44°S of E
7. 35°S of W
8. 20°S of W
9. 70°S of E
10. 9°S of E
11. 120°
12. 105°
13. 290°
14. 215°
15. 227°
16. 307°
17. 10°
18. 231°
19. $R_x = 257$ m
    $R_y = 306$ m

20. $R_x = 58.0$ lb
    $R_y = 15.5$ lb
21. $R_x = -100$ N
    $R_y = 173$ N
22. $R_x = -498$
    mi/h
    $R_y = -418$
    mi/h
23. $R_x = 80$ m
    $R_y = -139$ m
24. $R_x = -139$ N
    $R_y = 144$ N
25. $R_x = 10.9$ lb
    $R_y = -21.4$ lb
26. $R_x = 433$ km
    $R_y = 250$ km
27. $R_x = -1230$ mi
    $R_y = 1030$ mi
28. $R_x = -15.6$
    m/s
    $R_y = -9.00$
    m/s

29. $R = 500$ ft
    $\theta = 126.9°$
30. $R = 44.7$ m
    $\theta = 296.6°$
31. $R = 424$ m
    $\theta = 70.7°$
32. $R = 721$ lb
    $\theta = 213.7°$
33. $R = 50.6$ N
    $\theta = 341.6°$
34. $R = 608$ m
    $\theta = 99.5°$
35. $R = 64.6$ lb
    $\theta = 21.8$ N
    of E
36. $R = 215$ lb
    $\theta = 68.2°$S
    of E
37. $R = 532$ m
    $\theta = 48.8°$S
    of W
38. $R = 30.0$ N

$\theta = 53.1°$N
of W
39. 68.9, left
    57.9,
    downward
40. 530 lb, east
    282 lb, south
41. 10.6 lb
    17.0 lb
42. -80 N
    139 N
43. 297 lb
    268 lb
44. $R = 361$ lb
    $\theta = 33.7°$
45. $\theta = 73.9$
    $R = 72.1$ lb
46. $F = 62.2$ lb
47. $R = 302$ N,
    41.4°
    $\theta = 318.6°$
48. $R = 85.6$ lb

INDEX